Polynomial Signal Processing

V. John Mathews
University of Utah

Giovanni L. Sicuranza
University of Trieste

A Wiley-Interscience Publication
JOHN WILEY & SONS, INC.
New York • Chichester • Weinheim • Brisbane • Singapore • Toronto

This book is printed on acid-free paper. ∞

Copyright © 2000 by John Wiley & Sons, Inc. All rights reserved.

Published simultaneously in Canada.

For ordering and customer service, call 1-800-CALL-WILEY.

Library of Congress Cataloging-in-Publication Data:

Mathews, V. John.
    Polynomial signal processing / V. John Mathews, Giovanni L. Sicuranza.
        p.   cm.
        "A Wiley-Interscience publication."
        ISBN 0-471-03414-2 (alk. paper)
        1. Signal processing. 2. Nonlinear theories. 3. Filters (Mathematics) I. Sicuranza, Giovanni L. II. Title.
    TK5102.9.M385 2000   99-22018
    621.382'2– dc21

Printed in the United States of America.
10 9 8 7 6 5 4 3 2 1

Polynomial Signal Processing

To Suja
—VJM

To Maria Lucia
—GLS

CONTENTS

PREFACE

Linear filters have had enormous impact on the development of various techniques for processing stationary and nonstationary signals. However, there are a large number of practical applications in which linear filters perform poorly. Even though nonlinear systems and nonlinear models have been studied for many centuries, engineers have often shied away from using such techniques mainly for two reasons: (1) nonlinear filters are often difficult to implement because of their computational complexity and (2) they are also often conceptually difficult to understand. The ever-increasing capabilities of digital computers have significantly reduced the magnitude of the first problem in recent years. Furthermore, many new insights into the properties and capabilities of nonlinear filters have become available recently. As a result, there has been a large amount of research activity in the last two decades in the area of nonlinear filtering, and this in turn, has resulted in a large number of applications of such techniques.

Our objective in writing this book was to present the readers with a relatively easy-to-read exposition of the state of the art in one area of nonlinear signal processing known as *polynomial signal processing*. Since all systems that are not linear are nonlinear, it is impossible to find a meaningful or useful unifying theory that is applicable to all nonlinear systems. Consequently, studies of nonlinear systems often restrict themselves to special classes of nonlinear systems for which unifying theories do exist. Polynomial systems form one such class. Polynomial systems may be interpreted as conceptually straightforward extensions of linear systems to the nonlinear case. Therefore, many techniques for realization and design of linear systems as well as those for the adaptation of the parameters of linear system models can be extended to the case of polynomial systems in relatively easy ways. Wherever possible, we have tried to make use of this relationship among linear and polynomial systems to simplify our presentation.

We have attempted to present a comprehensive description of Volterra series expansion, which forms the basis of polynomial systems theory, realization theory for polynomial filters both in the time and frequency domains, multidimensional extensions, adaptive and nonadaptive estimation of the parameters of a polynomial system model, signal modeling using polynomial time series analysis, and a large number of applications of polynomial filtering in this book. One of the characteristics that the reader should become aware of while going through the book is that there are many applications in which polynomial system models can improve the performance over that of linear system models with minimal increase in the computational complexity. What we hope will distinguish this book from several others available in this area are the following factors:

1. We have attempted to write this book at a level that is suitable for first-year graduate students and practicing engineers who have an interest in this area. Consequently, we have tried to use no more than a basic understanding of digital signal processing, linear algebra, and probability theory in the development of this book. While we do not believe that mathematical rigor has been compromised in the descriptions, we have not presented highly mathematical derivations except in cases where they are absolutely necessary for the understanding of the concepts presented in the book.

2. We have emphasized practical applications and implementation issues in our descriptions.

3. There have been many new developments in both theory and applications of polynomial signal processing since the publication of the books by Schetzen [274] and Rugh [263], which are by far the books most commonly referred to on polynomial signal processing at this time. We have described many such topics in this book. The chapters on adaptive polynomial filters, multidimensional systems, time series analysis, and applications are examples of such discussions. It is our hope that we have been successful in presenting an easy-to-read and reasonably comprehensive discussion of the state of the art in polynomial signal processing in this book.

This book is suitable for use as a textbook in a semester-long or a quarter-long course at the first-year graduate level. At the University of Utah, we have used Chapters 1, 2, 3, 5, 6, 7, 8, and portions of 10 for a quarter-long course. We have provided several exercises and computing assignments that may be used as homework problems at the end of all chapters except the first and the last ones. We have also provided a relatively comprehensive list of references at the end of the book.

Financial support from NATO through Grant CRG 950379 enabled the authors to travel to Salt Lake City and Trieste several times during the preparation of this book. This support is gratefully acknowledged. These trips not only helped make the long-distance collaboration much easier than it would otherwise have been, but also allowed us to get to know each other's families very well.

There are several individuals who helped us in the preparation of this manuscript in different ways. We have learned much from our students, and we have tried to present many of the things we learned from them in this book. We would like to especially thank Alberto Carini, Junghsi Lee, Stefano Marsi, Shan Mo, Thomas

Panicker, Vittorio Rochelli, and Mushtaq Syed for their contributions and their friendships. We owe a special debt of gratitude to Gianni Ramponi for allowing us to incorporate his work on the bi-impulse response and the rational filters into this book and for the very useful discussions we had with him on many aspects of polynomial signal processing. We also wish to acknowledge the valuable contributions of Enzo Mumolo on applications of polynomial filters in speech modeling and computer engineering and the helpful comments of Sabine Kröner on higher-order neural networks and polynomial invariants. Richard Fay went through the entire manuscript and made several valuable suggestions for improving the book. His interest in this material has been tremendous, and his ability to catch even the smallest of problems is amazing. Using the manuscript as the text for a first-year graduate level course on nonlinear signal processing has helped us revise the book with improved pedagogy in mind. In particular, Shy Shoham reviewed almost all the chapters. Bill Pohlchuck also reviewed several chapters. They both provided many useful suggestions for change. We have been fortunate to have the prepublished versions of the book reviewed by several experts in the general area of nonlinear filtering. We especially thank Alberto Carini, Moncef Gabbouj, Veit Kafka, Sanjit Mitra, Rob Nowak, Thomas Panicker, and Gianni Ramponi for their willingness to devote considerable time and energy for reviewing the book, and for their many valuable suggestions. Lynn Kirlin taught from parts of the manuscript at the University of Victoria and provided valuable feedback. John Proakis, our series editor, and George Telecki at Wiley Interscience have been extremely patient with us through many years of delays in the completion of the book. We also thank Angioline Loredo and Andrew Prince at Wiley for helping us through the publication process.

Our departments, the Department of Electrical Engineering at the University of Utah and the Department of Electrical, Electronic and Computer Engineering at the University of Trieste, have been very supportive of our activities. We are honored to be part of these departments. The environments provided by our departments have been extremely conducive to the collaborative production of this book. We also thank Doris Marx and Wilma Johnson for typing parts of the manuscript. They had to learn variations of LATEX to help us in this endeavor.

Finally, the preparation of this book has caused significant inconveniences to our family members. Suja, Maria Lucia, Kiran, and Ella have suffered through countless hours when we were not accessible to them because of our efforts to write this book. Throughout this process, they have been patient, encouraging, and helpful. Without their active support, this book could not have been written. We are grateful to them, and we are thankful that we belong to our respective families.

V. JOHN MATHEWS

Salt Lake City

GIOVANNI L. SICURANZA

Trieste

Polynomial Signal Processing

1

INTRODUCTION

Nonlinear systems are all systems that are not linear. Linear systems are well understood, and have found a variety of applications in all walks of life. However, as we will soon see, there are many situations in which signals are generated by nonlinear phenomena, or in which signals are subject to nonlinear processing. In most such situations, linear filters will not perform adequately, and nonlinear filters are required to provide the required levels of performance. This book aims to present a comprehensive description of the theory and applications of a class of nonlinear systems called *polynomial systems.*

All linear systems obey the *superposition principle,* which implies that the output of a linear combination of input signals to a linear system is the same linear combination of the outputs of the system corresponding to the individual components. That is,

$$\mathcal{L}\{\alpha x_1(n) + \beta x_2(n)\} = \alpha\mathcal{L}\{x_1(n)\} + \beta\mathcal{L}\{x_2(n)\}, \tag{1.1}$$

where $\mathcal{L}\{(\cdot)\}$ denotes the output of the linear system when its input is $(\cdot)$, $x_1(n)$ and $x_2(n)$ are two different input signals, and α and β are two arbitrary constants. The superposition principle is a powerful mechanism that allows us to study all linear systems in a unified manner. Nonlinear systems do not satisfy the superposition principle. Furthermore, since every system that does not satisfy the superposition principle is nonlinear, it is impossible to develop and present a theory for all nonlinear systems in a unified manner. Consequently, the traditional approach for studying nonlinear systems is to consider one or more classes of such systems and to develop the theory for analysis, design, and realization as well as applications of such classes individually. Polynomial systems form one such class of nonlinear systems. This class of systems is defined by input–output relationships of the form

$$y(n) = \sum_{i=0}^{P} f_i\{x(n), x(n-1), \ldots, x(n-N), y(n-1), \ldots, y(n-M)\}, \tag{1.2}$$

1

where $x(n)$ and $y(n)$ represent the input and output signals, respectively, $f_i(\cdots)$ is a polynomial of order i in the variables within the parentheses, and P is the maximum order of the polynomials employed in the model. Commonly used polynomial system models include the *quadratic filter* described by the input–output relationship

$$y(n) = \sum_{m_1=0}^{N_1-1} h_1(m_1)x(n-m_1) + \sum_{m_1=0}^{N_2-1}\sum_{m_2=0}^{N_2-1} h_2(m_1, m_2)x(n-m_1)x(n-m_2) \quad (1.3)$$

and the *bilinear filter* whose input–output relationship is given by

$$y(n) = \sum_{i=0}^{N_1} a_i x(n-i) + \sum_{j=1}^{N_2} b_j y(n-j) + \sum_{i=0}^{N_3}\sum_{j=1}^{N_4} c_{ij} x(n-i)y(n-j). \quad (1.4)$$

In these expressions, $h_1(m_1)$, $h_2(m_1, m_2)$, a_i, b_j, and c_{ij} represent the coefficients or parameters that completely characterize the system. The quadratic system model has only feedforward terms in the output, and may be considered as an extension of the finite impulse response (FIR) linear filters. The bilinear filter employs feedback of its outputs to generate the output signal, and thus may be thought of as an extension of the infinite impulse response (IIR) linear filters.

The primary objective of this book is to study the properties of systems described by (1.2). We will employ (1.3) and (1.4) and their variations as specific examples on which we base our study. In the next few chapters, we will learn about analysis techniques for characterizing such systems, certain design techniques, estimation of the parameters of the system models, adaptation of the parameters, and several applications. However, before getting into a detailed study of polynomial systems and polynomial signal processing, we motivate such studies by considering a variety of applications in which nonlinear filters are required for adequate performance.

1.1 EXAMPLES IN APPLICATIONS OF NONLINEAR FILTERS

In this section, we consider several engineering applications of nonlinear filters. The objective is not to provide the details, but simply to show where and how nonlinear filters are used. Polynomial system models have been used in most of the applications discussed below. The list of applications in this section is not exhaustive, and the details associated with many of these as well as several other applications are provided in Chapter 10.

1.1.1 Communication Systems

In satellite communication systems, the amplifiers located in the satellites usually operate at or near the saturation region in order to conserve energy. The saturation nonlinearities of the amplifiers introduce nonlinear distortions in the signals they process. The satellite channel is typically modeled using three distinct components as shown in Figure 1.1 [232]. The path from the earth to the satellite as well as from the satellite to the earth may be modeled as linear dispersive systems. The amplifier characteristics are modeled usually using memoryless nonlinearities. The equalizers at the receiver must be able to compensate for the nonlinear distortions so that the full capacity of the channel can be utilized.

1.1.2 Perceptually Tuned Signal Processing

In a large number of applications involving the processing of audio signals, images, and video signals, the final judges of quality of the processed signals are human beings. Examples of such applications include audio and video compression and image enhancement. In such situations, our processor should be tuned to the perceptual properties of the human brain; specifically, they should make use of known properties of vision or hearing depending on the application. Perceptual quality measurement is a highly nonlinear and complex process, and any system that hopes to match the subjective measures of quality employed by human beings must be nonlinear.

It is typical to employ models of the early portions of the human visual system in image processing applications. Figure 1.2 shows a block diagram of the *homomorphic* model of the eye developed by Stockham [301]. The logarithmic nonlinearity models the lens in the eye. The existence of such a nonlinearity explains the ability of the human eye to adapt to environments in which light intensities differ by seven or eight orders of magnitude. The early portion of the retina after the lens can be modeled by a highpass filter in the spatial domain followed by a saturation device. We can see that even in this simple model, we employ highly nonlinear devices to represent the behavior of the eye. Image compression systems that explicitly make use of the properties of the human eye provide compressed images that fare distinctly better in subjective quality evaluations when compared with systems that do not make use of perceptual criteria [102].

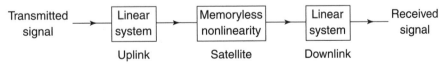

Figure 1.1 A model of a satellite communication channel.

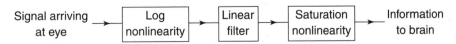

Figure 1.2 Homomorphic model of the human visual system due to Stockham [301].

1.1.3 Harmonic Distortion in Loudspeakers

The harmonic distortions introduced by loudspeakers into the audio signals are caused by nonlinearities in the loudspeaker characteristics. The main cause of the nonlinearities are the nonuniform flux of the permanent magnet and the nonlinear response of the suspensions [114]. Several methods have been devised to characterize and compensate for such distortions. One commonly employed model for loudspeakers employ a low-order nonlinearity in the form of a *truncated Volterra* system model with input–output relationship [70,76]

$$y(n) = \sum_{m_1=0}^{N_1-1} h_1(m_1)x(n-m_1) + \sum_{m_1=0}^{N_2-1}\sum_{m_2=0}^{N_2-1} h_2(m_1,m_2)x(n-m_1)x(n-m_2)$$

$$+ \sum_{m_1=0}^{N_3-1}\sum_{m_2=0}^{N_3-1}\sum_{m_3=0}^{N_3-1} h_3(m_1,m_2,m_3)x(n-m_1)x(n-m_2)x(n-m_3), \qquad (1.5)$$

where $h_1(m_1)$, $h_2(m_1,m_2)$, and $h_3(m_1,m_2,m_3)$ represent the first, second, and third-order kernels, respectively, of the model. Figure 1.3 displays the plots of the second-order kernel of a loudspeaker as measured by Frank [70]. Compensation for the nonlinear distortions is typically achieved by predistorting the audio signals prior to introducing them to the loudspeakers [70,76].

1.1.4 Enhancement of Noisy Images

Sharpening the edges in an image often produces an image that is subjectively more pleasing to human viewers. The original image may also contain noise that we wish to remove. If we were to use a linear filter in this application, we would need a highpass filter to perform the task of edge sharpening. However, highpass filters

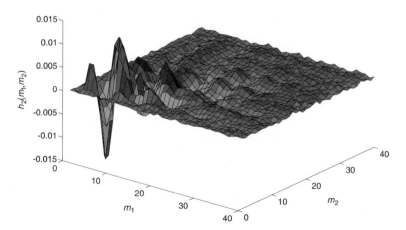

Figure 1.3 The second-order kernel of a loudspeaker as measured by Frank [70].

amplify wideband noise. On the other hand, lowpass filters reduce the noise present in the input images, but blur or smooth the edge information. Clearly, a linear filter is unable to accomplish both the objectives of edge enhancement *and* noise removal at the same time. A nonlinear filter is needed to obtain satisfactory performance in this situation.

Figure 1.4*a* shows a noisy image on which we wish to perform edge enhancement. The result of applying a *linear unsharp masking*[1] algorithm [252] to this image is shown in Figure 1.4*b*. Even though we can clearly see the improved rendition of the edges, the noise level in this image is unacceptably high. The result of a nonlinear processor for edge enhancement [252] is depicted in Figure 1.4*c*. The noise level is significantly lower in this image than in the output of the linear filter. At the same time, we can see improved edge rendition in this figure, demonstrating the usefulness of the nonlinear processor in this application.

1.1.5 Motion of Moored Ships in Ocean Waves

The force acting on ships due to the waves has been modeled as consisting of two components [14]. The first component is a linear force term that is proportional to the instantaneous elevation of the waves. The second term is nonlinear, and is usually modeled as proportional to the squared value of the *envelope* of the wave. In order to stabilize a moored vessel in the ocean, one must take into account the nonlinear relationship between the waves and the forces that the waves apply on the vessels.

1.1.6 Distortions in Magnetic Recording Systems

The nonlinear distortions in digital magnetic recording systems are caused by the nonlinearities in the magnetoresistive read transducers and by interactions between adjacent transitions which increase with increasing linear density of the channel. Consequently, it is necessary to compensate for the nonlinearities in high-density magnetic recording systems. Hermann [91] has developed a model for the nonlinearities in magnetic recording systems using a third-order Volterra series expansion. A discrete-time approximation of Hermann's model has input–output relationship [21]

$$
y(n + D) = \sum_{i=0}^{M} h_i x(n + i) + \sum_{i=0}^{M} c_i^{(1)} x(n + i) x(n + i - 1)
$$

$$
+ \sum_{i=0}^{M} c_i^{(2)} x(n + i) x(n + i - 2) + \sum_{i=0}^{M} c_i^{(3)} x(n + i) x(n + i - 3)
$$

$$
+ \sum_{i=0}^{M} c_i^{(1,2)} x(n + i) x(n + i - 1) x(n + i - 2), \tag{1.6}
$$

---

[1]Linear and nonlinear edge enhancement algorithms are described in detail in Chapter 10.

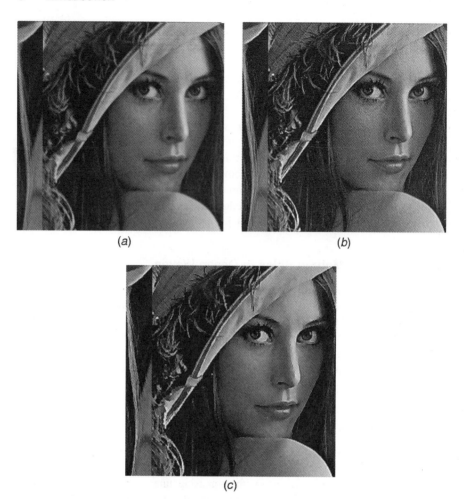

Figure 1.4 Comparison of a linear and a nonlinear edge enhancement procedure for noisy input images: (*a*) noisy input image; (*b*) output of the linear processor; (*c*) output of the nonlinear processor. (Courtesy of G. Ramponi. Copyright © 1996 SPIE *J. Electronic Imaging*.)

where $y(n)$ are the observed values at the output of the read process and $x(n)$ are the stored bits that take on values ± 1. The parameter D is a delay associated with the read and detection process, and the system is assumed to have a memory of $M + 3$ samples. A method for compensating for this type of nonlinearity in magnetic recording systems is described in Biglieri et al. [21].

1.2 CLASSES OF NONLINEAR SYSTEMS

As described in the previous section, it is in general impossible to develop a framework that is applicable to all nonlinear systems. In the following subsections, we briefly describe several classes of nonlinear systems that have found applications in many engineering problems.

1.2.1 Homomorphic Systems

We recall that linear systems satisfy the superposition principle. There are many applications in which the signal is formed as a multiplicative combination of two different signals. For example, the intensity of light reflected from an object can be modeled as the product of the intensity of the light that shines on the object at each location and the reflection coefficient of the object, i.e., an image $x(n_1, n_2)$ that the eye views may be expressed as [301]

$$x(n_1, n_2) = I(n_1, n_2)r(n_1, n_2), \tag{1.7}$$

where $I(n_1, n_2)$ is the light intensity that falls on the object at location (n_1, n_2) and $r(n_1, n_2)$ denotes the reflection coefficient of the object at the same location. *Homomorphic filters* are particularly useful in situations such as this, where we would like to separate the information-bearing function $r(n_1, n_2)$ from the image, or we would like to process the two components in different ways using an appropriate nonlinear transformation and a linear filter following the nonlinearity. As an example, we consider the block diagram of a homomorphic filter given in Figure 1.5. In this figure, the first block transforms the input signal using a logarithmic nonlinearity. The transformed signal is then processed by a linear filter. Suppose now that the signal $x(n_1, n_2)$ is the input to this system. The output of the logarithmic nonlinearity is

$$x'(n_1, n_2) = \ln(I(n_1, n_2)) + \ln(r(n_1, n_2)). \tag{1.8}$$

Thus, the linear filter will see the multiplicative components as additive components at its input. Consequently, we can use linear filtering techniques to discriminate the two components. The output of the linear filter may then be exponentiated to get a signal in the domain of the original signal. The basic structure of the system in Figure 1.5 can be used to enhance photographic images shot in poor lighting conditions since the signals $I(n_1, n_2)$ and $r(n_1, n_2)$ typically have different spectral contents. This system can also be used to remove multiplicative noise from signals.

The class of homomorphic filters is a generalization of the ideas described above. Let $*$ and $+$ be two operations defined on the input signal and the output signal,

Figure 1.5   A homomorphic filter employing a logarithmic nonlinearity.

respectively. Then a homomorphic system satisfies the generalized superposition principle stated as

$$\mathcal{H}\{x_1(n) * x_2(n)\} = \mathcal{H}\{x_1(n)\} + \mathcal{H}\{x_2(n)\}. \tag{1.9}$$

In the previous example, the operation $*$ denoted sample by sample multiplications while the operation $+$ denoted sample by sample addition. More detailed descriptions of homomorphic filters and their applications can be found in Oppenheim et al. [206] and Stockham [301].

1.2.2 Order Statistic Filters

Order statistic filters are employed mostly in applications in which the input signal is corrupted by impulsive noise. Median filters belong to this class. A one-dimensional, $(2K + 1)$-point median filter is defined by the input–output relationship

$$y(n) = \text{median}\{x(n + K), x(n + K - 1), \ldots, x(n - K)\}, \tag{1.10}$$

where the median of the $2K + 1$ samples within the curly brackets is the $(K + 1)$th value among the samples rearranged in the ascending or descending order of magnitude. One advantage of median filters is that unlike linear lowpass filters they are capable of removing impulsive noises without distorting the edges in the input signals. As an example, consider the three sequences shown in Figure 1.6. The noise-free input signal and its noisy version are shown in Figure 1.6a and 1.6b, respectively. We note that the noise is impulsive, and the samples that are affected by the noise differ significantly from the noise-free version. The output of a three-point median filter when its input is the noisy signal of Figure 1.6b is plotted in Figure 1.6c. We can see from this figure that the median filter has removed the impulsive noise successfully in this example, and that the system also preserved the sharp edge that was present in the signal without distortion. It is not difficult to show that a three-point median filter can remove impulsive noise samples as long as they do not occur more than once in any interval of length three samples.

The class of order statistic filters includes the median filter and its generalizations. One such generalization computes the output signal as a linear combination of samples arranged in the ascending or descending order of magnitude. The output of such a filter may be expressed as

$$y(n) = \sum_{i=0}^{N-1} h_i x_i(n), \tag{1.11}$$

where $x_i(n)$ is the ith sample in a rearrangement of the set $x(n)$, $x(n - 1)$, $\ldots, x(n - N + 1)$ in the ascending order of magnitude and h_i denotes the coefficient of $x_i(n)$. This type of a generalization provides the filter with a combination of properties associated with the linear filters and the median filters. Additional details of this and other types of order statistic filters such as weighted median filters, FIR-median hybrid filters, weighted order statistic filters, rank selection

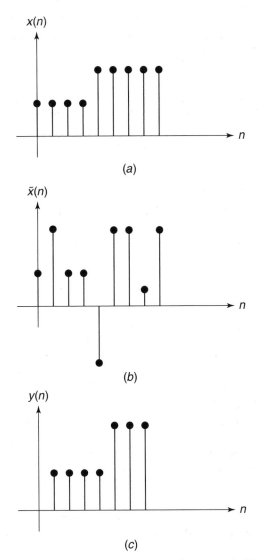

Figure 1.6 Operation of a median filter: (*a*) noise-free input signal; (*b*) noisy input signal; (*c*) output of a three-point median filter.

filters, and stack filters may be found in Astola and Kuosmanen [5], Yin et al. [334] and Pitas and Venetsanopoulos [228].

1.2.3 Morphological Filters

In many applications of pattern recognition and robotics, it is necessary to partition an image into segments on the basis of the geometric properties of the objects depicted in the image. In these and other applications it may also be necessary to obtain skeletal sketches of these objects. These objectives can be accomplished by the use of mathematical transformations of the input images using operators known as *morphological transforms*. Signal processing systems that employ morphological transforms are known as *morphological filters*. It is difficult to provide a description of morphological filters without introducing additional terminology and additional mathematical concepts. Consequently we refer the interested readers to Serra [278,279] and Maragos and Schafer [154]. In addition to pattern and shape recognition and image decomposition using skeletal representations, morphological filters have also been used in image compression systems [153].

1.2.4 Neural Networks

Neural networks attempt to model nonlinear systems using interconnections of simple nonlinear devices called *artificial neurons*. The artificial neurons are typically multiple-input, single-output systems with input– output relationship in the form

$$y(n) = f\left\{ \sum_{i=1}^{N} w_i x_i(n) - \theta \right\}, \tag{1.14}$$

where w_i is the weight associated with the ith input $x_i(n)$ to the device and θ is a constant term that controls the operating point of the nonlinearity f. An artificial neuron with this characteristic is depicted in Figure 1.7. Many different types of nonlinear functions have been employed, and some of the common ones are shown in Figure 1.8. It can be shown that appropriate interconnections of cells similar to the one depicted in Figure 1.7 is capable of approximating a large class of nonlinear systems accurately. In typical applications, the weights w_i used in the network are selected by training the neural network on data that are representative of what the network will encounter in normal applications. The training is accomplished using an adaptation algorithm such as the *backpropagation* algorithm. More details on algorithms for neural network design and their applications can be obtained from other sources [88,146].

The basic approach to the design of neural networks is an attempt to mimic the operation of the nervous system. The nervous system is made up of massively

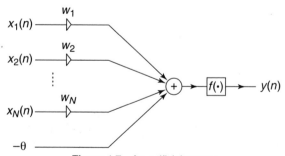

Figure 1.7 An artificial neuron.

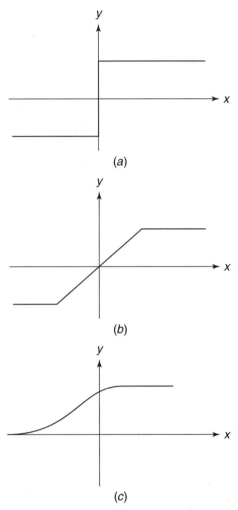

Figure 1.8 Nonlinearities commonly employed in neural networks: (*a*) hard limiter; (*b*) threshold logic; (*c*) sigmoid.

parallel interconnections of a very large number of neurons, and this interconnection is capable of performing recognition and decisionmaking tasks that cannot be matched by even the most powerful supercomputers that are available today. The advantage of artificial neural networks is their ability to model most nonlinear systems. However, in order to perform the modeling adequately, the network might require a very large number of artificial neurons. Another disadvantage of neural networks is that global convergence of training algorithms such as back propagation is not guaranteed.

1.2.5   Polynomial Filters

Causal, discrete-time polynomial filters satisfy the input–output relationship of the form

$$y(n) = \sum_{i=0}^{P} f_i\{x(n), x(n-1), \ldots, x(n-N), y(n-1), \ldots, y(n-M)\}, \quad (1.12)$$

where the function $f_i(\cdots)$ is an ith-order polynomial in the variables within the parenthesis. This definition contains the linear filters as a special case since the system of (1.12) is linear if $f_i(\cdots) = 0$ for all $i \neq 1$. If the system of (1.12) is stable in the bounded-input bounded-output sense, it admits a convergent *Volterra series expansion* of the form

$$y(n) = h_0 + \sum_{m_1=0}^{N_1-1} h_1(m_1)x(n-m_1) + \sum_{m_1=0}^{N_2-1}\sum_{m_2=0}^{N_2-1} h_2(m_1, m_2)x(n-m_1)x(n-m_2) + \cdots$$

$$+ \sum_{m_1=0}^{N_r-1}\sum_{m_2=0}^{N_r-1} \cdots \sum_{m_r=0}^{N_r-1} h_r(m_1, m_2, \ldots, m_r)x(n-m_1)x(n-m_2)\ldots x(n-m_r)$$

$$+ \cdots, \quad (1.13)$$

where $h_r(m_1, m_2, \ldots, m_r)$ denotes the rth-order *Volterra kernel* of the nonlinear system. We note from this expression that the term involving the rth-order kernel looks like an r-dimensional convolution. Consequently, we can consider polynomial systems as generalizations of linear systems.

Polynomial systems can be broadly classified into recursive and nonrecursive systems. Nonrecursive polynomial systems are characterized by input–output relationships that take the form of a truncated Volterra series expansion given by

$$y(n) = h_0 + \sum_{m_1=0}^{N_1-1} h_1(m_1)x(n-m_1) + \sum_{m_1=0}^{N_2-1}\sum_{m_2=0}^{N_2-1} h_2(m_1, m_2)x(n-m_1, n-m_2)$$

$$\cdots + \sum_{m_1=0}^{N_p-1}\sum_{m_2=0}^{N_p-1} \cdots \sum_{m_p=0}^{N_p-1} h_p(m_1, m_2, \ldots, m_p)x(n-m_1, n-m_2, \ldots, n-m_p).$$

$$(1.15)$$

The bilinear system of (1.4) is an example of a recursive polynomial system. Such systems are characterized by (possibly nonlinear) feedback terms in their input–output relationships. Since this book covers polynomial systems comprehensively, we state some advantages and disadvantages of polynomial systems and defer the detailed study of such systems to later chapters.

Some of the main advantages of polynomial systems are the following.

1. The theory of polynomial systems can be viewed as an extension to the theory of linear systems. Consequently, many results for the analysis and design of linear systems can be extended in a relatively straightforward manner to the case of polynomial systems.

2. It can be shown under relatively mild conditions that polynomial system models are capable of approximating a large class of nonlinear systems with a finite number of coefficients.

3. There are many processes in the real world that can be modeled using physical considerations as polynomial systems. An example is the interaction between ship motion and the ocean waves described earlier. There are also several applications in which parsimonious polynomial system models are able to provide significant performance improvements over linear system models. Many such applications in image processing are described in Chapter 10.

Polynomial system models also suffer from some disadvantages. The main disadvantages of polynomial models are listed below.

1. It requires a large number of coefficients to adequately model many real-world nonlinear systems using polynomial models. We can see from the input–output relationship in (1.12) that the complexity of a generic pth-order polynomial is a function of the pth power of the memory span of the system model. Thus as the memory span and the order of nonlinearity increase to even moderately large values, the complexity of implementing such systems becomes overwhelming. Even though the power of modern digital computers has alleviated the problems due to system complexity to some extent, most of the applications of polynomial systems until now have involved system models employing lower-order polynomials.

2. The stability properties of recursive polynomial systems are not completely understood at this time. For many system models, we can only develop sufficient conditions for stability that are often very restrictive. In many cases, it is also necessary to restrict the amplitude range of input signals to the system to ensure stable operation.

Despite these disadvantages, the performance improvements possible with polynomial system models are significant in many applications, and therefore there is a large amount of interest among researchers and applied engineers in studying such systems at present. The remaining chapters of this book describe the state of the art in the analysis and application of such systems.

1.3 ORGANIZATION OF THE BOOK

This book contains ten chapters in addition to this introductory chapter, which sets the framework for our studies. In Chapter 2, we review the theory of Volterra series expansion. A good understanding of the material in this chapter is required for reading the rest of the book. Rather than presenting a highly mathematical description of the theory of Volterra series expansion, we have chosen to collect together in a compact form several theoretical results that deal with the existence and convergence of the Volterra series expansion, and refer the student to references in the literature for detailed proofs. In this way, the reader will be able to advance more quickly to the subsequent chapters. As stated earlier, polynomial systems can be broadly classified as recursive and nonrecursive systems. Chapters 3–7 deal with truncated Volterra systems. Most of the current applications of polynomial systems employ truncated Volterra system models, and therefore we emphasize such models much more in this book than the recursive polynomial models.

Chapter 3 deals with different structures for realizing truncated Volterra systems. Our discussions involve time-domain and frequency-domain realizations. We also present alternatives to direct form realization of truncated Volterra systems. Such realizations include lattice filters and parallel-cascade systems. The latter class of systems is of particular importance because they realize higher-order systems using multiplicative connections of lower-order systems. They also provide us with a systematic method for approximating truncated Volterra systems in an efficient manner.

The results of Chapters 2 and 3 are extended to the case of multidimensional systems in Chapter 4. Because the complexity of truncated Volterra systems becomes exorbitant as the dimensionality of the signal and the order of nonlinearity increase, most of the applications of such filters to date have involved two- or three-dimensional signals and low nonlinearity orders such as two or three. It will be shown that computationally efficient, multidimensional Volterra systems with small regions of support are capable of providing significant performance improvement over linear systems in many applications. In this chapter, we concentrate primarily on two-dimensional quadratic filters even though their extensions to higher dimensions and orders of nonlinearity are conceptually straightforward.

Techniques for estimating the parameters of truncated Volterra systems are described in Chapters 5–7. Chapter 5 describes time-domain methods. Frequency-domain parameter estimation techniques are presented in Chapter 6. Adaptive filters employing truncated Volterra series expansions form the subject of Chapter 7.

Recursive polynomial systems are discussed in Chapter 8. Since much of the recent work on such models has involved bilinear systems, we concentrate primarily on bilinear system models in this chapter. One difficulty that arises with recursive system models is that of its stability. We present some relatively simple checks for the bounded-input/bounded-output stability of bilinear systems in this chapter. Adaptive filters employing bilinear system models as well as an experimental comparison of the bilinear and quadratic system models in a nonlinear channel equalization problem are also described in this chapter.

Chapter 9 deals with inversion of polynomial systems and time series modeling using polynomial system models. The first part of the chapter derives some results on exact inversion of certain nonlinear system models. Approximate inverses are also described in this part since the exact inverses are often not realizable. In the second part, it is shown using experimental results that it is often possible to provide better prediction of signals using a nonlinear model for its generation. In Chapter 10, we describe a large number of applications of polynomial signal processing involving one-, two-, and three-dimensional signals and systems. We will see in these applications that often very simple nonlinear system models are able to provide significant performance improvements over linear system models. The final chapter of the book provides a brief overview of some recent work in polynomial and related signal processing techniques.

1.4   A BRIEF HISTORY OF POLYNOMIAL SIGNAL PROCESSING

In this section we provide a brief history of polynomial signal processing. Even though the emphasis of this book is on discrete systems, this section discusses developments in discrete and continuous systems. The authors have relied heavily on the historical accounts in two other sources [264,275] to create this section. They are aware that the bibliography is far from complete. Rather than providing a complete set of references, we have attempted to refer to the early works in each subsection. Newer works are cited more often in the later chapters. The books by Schetzen [275] and Rugh [264] and the survey papers [22,165,274,290] are also good sources for additional references.

1.4.1   Volterra's Work

Vito Volterra, an Italian mathematician, first studied the functional series named after him as an extension of the Taylor series expansion. His first accounts of the theory of functionals were published in 1887 when he was in his twenties [324]. Volterra was born in 1860 and almost lost his life when he was only three months old [328]. The town of Ancona, where he was born, was besieged by the Italian army, and a bomb fell near him, actually destroying his cradle. He started studying mathematics when he was 11, received his doctor of physics degree from the University of Pisa when he was 22, and became full professor of Mechanics in the University of Pisa when he was only 23 years old. Besides the theory of functionals, he has made many other contributions to science, including in integral and integrodifferential equations, hydrodynamics, elasticity, submarine and aerial warfare during World War I, and in biological sciences. His works on the theory of functionals are available in English as a book [326]. A brief biography of Vito Volterra as published by Whittaker [328] was reproduced with a complete list of his publications (270 in all) in [326].

Fréchet, in a 1910 publication [72], showed that the set of Volterra functionals is complete. Fréchet's theorem implies that every continuous functional of a signal $x(t)$ can be approximated with arbitrary precision as a sum of a finite number of Volterra functionals in $x(t)$. This result was a generalization of the Weierstrass theorem, which states that every continuous function of a variable x can be approximated with arbitrary precision as a sum of a finite number of polynomials in x.

1.4.2   Wiener and His Influence

The first use of Volterra's theory in nonlinear system theory occurs in the work of Norbert Wiener in the early 1940s. Wiener, who started studying nonlinear systems in the 1930s, is one of the most famous prodigies of the early twentieth century. He finished high school at the age of 11, entered graduate school at the age of 14 and received his doctorate degree in the philosophy of mathematics from Harvard University when he was 18 years old. He is perhaps best known for developing the Wiener filter. He is also known for coining and promoting the term *cybernetics*. Some of his greatest mathematical achievements were in generalized harmonic analysis, in which he extended the Fourier transform to functions of finite power. Wiener's method of analyzing nonlinear systems was to determine the coefficients of the Wiener model, which employs the so-called *G*-functionals. The *G*-functionals are mutually orthogonal when the input signal to the system is white and Gaussian. The basic ideas related to Wiener's techniques are described in Chapter 5. Much of Wiener's work in this area is available in the form of transcribed lecture notes [332].

Wiener's work resulted in a flurry of activities in the late 1950s and the early 1960s at the Massachusetts Institute of Technology (MIT) under the direction of Y. W. Lee. Most of the results obtained at the MIT are contained in *Research Laboratory for Electronics Quarterly Progress Reports* during that time. Some of the most referred to reports among these publications are those by Brilliant [28] and George [78]. Among the results obtained during the 1960s and the early 1970s are the method developed by Lee and Schetzen for measuring the Wiener kernels using cross-correlation functions [133], extension of these results to colored Gaussian signals [273], development of orthogonal functionals for non-Gaussian input signals [205,277], and extensions of Wiener theory to systems with multiple inputs [50].

Easily readable accounts of the fundamentals of Volterra system theory and the developments that occurred till the late seventies are the books by Schetzen [275], Rugh [264], and Marmarelis and Marmarelis [161]. The latter book deals primarily with the applications of Volterra and Wiener theory in the study of biological systems.

1.4.3 Early Work on Discrete-Time Systems

It appears that the first published study of Volterra series expansions for discrete systems was by Alper [3]. Two other early contributions to the theory of discrete-time Volterra systems were due to Bush [37] and Barker and Ambati [12].

Even though there was considerable interest in the analysis and applications of nonlinear systems, the activities soon faded, primarily because of the recognition that system realizations using Volterra series models were computationally complex, and the computers in the 1960s and 1970s were not powerful enough to implement such systems. Analysis and applications of polynomial systems saw a revival in the 1980s with the arrival of fast computers and the increasing ability to digitally implement nonlinear systems.

1.4.4 Adaptive Polynomial Filters

Billings has done considerable work on recursive identification of nonlinear systems. The survey paper [22] describes the work done prior to 1980 in this area. Thomas described an adaptive echo canceller using Volterra system representation in 1971 [312]. The first work that employs a least-mean-square (LMS) type adaptive algorithm for Volterra filtering appears to be that of Coker and Simkins [58]. They applied their adaptive filter for noise cancellation in a biological signal processing problem. Related developments in the early 1980s include those reported by Sicuranza et al. [283], Casar-Corredera et al. [45], and Koh and Powers [117]. The latter paper [117] is perhaps the most referenced work in this area. Koh and Powers [115] were also the first to publish lattice algorithms for adaptive Volterra filtering. Mathews and Lee [164] derived a computationally efficient algorithm for recursive least-squares (RLS) adaptive Volterra filtering. RLS algorithms have the advantage of fast convergence over LMS algorithms, but even the most efficient RLS adaptive Volterra filter available today is considerably more complex than the LMS algorithms. Recursive least-squares adaptive lattice Volterra filters were first derived by Zarzycki [339].

1.4.5   Frequency-Domain Methods

One of the earliest uses of multivariable Laplace transform theory to analyze nonlinear systems was by George [78]. The works of Brilliant [28] and Barrett [13] also deal with the use of multivariable transform theory to the analysis of nonlinear systems. Other useful sources that describe the frequency-domain techniques for analyzing continuous-time nonlinear systems include [52,53,130,219]. Similar analysis of discrete-time nonlinear systems using multidimensional z-transform has also been described [12,98,131]. Tick [313] appears to be the first to use frequency-domain ideas to identify a quadratic system. Brillinger [30] used higher-order spectral analysis to identify polynomial systems. Efficient frequency-domain algorithms to identify nonlinear systems, including techniques employing the fast Fourier transform, were developed by French and Butz [73] and French [74]. Adaptive Volterra filters using frequency-domain techniques were first developed by Mansour and Gray [152].

1.4.6   Recursive Polynomial Systems

George's work [78] involved the use of multidimensional Laplace transform for the study of nonlinear feedback systems. Important early contributions to the analysis of recursive nonlinear systems include the results presented in Zames [336,337] and Sandberg [268,269]. Bilinear system models have been studied in great detail. Fundamental results related to bilinear system models can be found in several other sources [32,86,185,302,303].

1.4.7   Some Early Applications

Several applications of polynomial signal processing are described in Chapter 10. The objective of this section is to cite some early work involving practical applications of polynomial signal processing. It appears that the first one to use Volterra theory to the analysis of nonlinear devices was Wiener [331]. In this paper,

Wiener analyzed the response of a series *RLC* circuit with a nonlinear resistor to a white Gaussian input signal.

Several researchers have studied biological systems using Volterra models. A comprehensive description of the developments in the analysis of physiological systems using nonlinear system models can be found in the book by Marmarelis and Marmarelis [161]. This book also contains a large number of references to related work. Identification of biological systems using cascade nonlinear systems have been studied by Korenberg and his colleagues [94,118,119,121]. Other examples of early application of polynomial signal processing in biological sciences include characterization of human pupils [267], analysis of neuron behavior [160], identification of biological control systems [327], and immunology [187].

Some of the earliest applications of Volterra series modeling involved the characterization of nonlinear properties of semiconductor devices and circuits [100,179,193]. The need for compensating for the nonlinearities in communication systems was recognized by researchers very early. Lucky conjectured in 1975 that the error probability performance of data transmission systems operating at rates higher than 4800 bits per second (bits/s) is almost entirely due to nonlinear distortion [150]. Thomas [312] employed the Volterra series representation to develop an echo canceller in 1971. Digital satellite channels were modeled using the Volterra series approach described by Benedetto et al. [15]. Goldman [84] modeled crosstalk interference in communication systems using Volterra system models.

Siegel, Imamura and Meecham used Wiener's approach to model turbulence [292]. The motion of ships in random seas was studied using Volterra series analysis by Vassilopoulos [321]. Another early reference in this area is the paper by Burcher et al. [35]. Polynomial models have also been used in hydrology [4,338], population studies [31,314], nuclear fission [184], and several other widely different applications.

2

VOLTERRA SERIES EXPANSIONS

Volterra series expansions form the basis of the theory of polynomial nonlinear systems. We introduce the Volterra series expansions and discuss their properties in this chapter. Both continuous-time and discrete-time systems are considered here. In order to obtain a good understanding of the characteristics of such systems, we study time-domain and frequency-domain representations of polynomial systems. Several simplified structures for facilitating our analysis are also introduced in this chapter.

2.1 CONTINUOUS-TIME SYSTEMS

A *system* is defined mathematically as a rule for transformation of an input signal x into another signal y by means of an operator S so that

$$y = S[x]. \tag{2.1}$$

The input and output signals are usually functions of one or more independent variables such as position or time. If they are functions only of time t, and the variable t is defined over a continuous range of values, (2.1) defines a continuous one-dimensional system whose input–output relationship can be represented as

$$y(t) = S[x(t)]. \tag{2.2}$$

Figure 2.1 displays a block diagram for an arbitrary, one-dimensional system. The terms system, filter, and algorithm are used interchangeably in this book.

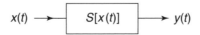

Figure 2.1 An arbitrary, continuous one-dimensional system.

2.1.1   Linear Shift-Invariant Systems

A *shift-invariant* system is characterized by the invariance of its output with respect to a shift in the independent variable. For example, a *time*-invariant (TI) system satisfies the relationship

$$y(t + \tau) = S[x(t + \tau)] \tag{2.3}$$

for all values of τ. A *linear* system is one that satisfies the superposition principle. This implies that

$$S[\alpha x_1(t) + \beta x_2(t)] = \alpha S[x_1(t)] + \beta S[x_2(t)] \tag{2.4}$$

for all arbitrary constants α and β and arbitrary inputs $x_1(t)$ and $x_2(t)$.

It is well-known that the output of a continuous-time, linear, and time-invariant (LTI) system is related to the input signal through the convolution integral

$$y(t) = \int_{-\infty}^{\infty} h(\tau)x(t - \tau)d\tau, \tag{2.5}$$

where $h(t)$ is the response of the system to a unit impulse signal. Equation (2.5) implies that the unit impulse response signal completely characterizes an LTI system. A system is said to be *causal* if its output at any given time does not depend on the future values of its input. An LTI system is causal if and only if

$$h(t) = 0 \quad \text{for} \quad t < 0. \tag{2.6}$$

The unit impulse response signal represents the *memory* of the LTI system since the contribution to the current value of the output signal from the value of the input signal T seconds prior to the present time is determined by $h(T)$. An LTI system without memory is characterized by an impulse response that is also an impulse at time $t = 0$, i.e., $h(t) = cu_0(t)$, where $u_0(t)$ represents the unit impulse signal and c is a constant that determines the extent of amplification or attenuation of the input signal at the output of the system. The output of a memoryless LTI system is given by

$$y(t) = cx(t), \tag{2.7}$$

which is obtained by convolving the input $x(t)$ with $cu_0(t)$. The preceding description is intended only to provide the necessary definitions to facilitate our study of nonlinear systems. A large number of books are available on linear systems (see, e.g., Oppenheim and Schafer [208]), and the reader may refer to any one of them for more details.

2.1.2 Volterra Series Expansion for Nonlinear Systems

A nonlinear system without memory can be often described by means of an appropriate series expansion. Within some restrictions, a power series expansion such as the Taylor series expansion may be used to describe the output of such systems as

$$y(t) = \sum_{p=0}^{\infty} c_p x^p(t). \tag{2.8}$$

A nonlinear system with memory can be represented by means of an extension of this expression. Such an extension, known as the *Volterra series expansion* [264,275,324–326], relates the input and output signals of the system as

$$
y(t) = h_0 + \int_{-\infty}^{\infty} h_1(\tau_1)x(t - \tau_1)d\tau_1
$$

$$
+ \int_{-\infty}^{\infty} \int_{-\infty}^{\infty} h_2(\tau_1, \tau_2)x(t - \tau_1)x(t - \tau_2)d\tau_1 d\tau_2 + \cdots
$$

$$
+ \int_{-\infty}^{\infty} \cdots \int_{-\infty}^{\infty} h_p(\tau_1, \ldots, \tau_p)x(t - \tau_1) \cdots x(t - \tau_p)d\tau_1 \cdots d\tau_p + \cdots. \tag{2.9}
$$

The nonlinear system represented by a Volterra series expansion is completely characterized by the multidimensional functions $h_p(t_1, \ldots, t_p)$, called the *Volterra kernels*. The zeroth-order kernel h_0 is a constant. The higher-order kernels can be assumed, without loss in generality, as symmetric functions of their arguments so that any of the $p!$ possible permutations of $t_1, \ldots, t_p$ leaves $h_p(t_1, \ldots, t_p)$ unchanged. This symmetry is a direct result of the invariance of the products of the delayed input functions $\{x(t - \tau_i), i = 1, \ldots, p\}$ with respect to their permutations. The system is *causal* if and only if

$$h_p(t_1, \ldots, t_p) = 0 \quad \text{for any} \quad t_i < 0 \quad \text{and} \quad i = 1, \ldots, p. \tag{2.10}$$

The lower limits in the integrals in (2.9) are therefore set to zero for causal nonlinear systems. The upper limits of the integrals in (2.9) given as ∞ indicate that the system may have infinite memory. If the upper limits are all finite, the system possesses finite memory. Note that each integral in (2.9) has the form of a multidimensional convolution. This property will be discussed in detail later in this chapter. By defining the pth order Volterra operator $\overline{h}_p[x(t)]$ as

$$\overline{h}_p[x(t)] = \int_{-\infty}^{\infty} \cdots \int_{-\infty}^{\infty} h_p(\tau_1, \ldots, \tau_p)x(t - \tau_1) \cdots x(t - \tau_p)d\tau_1 \cdots d\tau_p, \tag{2.11}$$

(2.9) may be written more compactly as

$$y(t) = h_0 + \sum_{p=1}^{\infty} \bar{h}_p[x(t)]. \tag{2.12}$$

A truncated Volterra series expansion is obtained by setting the upper limit of the summation in (2.12) to a finite integer value P. The parameter P is called the *order,* or the *degree,* of the Volterra series expansion. It is known [72] that any time-invariant, finite-memory system that is a continuous functional of its input can be uniformly approximated over a uniformly bounded and continuous set of input signals by a Volterra series expansion of appropriate finite order P. Moreover, (2.12) reveals the similarity of Volterra series expansions with the Taylor series expansions. When the input signal is multiplied by a constant factor c, the output $y(t)$ becomes

$$y(t) = h_0 + \sum_{p=1}^{\infty} \bar{h}_p[cx(t)] = h_0 + \sum_{p=1}^{\infty} c^p \bar{h}_p[x(t)], \tag{2.13}$$

which is a power series expansion in c.

2.1.3 Limitations of Volterra Series Expansions

As a consequence of its power series characteristic, there are some limitations associated with the application of the Volterra series expansion to nonlinear system modeling. We briefly discuss a few such limitations below.

2.1.3.1 Convergence Issues Because of its close relationship with Taylor series expansions, Volterra series expansions exhibit convergence problems when the nonlinear systems to be modeled include strong nonlinearities such as saturating elements. An example is shown in Figure 2.2, which shows three types of saturation nonlinearities for memoryless systems. A convergent Taylor series expansion exists only for the nonlinearity depicted in Figure 2.2b since the function in Figure 2.2a has no derivatives at $\pm\tau$ and the function in Figure 2.2c has a discontinuity at the origin. It is straightforward to infer that no convergent Volterra series expansions exists for systems that possess similar types of nonlinearity. Therefore, the Volterra series approach can be applied with good results only to systems with mild nonlinearities. Convergence of the Volterra series for both deterministic and stochastic inputs has been studied in the literature [80,130,213,270].

2.1.3.2 The Impulse Response Function Does Not Completely Characterize a Nonlinear System In contrast to the linear time-invariant systems, the impulse response function does not fully characterize a nonlinear system. To see this, let $x(t) = cu_0(t)$. The response of the nonlinear system to this

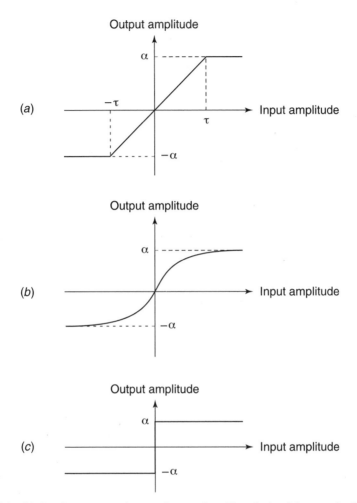

Figure 2.2 Of the three types of saturation nonlinearities depicted here, only *(b)* has a convergent Taylor series expansion. (Copyright © 1991 IEEE.)

impulse is given by

$$h(t) = h_0 + ch_1(t) + c^2h_2(t, t) + \cdots + c^P h_p(t, \ldots, t) + \cdots. \qquad (2.14)$$

This indicates that the impulse response is determined only by the diagonal values of the kernels h_p, that is for $t = t_1 = t_2 = \cdots = t_p$, and therefore it does not completely specify the pth-order Volterra kernels of order higher than 1.

2.1.3.3 Multiple-Valued Nonlinearities

Since the Volterra series expansion is a single-valued representation, it cannot be used to represent multiple-valued nonlinearities. Therefore, important physical phenomena such as hysteresis cannot

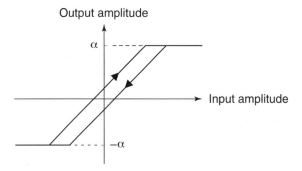

Figure 2.3   An example of a double-valued nonlinearity.

be studied using Volterra series expansions. An example of the input–output relation of a system exhibiting hysteresis is shown in Figure 2.3.

2.1.3.4   Measurement of Volterra Kernels   The measurement of the Volterra kernels is possible only if the contribution of each Volterra operator can be separated from the total system response. In general, if the order of the system is not finite, such contributions cannot be separated from the total system response. Wiener overcame this problem in his classical theory of nonlinear systems by using the so-called *G*-functionals, which are orthogonal when the input is a white Gaussian process [275,332]. *G*-functionals are described in Chapter 5.

2.2   DISCRETE-TIME SYSTEMS

Consider a system for which the input and output values are known only at given time instants $t = nT$, where n is an integer variable. By assuming a normalized time interval $T = 1$, (2.2) becomes

$$y(n) = S[x(n)]. \tag{2.15}$$

This equation describes a general one-dimensional discrete-time system as depicted in Figure 2.4. Many classifications of discrete-time systems can be made in a manner similar to those of continuous-time systems. The following paragraphs briefly describe some of the most important classifications.

Shift-Invariant Systems.   A discrete, shift-invariant (DSI) system is characterized by the property that it is invariant to shifts in the independent variable:

$$y(n + m) = S[x(n + m)], \tag{2.16}$$

where m is an integer shift. A DSI system is said to be *discrete, time-invariant* (DTI) when the independent variable is time.

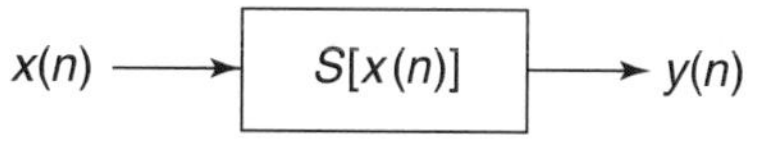

$$x(n) \longrightarrow \boxed{S[x(n)]} \longrightarrow y(n)$$

Figure 2.4 A discrete-time, one-dimensional system.

Linear Systems. A discrete-time system is said to be *linear* if it obeys the superposition principle:

$$S[\alpha x_1(n) + \beta x_2(n)] = \alpha S[x_1(n)] + \beta S[x_2(n)], \tag{2.17}$$

for all arbitrary constants α and β and arbitrary input signals $x_1(n)$ and $x_2(n)$. If a discrete-time system is both linear and time-invariant, its output can be evaluated by convolving its input signal with its unit impulse response function $h(n)$ as

$$y(n) = \sum_{m=-\infty}^{\infty} h(m)x(n-m). \tag{2.18}$$

A discrete-time LTI system is *causal* if and only if

$$h(n) = 0 \quad \text{for any} \quad n < 0. \tag{2.19}$$

2.2.1 DTI Nonlinear Systems

In a manner similar to the continuous case, it is now possible to describe DTI nonlinear systems with memory by means of the discrete-time Volterra series expansion

$$y(n) = h_0 + \sum_{p=1}^{\infty} \overline{h}_p[x(n)], \tag{2.20}$$

where $y(n)$ and $x(n)$ are the output and input signals, respectively, and

$$\overline{h}_p[x(n)] = \sum_{m_1=-\infty}^{\infty} \cdots \sum_{m_p=-\infty}^{\infty} h_p(m_1, \ldots, m_p)x(n-m_1)\cdots x(n-m_p), \tag{2.21}$$

where $h_p(m_1, \ldots, m_p)$ is the pth-order Volterra kernel of the system. If

$$h_p(m_1, \ldots, m_p) = 0 \quad \text{for all} \quad m_i < 0, \quad \text{and} \quad i = 1, \ldots, p, \tag{2.22}$$

the DTI nonlinear system is causal, and (2.21) becomes

$$\overline{h}_p[x(n)] = \sum_{m_1=0}^{\infty} \cdots \sum_{m_p=0}^{\infty} h_p(m_1, \ldots, m_p)x(n-m_1)\cdots x(n-m_p). \tag{2.23}$$

We can interpret the discrete-time Volterra kernels in a manner similar to our interpretation of the continuous-time systems. The constant h_0 is an offset term, $h_1(m_1)$ is the impulse response of a discrete-time LTI filter, and the pth-order kernel $h_p(m_1, \ldots, m_p)$ can be considered as a generalized pth-order impulse response characterizing the nonlinear behavior of the system. The upper limit in the summations in (2.23) given as infinity indicates that the discrete system may have infinite memory. The difficulties that arise because of the infinite summations in (2.23) may be avoided by using recursive polynomial system models. This approach is similar to the use of IIR filters in applications involving linear system models. In recursive polynomial system models, the relationship between the input and output signals is described using a nonlinear difference equation of finite order involving delayed values of the output signal as well as the current and delayed values of the input signal as

$$y(n) = f_i\{x(n), x(n-1), \ldots, x(n-N), y(n-1), y(n-2), \ldots, y(n-M)\}, \quad (2.24)$$

where $f_i\{\cdot\}$ is an ith-order polynomial in the variables within the parentheses. Such systems are described in Chapter 8. Very often, however, the memory required to adequately approximate a nonlinear system is finite. In such situations, a Volterra series expansion involving only the input signal is sufficient to model the system. This simpler but still useful class of nonlinear system models is derived by limiting all the summations in (2.23) to finite values, i.e., by defining the discrete independent variables $m_1, \ldots, m_p$ on a finite domain. In this, case, $h_1(m_1)$ represents the impulse response of an FIR filter, and the effect of the nonlinearity on the output depends only on the present and past values of the input signals defined on the extent of the filter support. If the discrete Volterra series expansion is truncated by limiting the summation in (2.20) to a finite value P, a finite-memory, finite-order expansion is obtained as

$$y(n) = h_0 + \sum_{p=1}^{P} \bar{h}_p[x(n)], \quad (2.25)$$

where

$$\bar{h}_p[x(n)] = \sum_{m_1=0}^{N-1} \cdots \sum_{m_p=0}^{N-1} h_p(m_1, \ldots, m_p)x(n-m_1)\cdots x(n-m_p). \quad (2.26)$$

The upper limits in all the summations of (2.26) are made identical only for convenience. They may be set to arbitrary values to obtain a more general expression.

The nonrecursive models described by truncated Volterra systems have been studied extensively because of the relative simplicity of the input–output relationship of such models. The simplest polynomial filter of this class is the quadratic filter obtained by choosing $P = 2$ in (2.25). The use of a quadratic term, alone or in addition to a linear filter, often offers very interesting effects.

2.2.1.1 Homogeneous Systems   The system represented by (2.21) is called a *homogeneous* system of order p, since the output corresponding to an input $cx(n)$, where c is a scalar, is given by $c^p y(n)$, where $y(n)$ is the response to $x(n)$. In traditional system theory, a system S is said to be homogeneous if

$$S[cx(n)] = cS[x(n)], \tag{2.27}$$

where c is a constant. In the case of the Volterra system of (2.21), scaling the input signal by c results in $c^p y_p(n)$. Consequently, the definition of homogeneous Volterra systems is an extension of the traditional definition.

2.2.1.2 Examples of Computation of Volterra Kernels   The computation of the Volterra kernels of arbitrary nonlinear systems is often a difficult problem. However, there are a few situations in which it is possible to analytically derive the Volterra kernels. In the following, we consider some practical situations in which such a derivation is possible. These situations involve interconnections of LTI systems and memoryless nonlinear systems.

Example 2.1   Consider the cascade connection of an FIR system followed by a static, cubic nonlinearity as shown in Figure 2.5. The input–output relationship of this system is given by

$$y(n) = \left[\sum_{m=0}^{N-1} h(m)x(n-m) \right]^3.$$

The elements of the third-order kernel $h_3(n_1, n_2, n_3)$ of the Volterra series expansion can be easily derived from the preceding equation by means of algebraic manipulations. Since

$$y(n) = \sum_{m_1=0}^{N-1} \sum_{m_2=0}^{N-1} \sum_{m_3=0}^{N-1} h(m_1)h(m_2)h(m_3)x(n-m_1)x(n-m_2)x(n-m_3),$$

it immediately follows that

$$h_3(m_1, m_2, m_3) = h(m_1)h(m_2)h(m_3); \quad 0 \le m_1, m_2, m_3 \le N-1,$$

Figure 2.5   The cascade nonlinear system of Example 2.1.

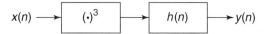

Figure 2.6 The cascade nonlinear system of Example 2.2.

and that all other Volterra kernels are zero. A cascade, nonlinear model such as this, in which a static nonlinearity follows a linear system, is also known as the *Wiener model*.

Example 2.2 Consider the cascade connection of Figure 2.6, in which a static, cubic nonlinearity is followed by an FIR system. The input–output relationship of this system is given by

$$y(n) = \sum_{m=0}^{N-1} h(m)x^3(n-m).$$

The elements of the third-order kernel $h_3(m_1, m_2, m_3)$ of the Volterra series expansion for this system can be easily derived as

$$h_3(m_1, m_2, m_3) = \begin{cases} 0; & m_1 \neq m_2 \neq m_3 \\ h(m); & m_1 = m_2 = m_3 = m, 0 \leq m \leq N-1. \end{cases}$$

A cascade, nonlinear model such as this, in which a linear system follows a static nonlinearity is also known as the *Hammerstein model*.

Example 2.3 Consider a multiplicative interconnection of three linear, time-invariant systems shown in Figure 2.7. Let the impulse responses of the three systems be $h_a(n)$, $h_b(n)$ and $h_c(n)$, respectively. Then, the resultant nonlinear system has Volterra kernels given by

$$h_3(m_1, m_2, m_3) = h_a(m_1)h_b(m_2)h_c(m_3)$$

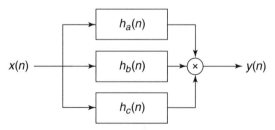

Figure 2.7 A multiplicative interconnection of LTI systems described by a third-order Volterra kernel.

since

$$y(n) = \left[\sum_{m_1=-\infty}^{\infty} h_a(m_1)x(n-m_1) \right] \left[\sum_{m_2=-\infty}^{\infty} h_b(m_2)x(n-m_2) \right]$$

$$\times \left[\sum_{m_3=-\infty}^{\infty} h_c(m_3)x(n-m_3) \right]$$

$$= \sum_{m_1=-\infty}^{\infty} \sum_{m_2=-\infty}^{\infty} \sum_{m_3=-\infty}^{\infty} h_a(m_1)h_b(m_2)h_c(m_3)x(n-m_1)x(n-m_2)x(n-m_3).$$

Volterra kernels of all orders other than three are zero in this example.

2.3   PROPERTIES OF VOLTERRA SERIES EXPANSIONS

Volterra series expansions possess some very interesting properties that are the main reasons for the popularity of Volterra series modeling in the analysis of nonlinear systems. For the sake of simplicity, we consider the properties of only discrete-time Volterra series expansions in this section. The continuous-time Volterra series expansions exhibit similar properties.

2.3.1   Linearity with Respect to the Kernel Coefficients

The linearity of the Volterra series expansions with respect to the kernel coefficients is evident from (2.18) and (2.19). In other words, the nonlinearity of the expansions is completely due to the multiple products of the delayed input values, while the filter coefficients appear linearly in the output expression. Because of this property, many classes of polynomial filters can be defined as conceptually straightforward extensions of linear system models. Recall that the output of a linear FIR filter with N-sample memory is computed as a linear combination of a set of N input samples. The output of a finite-support, or finite-memory, Pth-order truncated Volterra filter with the same memory span is a linear combination of all the possible products of up to P samples belonging to the same N input samples. For example, when $P = 2$, the output signal is a linear combination of a bias term, the input samples, and products terms involving two input samples. This system corresponds to a quadratic filter. Similarly, for a cubic or third-order, truncated Volterra filter, $P = 3$, and the output can be expressed as a linear combination of the bias term, samples of the input signals, products of two input samples and products involving three samples of the input signals. This description can be extended to any order of nonlinearity.

Figure 2.8 shows an FIR filter defined on a support formed by three consecutive input samples, while Figure 2.9 shows a quadratic filter operating on the same set of input samples. These figures illustrate the linearity of the truncated Volterra systems with respect to its coefficients. The increase in complexity of the polynomial system is evident even in this simple example.

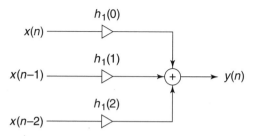

Figure 2.8 An FIR filter characterized by the kernel $h_1(m_1)$ for $m_1 = 0, 1, 2$.

The preceding examples correspond to nonrecursive system models. We can extend the notion of the recursive or IIR linear systems by constructing nonlinear system models that incorporate output signal feedback. In such cases the polynomial filters may have infinite memory. Figure 2.10 illustrates a simple recursive nonlinear filter whose input–output relationship is given by

$$y(n) = ax(n)x(n-1) + bx(n)y(n-1). \tag{2.28}$$

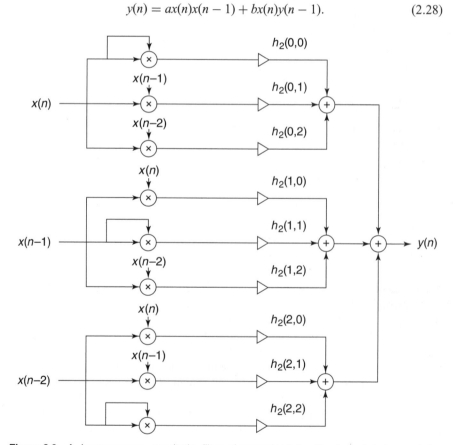

Figure 2.9 A homogeneous quadratic filter characterized by the kernel $h_2(m_1, m_2)$ for $m_1, m_2 = 0, 1, 2$.

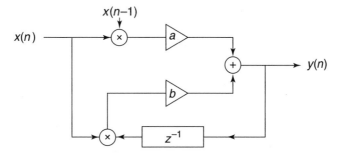

Figure 2.10   A simple recursive nonlinear filter.

Note that the system output is linear in the coefficients of this model also. The linearity of the output with respect to the coefficients of the system can be gainfully employed to extend many concepts in linear system theory to nonlinear systems. Examples of such extensions include frequency-domain representation of polynomial systems, optimum polynomial filter theory, and the derivation of adaptive polynomial filters. These concepts are considered in detail in later sections and chapters.

2.3.2 Multidimensional Convolution Property

Consider the pth-order term of the discrete-time Volterra series expansion given by

$$\bar{h}_p[x(n)] = \sum_{m_1=-\infty}^{\infty} \cdots \sum_{m_p=-\infty}^{\infty} h_p(m_1, \ldots, m_p)x(n-m_1)\cdots x(n-m_p), \qquad (2.29)$$

and define a p-dimensional signal

$$v(n_1, n_2, \ldots, n_p) = x(n_1)x(n_2)\cdots x(n_p). \qquad (2.30)$$

Now consider the p-dimensional convolution

$$w(n_1, \ldots, n_p) = \sum_{m_1=-\infty}^{\infty} \cdots \sum_{m_p=-\infty}^{\infty} h_p(m_1, \ldots, m_p)v(n_1-m_1, \ldots, n_p-m_p). \quad (2.31)$$

Comparing (2.31) with (2.29), we see that the pth-order terms of a Volterra filter can be evaluated by performing a p-dimensional convolution along the diagonal line defined by $n_1 = n_2 = \cdots n_p = n$. Even though such a realization is not very efficient, the characterization of Volterra systems using multidimensional convolutions is useful in understanding their properties. In particular, this characterization leads to the frequency-domain analysis presented in Section 2.5.

2.3.2.1 *Relationship of Volterra Kernels to Multidimensional Linear Systems*   The proof of the multidimensional convolution property is based on the interpretation of a pth-order Volterra kernel as the impulse response of a p-

dimensional linear system whose p-dimensional input signal is constrained to be the product of p identical one-dimensional signals. In other words, the nonlinearity in a one-dimensional Volterra system is mapped into a p-dimensional linear system using a constraint on the input signal. This correspondence between one-dimensional Volterra kernels and p-dimensional linear systems can be used to design Volterra filters using techniques employed for designing multidimensional linear filters.

2.3.2.2 Separable Kernels   The concept of *separability* plays an important role in the theory of multidimensional linear systems since it simplifies many problems associated with the design, realization, and stability of such systems. Because of the correspondence between one-dimensional Volterra kernels and p-dimensional linear systems, the concept of separability can be extended to the Volterra kernels also. We say that the pth-order Volterra kernel is *completely separable* if it can be expressed as a product of p first-order kernels. For discrete-time systems, this condition implies that the pth-order kernel can be factored as

$$h_p(m_1, m_2, \ldots, m_p) = \prod_{i=1}^{p} h_{1,i}(m_i). \tag{2.32}$$

The kernels given in Examples 2.1 and 2.3 are separable. While a separable representation may not necessarily be the most suitable one, it may provide for efficient analysis and realization of nonlinear systems.

2.3.3 Stability

We say that a system is stable in the *bounded-input bounded-output* (BIBO) sense if and only if every bounded-input signal results in a bounded-output signal. The BIBO stability criterion can be applied to higher-order Volterra operators. It is well-known that a sufficient and necessary condition for a linear and time-invariant system to be BIBO stable is

$$\sum_{-\infty}^{\infty} |h_1(m_1)| < \infty. \tag{2.33}$$

For a p-dimensional system, the corresponding condition is

$$\sum_{m_1=-\infty}^{\infty} \cdots \sum_{m_p=-\infty}^{\infty} |h_p(m_1, \ldots, m_p)| < \infty. \tag{2.34}$$

This condition can be shown to be a sufficient condition for the BIBO stability of homogeneous pth-order Volterra systems [275]. To prove the sufficiency of this condition for a Volterra system, consider a bounded input signal $x(n)$ such that $|x(n)| < M$. Then,

$$|y(n)| = |\bar{h}_p[x(n)]| = |\sum_{m_1=-\infty}^{\infty} \cdots \sum_{m_p=-\infty}^{\infty} h_p(m_1, \ldots, m_p) x(n - m_1) \cdots x(n - m_p)|$$

$$\leq \sum_{m_1=-\infty}^{\infty} \cdots \sum_{m_p=-\infty}^{\infty} |h_p(m_1, \ldots, m_p)| |x(n - m_1)| \cdots |x(n - m_p)|$$

$$\leq M^p \sum_{m_1=-\infty}^{\infty} \cdots \sum_{m_p=-\infty}^{\infty} |h_p(m_1, \ldots, m_p)| < \infty, \tag{2.35}$$

thus proving that the output is bounded whenever the input is bounded.

The condition of (2.34) is not necessary for the stability of arbitrary Volterra systems. Exercise 2.12 describes a system that is stable in the BIBO sense, but does not satisfy (2.34). However, there are subclasses of the Volterra system models for which condition (2.34) is also necessary. In particular, Volterra systems with separable kernels are BIBO stable if and only if (2.34) is satisfied. The proof is left as an exercise for the reader.

2.3.4 Symmetry of the Kernels and Equivalent Representations

The pth-order term of a generic Volterra series expansion, defined in (2.26), has N^p coefficients. This representation assumes that each permutation of the indices $m_1, m_2, \ldots, m_p$ results in a separate coefficient. However, since all such permutations multiply the same quantity, namely, $x(n - m_1), \ldots, x(n - m_p)$, only a fewer number of these coefficients can be uniquely determined. Consequently, a generic kernel $h_p(m_1, \ldots, m_p)$ can be replaced by a *symmetric* kernel $h_{p,sym}(m_1, \ldots, m_p)$ by defining its elements as

$$h_{p,sym}(m_1, \ldots, m_p) = \frac{1}{|\pi(m_1, m_2, \ldots, m_p)|} \sum_{\pi(\cdot)} h_p(m_{\pi(1)}, \ldots, m_{\pi(p)}), \tag{2.36}$$

where the summation is over all distinct permutations $\pi(\cdot)$ of the indices $m_1, m_2, \ldots, m_p$ and $|\pi(m_1, m_2, \ldots, m_p)|$ represents the number of such permutations. In order to evaluate $|\pi(m_1, m_2, \ldots, m_p)|$, let us denote the number of distinct values in a specific set of $(m_1, m_2, \ldots, m_p)$ as r. Let $k_1, k_2, \ldots, k_r$ denote the number of times these values appear in $(m_1, m_2, \ldots, m_p)$. Then

$$|\pi(m_1, m_2, \ldots, m_p)| = \frac{p!}{k_1! k_2! \cdots k_r!}. \tag{2.37}$$

For example, the number of distinct permutations of the indices $(1,3,3,4,4,4)$ is given by $6!/1!2!3! = 60$. Because of the redundancy of the coefficients in the generic

Volterra series expansions due to the symmetry, (2.26) can also be recast in the *triangular* form [56,220] as

$$\bar{h}_p[x(n)] = \sum_{m_1=0}^{N-1} \sum_{m_2=m_1}^{N-1} \cdots \sum_{m_p=m_{p-1}}^{N-1} h_{p,tri}(m_1, m_2, \ldots, m_p)$$

$$\times x(n - m_1)x(n - m_2)\cdots x(n - m_p). \tag{2.38}$$

The elements of the triangular kernel $h_{p,tri}(m_1, m_2, \ldots, m_p)$ for $m_1 \leq m_2 \leq \cdots \leq m_p$ can be computed as the sum of the generic kernel over all distinct permutations of the indices $(m_1, m_2, \ldots, m_p)$. Obviously, these elements can also be computed as the sum of the corresponding terms of the symmetric kernel:

$$h_{p,tri}(m_1, m_2, \ldots, m_p)$$
$$= \begin{cases} |\pi(m_1, m_2, \ldots, m_p)|h_{p,sym}(m_1, m_2, \ldots, m_p); & m_1 \leq m_2, \leq \cdots \leq m_p \\ 0; & \text{otherwise.} \end{cases} \tag{2.39}$$

Another way to make use of the kernel symmetry is to use the *regular* kernel representation $h_{p,reg}(k_1, \ldots, k_p)$ [56,220], which is derived from the triangular representation by means of a change of variable. By substituting $k_j = m_j - m_{j-1}$ for $j = 2, 3, \ldots, p$ and $k_1 = m_1$ in (2.37), we get

$$\bar{h}_p[x(n)] = \sum_{k_1=0}^{N-1} \sum_{k_2=0}^{N-1-k_1} \sum_{k_3=0}^{N-1-k_1-k_2} \cdots \sum_{k_p=0}^{N-1-k_1-\cdots-k_{p-1}} h_{p,reg}(k_1, k_2, \ldots, k_p)$$

$$\times x(n - k_1)x(n - k_1 - k_2)\cdots x(n - k_1 - k_2 - \cdots - k_p), \tag{2.40}$$

where

$$h_{p,reg}(k_1, k_2, \ldots, k_p) = h_{p,tri}(k_1, k_1 + k_2, \ldots, k_1 + k_2 + \cdots + k_p). \tag{2.41}$$

Figure 2.11 shows the supports of the symmetric, triangular, and regular kernels of a homogeneous quadratic Volterra operator.

Example 2.4   Consider a homogeneous quadratic operator with four-sample memory. Let the generic second-order Volterra kernel for this system be represented by the matrix

$$\mathbf{H}_2 = \begin{bmatrix} 0.5 & 0.8 & 0.7 & 0.6 \\ 0.3 & 1.0 & 0.4 & 0.2 \\ 1.0 & 0.5 & 1.0 & 0.4 \\ 2.0 & 1.0 & 0.5 & 2.0 \end{bmatrix}.$$

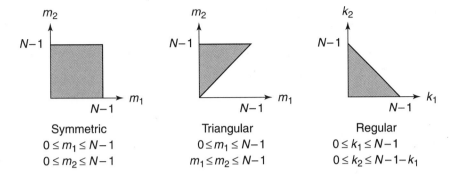

Figure 2.11 Supports of the symmetric, triangular, and regular kernels of a quadratic Volterra operator.

Then, the symmetric, triangular, and regular kernels of the system may be represented using the matrices $\mathbf{H}_{2,\mathrm{sym}}$, $\mathbf{H}_{2,\mathrm{tri}}$, and $\mathbf{H}_{2,\mathrm{reg}}$, respectively, as shown below:

$$
\mathbf{H}_{2,\mathrm{sym}} =
\begin{bmatrix}
0.50 & 0.55 & 0.85 & 1.30 \\
0.55 & 1.00 & 0.45 & 0.60 \\
0.85 & 0.45 & 1.00 & 0.45 \\
1.30 & 0.60 & 0.45 & 2.00
\end{bmatrix},
\qquad
\mathbf{H}_{2,\mathrm{tri}} =
\begin{bmatrix}
0.50 & 1.10 & 1.70 & 2.60 \\
0.00 & 1.00 & 0.90 & 1.20 \\
0.00 & 0.00 & 1.00 & 0.90 \\
0.00 & 0.00 & 0.00 & 2.00
\end{bmatrix},
$$

and

$$
\mathbf{H}_{2,\mathrm{reg}} =
\begin{bmatrix}
0.50 & 1.10 & 1.70 & 2.60 \\
1.00 & 0.90 & 1.20 & 0.00 \\
1.00 & 0.90 & 0.00 & 0.00 \\
2.00 & 0.00 & 0.00 & 0.00
\end{bmatrix}.
$$

Both the triangular and the regular representations reduce the computational complexity of the Volterra series expansions considerably. For example, by using the triangular kernel, the quadratic filter of Figure 2.9 will be simplified as shown in Figure 2.12.

2.3.5 Kernel Complexity

The huge number of the elements that are often present in the Volterra series expansion for many nonlinear systems is one of the drawbacks of such representations. The complexity of the model increases immensely with the length of the system memory and the order of the nonlinearity. In principle, the generic pth-order kernel of a Volterra series expansion with N-sample memory contains N^p coefficients. We have already seen that the number of independent coefficients can be greatly reduced because of the symmetry property. To find N_p, the number of

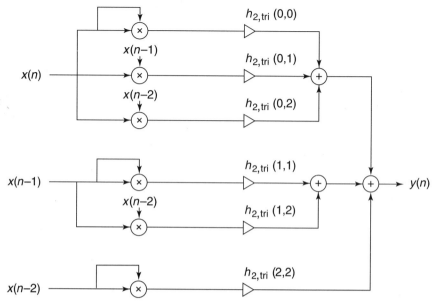

Figure 2.12 The quadratic filter of Figure 2.9 represented by the triangular kernel $h_{2,\text{tri}}(m_1, m_2)$ for $m_1, m_2 = 0, 1, 2$.

independent coefficients in the pth-order kernel of a Volterra series expansion with N-sample memory, let us consider a typical term of the form $h_p(m_1, \ldots, m_p)x(n - m_1)\ldots x(n - m_p)$. Let k_r denote the number of times $x(n - m_r)$ appears in this term. Note that

$$\sum_{r=1}^{N} k_r = p \qquad (2.42)$$

Then, the number of independent coefficients in the pth-order kernel is the number of distinct ways in which N nonnegative integers can add up to p. Many introductory-level textbooks on probability (for example, see Proposition 6.2 in Chapter 1 of Ross [261]) have solved this problem, and the solution is given by

$$N_p = \binom{N + p - 1}{p}. \qquad (2.43)$$

This result also gives the number of the nonzero elements in the triangular and regular representations of the Volterra kernels. The advantage of using the symmetry condition is measured by the ratio N^p / N_p. By taking the limit as N goes to ∞, we can show that this ratio tends to $1/p!$ asymptotically. Consequently, the reduction in the complexity of realizing higher-order Volterra filters is significant when the symmetry properties are utilized.

TABLE 2.1 Complexity of the Volterra Kernel as a Function of the Order of Nonlinearity for $N=8$

p	N^p	N_p
1	8	8
2	64	36
3	512	120
4	4,096	330
5	32,768	792
6	262,144	1,716
7	2,097,152	3,432
8	16,777,216	6,435

Example 2.5 Consider a quadratic kernel processing $N = 8$ input samples at each discrete time instant. The number of coefficients in the generic kernel is 64. When we apply the symmetry condition, the number of independent coefficients reduces to 36. When a homogeneous cubic kernel with only third-order terms is considered, the numbers of coefficients in the generic and symmetric kernels are 512 and 120, respectively. Table 2.1 tabulates N^p and N_p for several values of p for the case when $N = 8$. The advantage of using the symmetry conditions clearly increases with increases in the order of the nonlinearity.

2.3.6 Impulse Responses of Polynomial Filters

Unlike the case of linear, time-invariant systems, the unit impulse response signal is not sufficient to fully characterize most nonlinear systems. Let $h(n)$ represent the unit impulse response of the system. We can evaluate $h(n)$ by using (2.20) and (2.21) as

$$h(n) = h_0 + h_1(n) + h_2(n, n) + \cdots + h_p(n, \ldots, n) + \cdots. \qquad (2.44)$$

This result clearly shows that the impulse response function is determined only by the diagonal elements in the kernels, that is, by the samples of h_p at locations such that $n_1 = n_2 = \cdots = n_p$. Therefore, the impulse response alone is not sufficient to identify all the kernel elements. In order to identify all the kernel elements of a pth-order system, it is necessary in general to find its response to a set of p distinct impulse functions. We expand on this statement in the following example of second-order systems. Consider a homogeneous quadratic filter

$$y(n) = \overline{h}_2[x(n)]. \qquad (2.45)$$

Let the input to this system be decomposed into two arbitrary components as

$$x(n) = x_a(n) + x_b(n). \qquad (2.46)$$

The system output is now given by

$$y(n) = y_a(n) + y_b(n) + 2\overline{h}_{ab}[x_a(n), x_b(n)], \tag{2.47}$$

where $y_a(n)$ and $y_b(n)$ are the outputs of the filter when its input is $x_a(n)$ and $x_b(n)$, respectively, and

$$\overline{h}_{ab}[x_a(n), x_b(n)] = \sum_{m_1} \sum_{m_2} h_2(m_1, m_2) x_a(n - m_1) x_b(n - m_2). \tag{2.48}$$

To explore the response of this system to multiple impulses present at the input, let

$$x_a(n) = u_0(n - k_1) \tag{2.49}$$

and

$$x_b(n) = u_0(n - k_2). \tag{2.50}$$

The response $\overline{h}_{ab}[u_0(n - k_1), u_0(n - k_2)]$ of the quadratic system to a pair of distinct impulses is known as its *bi-impulse response*. Evaluating the bi-impulse response using (2.48), we find that

$$\overline{h}_{ab}[u_0(n - k_1), u_0(n - k_2)] = h_2(n - k_1, n - k_2). \tag{2.51}$$

This result provides us with one method for estimating the coefficients of a quadratic system.

2.3.6.1 A Deterministic Approach to Quadratic System Identification
Let us arrange the coefficients of the quadratic kernel with N-sample memory into an $N \times N$ matrix $\mathbf{H}$. It is easy to see from (2.51) that its diagonal corresponds to the response of the quadratic component to a single impulse. The off-diagonal elements of the form $h_2(m_1, m_1 - k)$ represents the coefficients that can be determined using two impulses located at $n = 0$ and $n = k$. The relation between the position of the input impulses and the quadratic coefficients which are determined as a result of the application of (2.51) is given in Table 2.2. A symmetric quadratic operator processing N samples at each time is uniquely defined by N bi-impulse responses.

TABLE 2.2   Relation between Positions of the Input Impulses and Quadratic Coefficients that Can Be Determined Using Them

$x_a(n)$	$x_b(n)$	Coefficients Identifiable Using $x_a(n)$ and $x_b(n)$			
$u_0(n)$	$u_0(n)$	$h_2(0, 0)$	$h_2(1, 1)$ $\cdots$	$h_2(N - 2, N - 2)$	$h_2(N - 1, N - 1)$
$u_0(n)$	$u_0(n - 1)$	$h_2(0, 1)$	$h_2(1, 2)$ $\cdots$	$h_2(N - 2, N - 1)$	
$\vdots$	$\vdots$		$\vdots$		
$u_0(n)$	$u_0(n - N + 1)$	$h_2(0, N - 1)$			

The filter has $N(N+1)/2$ independent coefficients, N of which are related to the response to a single impulse, $N-1$ to impulses separated by one sample, $N-2$ to impulses separated by two samples, and so on.

The preceding approach for determining the coefficients of a Volterra system is viable in particular for the finite-support Volterra filters of (2.26). This method was used by Schetzen [275] for the continuous-time case. It was reformulated for the discrete case by Ramponi [245], who also applied this formulation in the design of two-dimensional quadratic filters.

2.3.6.2 Generalization to pth-Order Systems
The interpretation of the second-order Volterra kernel as a two-dimensional unit impulse response function has been extended in the continuous-time case to higher-order Volterra kernels by Schetzen [275]. The pth-order kernel can be completely determined using p distinct impulses at the input of the system. Similar results can be derived for the discrete-time case also. However, because of the high complexity of such an approach, no practical applications have been devised up to now for identifying higher-order polynomial systems using this method.

2.4 EXISTENCE AND CONVERGENCE OF VOLTERRA SERIES EXPANSIONS

Much of the work on proving the existence and convergence of Volterra series expansions for nonlinear systems have been done for continuous-time systems. However, the basic approaches used for continuous-time Volterra series expansions can be directly extended to the discrete-time case. Although the issues of existence and convergence are essential from the mathematical point of view, they are in practice relatively less relevant in the context of many polynomial signal processing applications. This is a consequence of the fact that often nonlinear filters with specific and stable structures are used to process given sets of input signals to obtain the desired behavior at the output. Therefore, we will only state, without proof, some results related to the existence and convergence of Volterra series expansions of continuous nonlinear systems. More rigorous and detailed analysis may be found in Rugh [264] and Sandberg [271] and references cited therein.

The first two theorems deal with the approximation of continuous, time-invariant and nonlinear functions by time-invariant polynomial functions.

Theorem 2.1 (Weierstrass) Let $f(\lambda)$ be a continuous, real-valued function on the closed interval $[\lambda_1, \lambda_2]$. Then, for any $\epsilon > 0$, there exists a real polynomial $\mathcal{P}(\lambda)$ such that $|f(\lambda) - \mathcal{P}(\lambda)| < \epsilon$ for all $\lambda \in (\lambda_1, \lambda_2)$.

A generalization of this result is the following.

Theorem 2.2 (Stone–Weierstrass) Let X be a compact[1] space of signals and let F denote a continuous, time-invariant and causal mapping from X to a space of

---

[1]A compact space in $\mathcal{R}^n$ is one that is closed and bounded.

continuous, real-valued functions $C[0, T]$ on the interval $[0, T]$, specifically, $F : X \rightarrow C[0, T]$. Then, given any $\epsilon > 0$, there exists a continuous, time-invariant, and causal polynomial operator $\mathcal{P}$ such that for all $x(t) \in X$,

$$|F[x] - \mathcal{P}[x]| < \epsilon$$

for all t belonging to $[0, T]$. That is, there is a polynomial system that approximates F within ϵ in the range of t belonging to $[0, T]$.

   A proof of Theorem 2.2 can be found in Rugh [264]. Some remarks are in order here.

Remark 2.1   In order to prove Theorem 2.2, it is necessary to choose an appropriate algebra Φ of stationary, causal, and continuous operators from X to $C[0, T]$. Using the construction of such an algebra [264], it can be shown that there exist polynomial systems with separable kernels that satisfy the Stone–Weierstrass theorem.

Remark 2.2   Let us now consider a discrete-time, time-invariant, causal, finite-memory system that also has the property that *small* changes in the system input result in *small* changes in the system output. It follows from the Stone–Weierstrass theorem [64,212] that such a system can be uniformly approximated over a uniformly bounded set of input signals by a truncated Volterra series expansion obtained by limiting the summation in (2.20) to a sufficiently large, but finite, value P as in (2.25).

Remark 2.3   The preceding arguments indicate that a large class of nonlinear systems can be approximated with arbitrary precision using polynomial models. Consequently, the study of polynomial systems should enable us to characterize the properties of a significant class of nonlinear systems.

2.4.1   Special Classes of Polynomial Systems

The results described in the previous section guarantees the existence of general Volterra system models that can accurately represent large classes of nonlinear systems under relatively mild conditions. In many applications involving identification and realization of nonlinear systems, it is convenient to employ simpler, but less general system models. We now state, without proof, several theorems that describe the uniform approximation of nonlinear systems using simplified structures. The proofs of the results may be found in the references listed with the theorems. Even though some of the results have been derived for continuous-time systems, the proofs are valid for discrete-time systems also.

2.4.1.1   *The LNL Cascade Structure*   Consider a cascade structure as shown in Figure 2.13. This nonlinear system consists of a linear time-invariant system connected in series with a memoryless nonlinearity followed by another linear time-

Figure 2.13 An LNL cascade nonlinear system.

invariant system. This model is usually referred to as the *LNL cascade model.* Compared with the general Volterra series model, the LNL model is simpler. Consequently, approximating nonlinear systems using such simplified models is of interest. The following theorem shows that a large class of nonlinear systems can be uniformly approximated using a parallel combination of several LNL cascade systems.

Theorem 2.3 (Palm [212])   A discrete-time, time-invariant, causal, finite-memory system that is a continuous mapping of its input, in the sense that *small* changes in the system input result in *small* changes in the system output, can be uniformly approximated by a parallel-cascade system formed with a finite number of branches, each of which contains a dynamic linear system, a static nonlinear system, and a dynamic linear system in cascade as shown in Figure 2.14.

2.4.1.2 A Simpler Parallel-Cascade Structure The parallel-cascade nonlinear model depicted in Figure 2.15 is even simpler than the model of Figure 2.14. Each branch of this model consists of a linear time-invariant system connected in series with a memoryless nonlinearity. Korenberg [120,122] showed that such

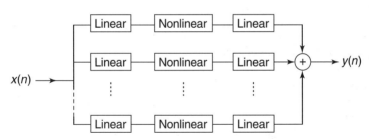

Figure 2.14 The parallel-cascade system of Theorem 2.3.

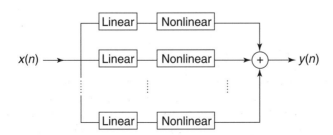

Figure 2.15 The parallel-cascade nonlinear model of Theorem 2.4.

models can also approximate a large class of nonlinear systems. The following theorem is due to Korenberg.

Theorem 2.4 (Korenberg [120,122])   The parallel-cascade form, in which each parallel channel contains a dynamic linear system followed by a static polynomial nonlinearity, can provide an arbitrarily close approximation, in the mean-square error sense, to any nonlinear, discrete-time, time-invariant, causal, finite-memory system that is a continuous mapping of its input.

2.4.1.3 Systems Represented by Linear–Analytic State Equations

Several other convergence conditions have been derived for different interconnection forms of Volterra systems [264] and for systems described by the so-called *linear–analytic* state equations. Such systems, studied mostly in the field of control theory, are special cases of the systems described by a set of controlled differential equations expressed in the form of a state model:

$$\dot{\mathbf{x}}(t) = a(\mathbf{x}(t), t) + b(\mathbf{x}(t), t)u(t), \qquad t \geq 0$$

$$y(t) = c(\mathbf{x}(t), t), \qquad \mathbf{x}(0) = \mathbf{x}_0, \tag{2.52}$$

where $\mathbf{x}(t)$ is the state vector, and the controlling input $u(t)$ and the output $y(t)$ are scalar signals. If the functions $a(\mathbf{x}(t), t)$, $b(\mathbf{x}(t), t)$ and $c(\mathbf{x}(t), t)$ are analytic in $\mathbf{x}$ and continuous in t, the model (2.52) describes a *linear–analytic* state equation. The system equations are analytic in the state and linear in the control signal u. It has been shown [32,126,143] that these systems admit a convergent Volterra series representation under general hypotheses. The following theorem describes the convergence characteristics of the Volterra series expansion of nonlinear systems represented by linear–analytic state equations.

Theorem 2.5 (Rugh [264])   Suppose that a solution to the unforced linear–analytic state equation exists for $t \in [0, T]$. Then, there exists an $\epsilon > 0$ such that for all continuous inputs satisfying $|u(t)| < \epsilon$ there is a Volterra series representation for the solution of the state equation that converges on $[0, T]$.

A particular case of linear–analytic state equations are the *bilinear* state equations. Such equations are given by

$$\dot{\mathbf{x}}(t) = A(t)\mathbf{x}(t) + B(t)\mathbf{x}(t)u(t), \qquad t \geq 0$$

$$y(t) = C(t)\mathbf{x}(t), \qquad \mathbf{x}(0) = \mathbf{x}_0, \tag{2.53}$$

where the system equations are linear with respect to both the state vector and the controlling input. This model has been independently studied because of its relevance and simplicity. It has been shown [33] that a system satisfying a bilinear state equation admits a Volterra series expansion. Furthermore, explicit formulas for the calculation of the Volterra kernels have been derived. A set of necessary and sufficient conditions for a continuous-time Volterra series to have a realization using bilinear state equations is given by d'Alessandro et al. [61]. Discrete-time bilinear systems are discussed in detail in Chapter 8.

2.5 TRANSFORM-DOMAIN REPRESENTATIONS OF VOLTERRA SYSTEMS

We have seen that the pth-order nonlinearity in the Volterra series expansions can be represented by a set of kernels that are p-dimensional arrays acting as filtering elements. A frequency-domain interpretation of the behavior of the Volterra filters is possible starting from the multidimensional convolution property. In the following, we make use of the frequency response of a p-dimensional linear time-invariant system to study the properties of Volterra systems characterized by a generic pth-order kernel.

2.5.1 Response to a Complex Sinusoid

Let us apply a p-dimensional, discrete complex sinusoid of the form

$$v(n_1, \ldots, n_p) = e^{j\omega_1 n_1} \cdots e^{j\omega_p n_p} \tag{2.54}$$

as the input sequence to a causal linear shift-invariant filter described by the p-dimensional impulse response function $h_p(n_1, \ldots, n_p)$. By convolving $v(n_1, n_2, \ldots, n_p)$ with $h_p(n_1, n_2, \ldots, n_p)$, the output sequence can be computed as

$$w(n_1, \ldots, n_p) = H_p(e^{j\omega_1}, \ldots, e^{j\omega_p})e^{j\omega_1 n_1} \cdots e^{j\omega_p n_p}, \tag{2.55}$$

where

$$H_p(e^{j\omega_1}, \ldots, e^{j\omega_p}) = \sum_{m_1=-\infty}^{\infty} \cdots \sum_{m_p=-\infty}^{\infty} h_p(m_1, \ldots, m_p)e^{-j\omega_1 m_1} \cdots e^{-j\omega_p m_p} \tag{2.56}$$

is the frequency response of the filter. The output of a Volterra filter characterized by the pth-order kernel to a single complex sinusoid $e^{j\omega_o n}$ can be derived using the multidimensional convolution property as

$$y(n) = w(n, \ldots, n) = H_p(e^{j\omega_0}, \ldots, e^{j\omega_0})e^{j(p\omega_0)n}. \tag{2.57}$$

This result immediately shows that frequency components that are not present at the input of the system are present at its output. This peculiarity is true of most nonlinear systems. Another interesting aspect of discrete-time Volterra systems is that if the input is not sampled fast enough, the output signal may exhibit aliasing even when the input signal does not. The ideas described above are illustrated in the following example.

Example 2.6   Consider a third-order system with input–output relationship

$$y(n) = x(n) + 0.5x(n-1) + 0.25x(n-2) + x(n)x(n-1)$$
$$+ x^2(n) + x(n-1)x(n-2) + x^3(n) + x^2(n-1)x(n-2).$$

Then

$$H_1(e^{j\omega_1}) = 1 + 0.5e^{-j\omega_1} + 0.25e^{-2j\omega_1},$$
$$H_2(e^{j\omega_1}, e^{j\omega_2}) = e^{-j\omega_2} + 1 + e^{-j\omega_1 - 2j\omega_2},$$

and

$$H_3(e^{j\omega_1}, e^{j\omega_2}, e^{j\omega_3}) = 1 + e^{-j\omega_1 - j\omega_2 - 2j\omega_3}.$$

Let us input a sinusoidal signal of the form $e^{j(\pi/6)n}$ to this system. The output of this system is

$$y(n) = H_1[e^{j(\pi/6)}]e^{j(\pi/6)n} + H_2[e^{j(\pi/6)}, e^{j(\pi/6)}]e^{j(\pi/3)n}$$
$$+ H_3[e^{j(\pi/6)}, e^{j(\pi/6)}, e^{j(\pi/6)}]e^{j(\pi/2)n},$$

indicating that the output contains three frequency components even though the input signal had only one frequency component. Now suppose that the input signal is $e^{j(3\pi/4)n}$. The output of this system can easily be calculated for this case as

$$y(n) = H_1[e^{j(3\pi/4)}]e^{j(3\pi/4)n} + H_2[e^{j(3\pi/4)}, e^{j(3\pi/4)}]e^{j(3\pi/2)n}$$
$$+ H_3[e^{j(3\pi/4)}, e^{j(3\pi/4)}, e^{j(3\pi/4)}]e^{j(9\pi/4)n}.$$

However, since discrete-time sinusoids are periodic in the frequency domain with period 2π radians per sample, the last component is aliased down to $\pi/4$ radians per sample from $9\pi/4$ radians per sample. The second component $H_2[e^{j(3\pi/4)}, e^{j(3\pi/4)}]e^{j(3\pi/2)n}$ can also be expressed as a sinusoid at the frequency $-\pi/2$ radians per sample for similar reasons. It should be clear from the preceding discussion that the output aliasing can be avoided by sampling the input signal at a rate higher than twice the maximum frequency component of the output signal.

2.5.2 Response to Multiple Complex Sinusoids

When the input signal is a combination of sampled complex sinusoids, the output of a Volterra system contains intermodulation terms in addition to the harmonics of the input frequencies. Let us assume, for example, that

$$x(n) = Ae^{j\omega_a n} + Be^{j\omega_b n} \tag{2.58}$$

is the input signal. The pth-order product term is given by

$$v(n_1, \ldots, n_p) = x(n_1) \cdots x(n_p) = (Ae^{j\omega_a n_1} + Be^{j\omega_b n_1}) \cdots (Ae^{j\omega_a n_p} + Be^{j\omega_b n_p}). \tag{2.59}$$

The term-by-term multiplications in (2.59) give all the intermodulation components. We demonstrate this idea in the following two examples.

Example 2.7 The response of a homogeneous quadratic filter to two distinct complex sinusoids can be obtained by setting $p = 2$ in (2.59). Then

$$v(n_1, n_2) = x(n_1)x(n_2)$$
$$= A^2 e^{j\omega_a(n_1+n_2)} + B^2 e^{j\omega_b(n_1+n_2)} + ABe^{j(\omega_a n_1 + \omega_b n_2)} + BAe^{j(\omega_b n_1 + \omega_a n_2)}$$

Because of the correspondence between a linear two-dimensional filter and the quadratic Volterra operator, we can show, by applying (2.55) with $n_1 = n_2 = n$ to each component of this equation, that the output signal is a superposition of three complex sinusoids at the angular frequencies $2\omega_a$, $2\omega_b$, and $\omega_a + \omega_b$. The output signal $y(n)$ may be evaluated as

$$y(n) = A^2 H_2(e^{j\omega_a}, e^{j\omega_a})e^{j2\omega_a n} + B^2 H_2(e^{j\omega_b}, e^{j\omega_b})e^{j2\omega_b n}$$
$$+ AB[H_2(e^{j\omega_a}, e^{j\omega_b}) + H_2(e^{j\omega_b}, e^{j\omega_a})]e^{j(\omega_a + w_b)n}.$$

Example 2.8 Assume that the input signal to a homogeneous quadratic filter is a sampled sinusoidal function of the type

$$x(n) = A\cos(\omega_a n) = \frac{A}{2}(e^{j\omega_a n} + e^{-j\omega_a n}).$$

The output signal for this input is

$$y(n) = \frac{A^2}{4} H_2(e^{j\omega_a}, e^{j\omega_a}) e^{j2\omega_a n} + \frac{A^2}{4} H_2(e^{-j\omega_a}, e^{-j\omega_a}) e^{-j2\omega_a n}$$

$$+ \frac{A^2}{4} [H_2(e^{j\omega_a}, e^{-j\omega_a}) + H_2(e^{-j\omega_a}, e^{j\omega_a})]$$

$$= \frac{A^2}{2} H_2(e^{j\omega_a}, e^{j\omega_a}) \cos(2\omega_a n) + \frac{A^2}{4} [H_2(e^{j\omega_a}, e^{-j\omega_a}) + H_2(e^{-j\omega_a}, e^{j\omega_a})].$$

In this case, a sinusoidal function at the harmonic frequency $2\omega_a$ and a sinusoidal function at the difference frequency $\omega = 0$, that is, a constant term appears at the output.

2.5.3 Response to a Signal with a Continuous Spectrum

We now extend the previous results for a finite number of sinusoids at the input of the Volterra system to the case when the input signal has a continuous spectrum. We first discuss the multidimensional shift-invariant linear systems and then derive the appropriate results for the Volterra systems.

2.5.3.1 *Response of a Multidimensional Linear, Shift-Invariant System*   When the input signal has a continuous spectrum, we can evaluate the Fourier transform of the output signal by means of an integral representation. It is well-known from the theory of linear multidimensional filters that the frequency response $H_p(e^{j\omega_1}, \ldots, e^{j\omega_p})$ is a continuous periodic function with period 2π in each one of the p dimensions. Because of the periodicity, the right-hand side of (2.56), which defines the frequency response of the kernel, can be thought of as the p-dimensional Fourier series expansion for $H(e^{j\omega_1}, e^{j\omega_2}, \ldots, e^{j\omega_p})$. Consequently, the samples of the unit impulse response function $h_p(n_1, n_2, \ldots, n_p)$ of the p-dimensional filter are the Fourier series coefficients of the frequency response. Those coefficients can be evaluated as

$$h_p(n_1, \ldots, n_p) = \frac{1}{(2\pi)^p} \int_{-\pi}^{\pi} \cdots \int_{-\pi}^{\pi} H_p(e^{j\omega_1}, \ldots, e^{j\omega_p}) e^{j\omega_1 n_1} \cdots e^{j\omega_p n_p} d\omega_1 \cdots d\omega_p.$$

$$(2.60)$$

Since the relations in (2.56) and (2.60) are valid for any sequence for which (2.56) exists, an arbitrary input sequence $v(n_1, \ldots, n_p)$ can also be represented in the form

$$v(n_1, \ldots, n_p) = \frac{1}{(2\pi)^p} \int_{-\pi}^{\pi} \cdots \int_{-\pi}^{\pi} V(e^{j\omega_1}, \ldots, e^{j\omega_p}) e^{j\omega_1 n_1} \cdots e^{j\omega_p n_p} d\omega_1 \cdots d\omega_p,$$

$$(2.61)$$

where $V(e^{j\omega_1}, \ldots, e^{j\omega_p})$ is the p-dimensional Fourier transform of $v(n_1, \ldots, n_2)$. The output sequence $W(n_1, \ldots, n_p)$ of a p-dimensional linear shift-invariant system can then be represented by the corresponding equation

$$w(n_1, \ldots, n_p) = \frac{1}{(2\pi)^p} \int_{-\pi}^{\pi} \cdots \int_{-\pi}^{\pi} W(e^{j\omega_1}, \ldots, e^{j\omega_p}) e^{j\omega_1 n_1} \cdots e^{j\omega_p n_p} d\omega_1 \cdots d\omega_p.$$

(2.62)

Because the system is linear and shift-invariant,

$$W(e^{j\omega_1}, \ldots, e^{j\omega_p}) = V(e^{j\omega_1}, \ldots, e^{j\omega_p}) H_p(e^{j\omega_1}, \ldots, e^{j\omega_p}).$$

(2.63)

Consequently, we can rewrite (2.62) as

$$w(n_1, \ldots, n_p) = \frac{1}{(2\pi)^p} \int_{-\pi}^{\pi} \cdots \int_{-\pi}^{\pi} H_p(e^{j\omega_1}, \ldots, e^{j\omega_p}) V(e^{j\omega_1}, \ldots, e^{j\omega_p}) e^{j\omega_1 n_1} \cdots$$
$$\times e^{j\omega_p n_p} d\omega_1 \cdots d\omega_p.$$

(2.64)

2.5.3.2 Response of Volterra Systems   Recall that

$$v(n_1, \ldots, n_p) = x(n_1) \cdots x(n_p)$$

(2.65)

is the multidimensional representation of the Volterra system. Consequently,

$$V(e^{j\omega_1}, \ldots, e^{j\omega_p}) = X(e^{j\omega_1}) \cdots X(e^{j\omega_p}).$$

(2.66)

We can immediately derive an expression for the Fourier transform of the output of the Volterra system with a pth-order kernel as follows:

$$y(n) = w(n, \ldots, n) = \frac{1}{(2\pi)^p} \int_{-\pi}^{\pi} \cdots \int_{-\pi}^{\pi} H_p(e^{j\omega_1}, \ldots, e^{j\omega_p})$$
$$\times X(e^{j\omega_1}) \cdots X(e^{j\omega_p}) e^{j(\omega_1 + \cdots + \omega_p)n} d\omega_1 \cdots d\omega_p.$$

(2.67)

We can extend this result in a straightforward manner to compute the output of a general pth-order Volterra filter. The output signal $y(n)$ of such a system can be expressed as

$$y(n) = h_0 + \sum_{i=1}^{P} \frac{1}{(2\pi)^i} \int_{-\pi}^{\pi} \cdots \int_{-\pi}^{\pi} H_i(e^{j\omega_1}, \ldots, e^{j\omega_i})$$
$$\times X(e^{j\omega_1}) \cdots X(e^{j\omega_i}) e^{j(\omega_1 + \omega_2 + \cdots + \omega_i)n} d\omega_1 \cdots d\omega_i.$$

(2.68)

This result also illustrates the presence of new frequencies in the output signal that are not present in the input signal. This characteristic is a significant difference of nonlinear systems from linear systems.

Remark 2.4   When the input signal is a single complex sinusoid described in the Fourier domain by the impulse function $2\pi A u_0(\omega - \omega_a)$, the application of (2.67) gives the same result as (2.57). Similarly, when the input signal consists of multiple sinusoids, the results obtained in Section 2.5.2 can still be derived from (2.67) by representing the input signal by means of appropriate impulse functions in the frequency domain. In particular, the result of Example 2.7 can be obtained from (2.67) by recognizing that

$$X(e^{j\omega}) = 2\pi A u_0(\omega - \omega_a) + 2\pi B u_0(\omega - \omega_b). \tag{2.69}$$

2.5.4 The z-Transform Representation

The Volterra kernels can also be described using their multidimensional z-transforms. Let the p-dimensional z-transform of the pth-order Volterra kernel be

$$H_p(z_1, \ldots, z_p) = \sum_{m_1=-\infty}^{\infty} \cdots \sum_{m_p=-\infty}^{\infty} h_p(m_1, \ldots, m_p) z_1^{-m_1} \cdots z_p^{-m_p}, \tag{2.70}$$

where we have assumed that the z-transform of $h_p(m_1, \ldots, m_p)$ converges in the region of interest. As before, we can represent the input–output relation

$$\bar{h}_p[x(n)] = \sum_{m_1=-\infty}^{\infty} \cdots \sum_{m_p=-\infty}^{\infty} h_p(m_1, \ldots, m_p) x(n - m_1) \cdots x(n - m_p) \tag{2.71}$$

using the following pair of equations:

$$w(n_1, \ldots, n_p) = \sum_{m_1=-\infty}^{\infty} \cdots \sum_{m_p=-\infty}^{\infty} h_p(m_1, \ldots, m_p) x(n_1 - m_1) \cdots x(n_p - m_p) \tag{2.72}$$

$$y(n) = w(n, \ldots, n). \tag{2.73}$$

Equation (2.72) can be rewritten in the z-transform domain in the form

$$W(z_1, \ldots, z_p) = H_p(z_1, \ldots, z_p) X(z_1) \cdots X(z_p). \tag{2.74}$$

The complete formulation of the input–output relationship using the transform domain requires the translation of (2.73) to the z-domain. Unfortunately, such a characterization requires use of multidimensional contour integrations [264], which are often difficult to evaluate in practice. Therefore, to identify the input–output relation of a system described by a given $H_p(z_1, \ldots, z_p)$, it is convenient to take the inverse transform of (2.74) and then directly apply (2.73). When the kernel $h_p(m_1, \ldots, m_p)$ is defined on a causal and finite support with N-sample memory, (2.70) becomes

$$H_p(z_1, \ldots, z_p) = \sum_{m_1=0}^{N-1} \cdots \sum_{m_p=0}^{N-1} h_p(m_1, \ldots, m_p)z_1^{-m_1} \cdots z_p^{-m_p}. \qquad (2.75)$$

In such situations, the z-transform converges at all points on the complex space (with the possible exceptions at zero and infinity). We now consider several examples illustrating the use of z-transform in analyzing Volterra systems.

Example 2.9   Consider a quadratic system described by the following equation:

$$y(n) = ax^2(n) + bx(n)x(n-1) + cx^2(n-1).$$

Since only three coefficients of the kernel are different from zero, namely, $h_2(0, 0) = a$, $h_2(0, 1) = b$ and $h_2(1, 1) = c$, the corresponding two-dimensional z-transform is

$$H_2(z_1, z_2) = a + bz_2^{-1} + cz_1^{-1}z_2^{-1}.$$

Substituting this result in (2.74) gives

$$W(z_1, z_2) = (a + bz_2^{-1} + cz_1^{-1}z_2^{-1})X(z_1)X(z_2).$$

We can take the two-dimensional inverse z-transform of both sides of this equation to obtain

$$w(n_1, n_2) = ax(n_1)x(n_2) + bx(n_1)x(n_2-1) + cx(n_1-1)x(n_2-1).$$

The original input–output relationship can be obtained from this equation by setting $n_1 = n_2 = n$, which gives the desired system representation:

$$y(n) = w(n, n) = ax^2(n) + bx(n)x(n-1) + cx^2(n-1).$$

Example 2.10   The cubic system described by the input–output relationship

$$y(n) = ax^3(n) + bx^3(n-1) + cx^3(n-2) + dx(n)x^2(n-1)$$

$$+ ex(n)x(n-1)x(n-2) + fx(n-1)x^2(n-2)$$

is equivalently represented by the following z-transform:

$$H_2(z_1, z_2, z_3) = a + bz_1^{-1}z_2^{-1}z_3^{-1} + cz_1^{-2}z_2^{-2}z_3^{-2} +$$
$$dz_2^{-1}z_3^{-1} + ez_2^{-1}z_3^{-2} + fz_1^{-1}z_2^{-2}z_3^{-2}.$$

2.5.4.1 *Relationship between z-Transform and Fourier Transform Representations*

Equation (2.56) can be obtained by substituting $z_1 = e^{j\omega_1}, \ldots, z_p = e^{j\omega_p}$ in (2.70). This result is just an illustration of the fact that the Fourier transform of a p-dimensional sequence is its z-transform evaluated on a p-dimensional sphere centered at the origin and having unit radius.

2.5.5 The Discrete Fourier Transform Representation

The z-transform and the Fourier transform are very useful in providing a frequency-domain interpretation for Volterra systems. However, they are less useful as tools for realizing truncated Volterra systems since the inverse transforms in both cases require the computation of multidimensional integrals. Such integrals are difficult to evaluate except for the simplest input signals and systems. The discrete Fourier transform (DFT) and its inverse can be computed using finite summations, and therefore can be used to implement truncated Volterra filters.

The p-dimensional DFT of the finite-support kernel $h_p(m_1, \ldots, m_p)$ is given by

$$H_p(k_1, \ldots, k_p) = \sum_{m_1=0}^{M-1} \cdots \sum_{m_p=0}^{M-1} h_p(m_1, \ldots, m_p) e^{-j(2\pi/M)m_1 k_1} \cdots e^{-j(2\pi/M)m_p k_p}, \quad (2.76)$$

where the indices k_i, $i = 1, \ldots, p$ range from 0 to $M-1$ and M is an integer greater than or equal to N. The pth-order kernel $h_p(m_1, \ldots, m_p)$ can be obtained from $H_p(k_1, \ldots, k_p)$ using the inverse DFT as

$$h_p(m_1, \ldots, m_p) = \begin{cases} \dfrac{1}{M^p} \sum_{k_1=0}^{M-1} \cdots \sum_{k_p=0}^{M-1} H_p(k_1, \ldots, k_p) e^{j(2\pi/M)k_1 m_1} \cdots e^{j(2\pi/M)k_p m_p}; \\ \qquad\qquad\qquad\qquad\qquad\qquad 0 \le m_1, \ldots, m_p \le M - 1 \\ 0; \qquad\qquad\qquad\qquad\qquad\qquad \text{otherwise.} \end{cases}$$
$$(2.77)$$

Example 2.11 Consider the simple system with input–output relationship

$$y(n) = ax^2(n) + bx^2(n - 1).$$

The two-dimensional DFT

$$H_2(k_1, k_2) = \begin{bmatrix} a+b & a+be^{-j(2\pi/M)} & \cdots & a+be^{-j(M-1)(2\pi/M)} \\ a+be^{-j(2\pi/M)} & a+be^{-j2(2\pi/M)} & \cdots & a+be^{-jM(2\pi/M)} \\ \vdots & \vdots & \vdots & \vdots \\ a+be^{-j(M-1)(2\pi/M)} & a+be^{-jM(2\pi/M)} & \cdots & a+be^{-j2(M-1)(2\pi/M)} \end{bmatrix}$$

provides a discrete frequency-domain representation for this system.

2.5.5.1 Relationships among the Transform-Domain Representations
It is straightforward to see from the definitions of the z-transform and the Fourier transform that

$$H_p(k_1, \ldots, k_p) = H_p(z_1, \ldots, z_p)\big|_{z_1 = e^{j(2\pi k_1/M)}, \ldots, z_p = e^{j(2\pi k_p/M)}}. \tag{2.78}$$

In other words, the DFT coefficients are uniformly sampled values of the z-transform on the p-dimensional unit sphere centered at the origin. Since the Fourier transform is the z-transform on the unit sphere, we can relate the DFT and Fourier transform in a similar manner.

Remark 2.5   Consider the expression

$$\tilde{h}_p(m_1, \ldots, m_p) = \frac{1}{M^p} \sum_{k_1=0}^{M-1} \cdots \sum_{k_p=0}^{M-1} H_p(k_1, \ldots, k_p) e^{j(2\pi/M)k_1 m_1} \cdots e^{j(2\pi/M)k_p m_p}. \tag{2.79}$$

It is easy to show that $\tilde{h}_p(m_1, \ldots, m_p)$ is periodic in all p dimensions, with period M along each dimension. The Volterra kernel $h_p(m_1, \ldots, m_p)$ is identical to $\tilde{h}_p(m_1, \ldots, m_p)$ in the first period defined by the region of support of the filter $0 \le m_1, \ldots, m_p \le M - 1$. In other words, $\tilde{h}_p(m_1, \ldots, m_p)$ is a periodic extension of $h_p(m_1, \ldots, m_p)$ with period M in each dimension. Consequently, $H_p(k_1, \ldots, k_p)$ may be thought of as the discrete Fourier series coefficients of a periodic extension of $h_p(m_1, \ldots, m_p)$.

2.5.5.2 Selection of the DFT Size
The parameter M in the definition of the DFT must be chosen carefully. If M is smaller than N, the memory span of the filter, the inverse DFT of $H_p(k_1, \ldots, k_p)$, will result in a sequence with smaller support than the actual filter. This happens because the inverse DFT exhibits time-domain aliasing in such situations. This idea is illustrated in the next example.

Example 2.12   Let $h_2(m_1, m_2)$ of a homogeneous quadratic system be given by

$$h_2(m_1, m_2) = \begin{cases} 1; & 0 \le m_1, m_2 \le 5 \\ 0; & \text{otherwise.} \end{cases}$$

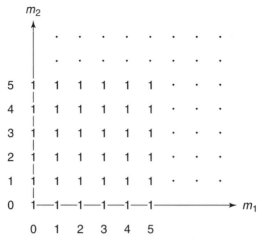

Figure 2.16 Quadratic kernel of Example 2.12.

Figure 2.16 shows this quadratic kernel. Suppose now that we compute the two-dimensional DFT $H_2(k_1, k_2)$ with $M = 4$. Then

$$H_2(k_1, k_2) = \sum_{m_1=0}^{5} \sum_{m_2=0}^{5} e^{-j(2\pi/4)k_1 m_1} e^{-j(2\pi/4)k_2 m_2}$$

Since $e^{j(2\pi/4)km}$ is periodic in m with period equal to four samples, we can rewrite this equation by combining the coefficients of all the exponential terms with identical values. This operation will result in

$$H_2(k_1, k_2) = \sum_{m_1=0}^{3} \sum_{m_2=0}^{3} h_2'(m_1, m_2) e^{-j(2\pi/4)k_1 m_1} e^{-j(2\pi/4)k_2 m_2},$$

where

$$h_2'(m_1, m_2) = h_2(m_1, m_2) + h_2(m_1 + 4, m_2)$$
$$+ h_2(m_1, m_2 + 4) + h_2(m_1 + 4, m_2 + 4).$$

Consequently, when the inverse DFT of $H_2(k_1, k_2)$ is computed, we obtain $h_2'(m_1, m_2)$ rather than $h_2(m_1, m_2)$. Figure 2.17 displays $h_2'(m_1, m_2)$. Note that $h_2'(m_1, m_2)$ is obtained by wrapping the portions of $h_2(m_1, m_2)$ that are outside the range $0 \leq m_1, m_2 \leq 3$ to within this range and adding up all the overlapped portions.

Remark 2.6 We note from (2.78) that the M-point DFT of a signal is its z-transform sampled on M equally-spaced points on the unit circle. In order to reconstruct the time-domain signal exactly from these samples on the complex plane, M must be greater than or equal to the number of time-domain samples of the

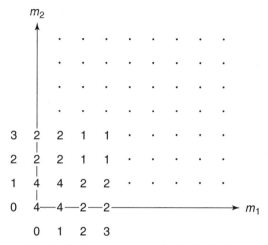

Figure 2.17 Illustration of time-domain aliasing in Example 2.12.

signal. If the sampling is done at a smaller rate than necessary, we will not be able to exactly reconstruct the original signal from the samples. Instead, we get time-domain aliasing. This phenomenon is a dual of the familiar aliasing phenomenon associated with undersampling of time-domain signals.

2.5.5.3 DFT of the Output of a Homogeneous pth-Order Volterra System
The DFT of the output of a Volterra system can be derived from the multidimensional convolution property of such systems. Referring to (2.29)–(2.31), we can see that

$$V(k_1, \ldots, k_p) = X(k_1)X(k_2)\cdots X(k_p) \tag{2.80}$$

and

$$W(k_1, \ldots, k_p) = H_p(k_1, \ldots, k_p)X(k_1)X(k_2)\cdots X(k_p), \tag{2.81}$$

where $V(k_1, \ldots, k_p)$ and $W(k_1, \ldots, k_p)$ are the DFTs of $v(n_1, \ldots, n_p)$ and $w(n_1, \ldots, n_p)$, respectively, and we have assumed that the parameter M is large enough not to cause any time-domain aliasing. The output $y(n)$ is the inverse DFT of $W(k_1, \ldots, k_p)$ at $n_1 = n_2 = \cdots = n_p = n$, and can be evaluated as

$$y(n) = \frac{1}{M^p} \sum_{k_1=0}^{M-1} \cdots \sum_{k_p=0}^{M-1} H_p(k_1, \ldots, k_p)X(k_1)\cdots X(k_p)e^{j(2\pi/M)(k_1+\cdots+k_p)n}. \tag{2.82}$$

We can compute the M-point DFT of $y(n)$ as

$$Y(k) = \sum_{n=0}^{M-1} y(n)e^{-j(2\pi/M)nk}$$

$$= \sum_{n=0}^{M-1} \left\{ \frac{1}{M^p} \sum_{k_1=0}^{M-1} \cdots \sum_{k_p=0}^{M-1} H_p(k_1, \ldots, k_p)X(k_1) \cdots X(k_p)e^{j(2\pi/M)(k_1+\cdots+k_p)n} \right\}$$

$$\times e^{-j(2\pi/M)nk}$$

$$= \frac{1}{M^p} \sum_{k_1=0}^{M-1} \cdots \sum_{k_p=0}^{M-1} H_p(k_1, \ldots k_p)X(k_1) \cdots X(k_p) \sum_{n=0}^{M-1} e^{j(2\pi/M)(k_1+\cdots+k_p-k)n}. \quad (2.83)$$

Since

$$\sum_{n=0}^{M-1} e^{j(2\pi/M)nk} = \begin{cases} M; & k = lM \\ 0; & \text{otherwise}, \end{cases} \quad (2.84)$$

where l is an integer than can take positive, negative, or zero values, we can rewrite $Y(k)$ as

$$Y(k) = \frac{1}{M^{p-1}} \sum_{k_1=0}^{M-1} \cdots \sum_{k_p=0}^{M-1} H_p(k_1, \ldots, k_p)X(k_1) \cdots X(k_p)$$

$$\times \sum_{l=-\infty}^{\infty} u_0[lM + k - (k_1 + \cdots + k_p)], \quad (2.85)$$

where $u_0(k)$ represents the unit impulse function as usual. That is, the DFT of the output sequence is obtained by summing all the product terms of the form $H_p(k_1, \ldots, k_p)X(k_1) \ldots X(k_p)$ over the values of $k_1, \ldots, k_p$ in the range $0 \leq k_i \leq M - 1$ such that they add up to k or $k + lM$ for some integer value of l. Note, however, that the maximum value of $k_1 + k_2 + \cdots + k_p$ in (2.85) is equal to $p(M - 1)$. Figure 2.18 shows the indices of the terms that are added together to evaluate each value of $Y(k)$ when $M = 4$ for a homogeneous quadratic filter. In this example $k_1 + k_2$ can fall in the range 0–3 or 4–6, indicating that only $l = 0$ and $l = 1$ are required in the summation. For a pth-order kernel, the sum $k_1 + k_2 + \cdots + k_p$ may fall in the range $[lM, (l+1)M - 1]$ for $l = 0, 1, \ldots, p - 1$ when $0 \leq k_1, \ldots, k_p \leq M - 1$.

2.5.5.4 Frequency-Domain Aliasing

It is important to point out again that components at frequencies higher than the Nyquist frequency associated with the input sequence may be present in the output signal. Consequently, the output signal may contain aliased frequency components. To avoid aliasing at the output of a pth-order Volterra system, the input signal must be band-limited to $1/2p$ cycles per sample. If this condition holds, we can approximate[2] the DFT $X(k)$ of the input signal to be zero when $(M/2p) \leq k < M - (M/2p)$. In such cases, we can simplify the system

---

[2]This is only an approximation since time-limited signals are not band-limited.

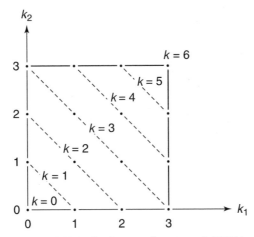

Figure 2.18   Frequency bins of $H_2(k_1, k_2)$ that contribute to each DFT bin of $Y(k)$ when $M = 4$.

representation using the DFT even more since only the non-zero terms in the product signal $H_p(k_1, \ldots, k_p)X(k_1) \cdots X(k_p)$ need to be evaluated. The shaded domain in Figure 2.19 indicates the range of (k_1, k_2) necessary to evaluate the output of a purely quadratic system when the input signal is band-limited as described above.

Example 2.13   Consider a homogeneous quadratic system described by the input–output relation $y(n) = x^2(n)$. Figure 2.20 shows 128 samples of a sinusoid of frequency $(2\pi)4$ radians per second, sampled at the rate of 128 samples per second. The 128-point DFT of this signal for k in the range $[-64, 63]$ is also shown in this figure. We see that the DFT consists of a pair of impulses at $k_a = \pm 4$. The system output as well as its DFT is displayed in Figure 2.21. The output signal consists of a constant term and a sinusoid at twice the frequency of the input signal. No aliasing effects appear at the output of the system since the maximum output frequency is still less than half of the Nyquist frequency.

Figure 2.22 displays the DFT of an input signal with frequency $(2\pi)60$ radians per second, sampled again at 128 samples per second, and the 128-point DFT of the corresponding output signal. In addition to the constant term, a sinusoidal signal at frequency 8 cycles/second appears in the output signal. This frequency component, rather than twice the input frequency given by 120 cycles/second, appears at the output as a result of aliasing. Thus, the discrete-time nonlinear filter produces the same signal for both the input signals. In order to obtain the sinusoid corresponding to the higher frequency of 120 cycles/second, we must sample the input signals at a rate higher than 240 samples per second. This idea is illustrated in Figure 2.23. This figure was obtained from the output signal that resulted from an input signal at frequency 60 cycles/second, and sampled at the rate of 2560 samples per second. It is worth noting that in all the figures in the time domain, no discontinuities appear at the end of the segments of data since there are an integer number of samples per period of the input sinusoids in all cases.

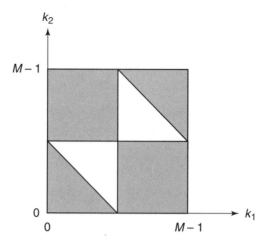

Figure 2.19 Domain of the terms used for the computation of the output signal when the input signal is band-limited.

2.5.6 Symmetry of $H_p(k_1, k_2, \ldots, k_p)$

There are two main sources of symmetry for $H_p(k_1, k_2, \ldots, k_p)$ when the pth-order Volterra kernel $h_p(m_1, m_2, \ldots, m_p)$ is real. The first one is due to the symmetry of the Volterra kernel, and the other one is because the kernel is real.

2.5.6.1 Symmetry of $H_p(k_1, k_2, \ldots, k_p)$ Due to the Symmetry of $h_p(m_1, m_2, \ldots, m_p)$ We assume that $h_p(m_1, m_2, \ldots, m_p)$ is a symmetric kernel. Since

$$h_p(m_1, m_2, \ldots, m_p) = h_p(\pi(m_1, m_2, \ldots, m_p)) \tag{2.86}$$

for any arbitrary permutation of the indices given by $\pi(m_1, m_2, \ldots, m_p)$, it follows immediately that

$$H_p(k_1, k_2, \ldots, k_p) = H_p(\pi(k_1, k_2, \ldots, k_p)). \tag{2.87}$$

That is, the DFT of a symmetric kernel is also invariant to the permutations of the frequency indices. For a symmetric quadratic kernel, this implies the symmetry of its DFT with respect to the $k_1 = k_2$ axis in the k_1, k_2 plane so that

$$H_2(k_1, k_2) = H_2(k_2, k_1). \tag{2.88}$$

The shaded portion of Figure 2.24 indicates the part of $H_2(k_1, k_2)$ that we must know for completely specifying the two-dimensional frequency response of a symmetric quadratic kernel.

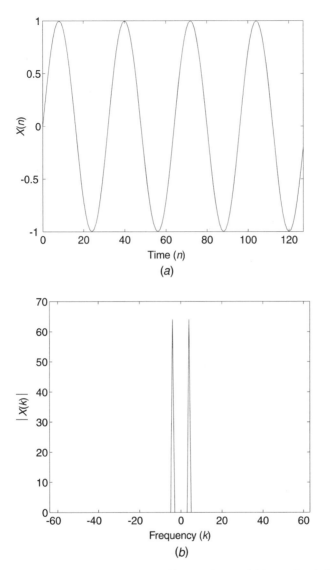

Figure 2.20 A sampled input sinusoid with $M = 128$, $k_a = 4$: (a) input signal; (b) magnitude spectrum of the input signal.

2.5.6.2 *Symmetry of* $H_p(k_1, k_2, \ldots, k_p)$ *for a Real Kernel* The DFT of a real kernel $h_p(m_1, m_2, \ldots, m_p)$ satisfies the conjugate symmetry property given by

$$H_p(k_1, k_2, \ldots, k_p) = H_p^*(-k_1, -k_2, \ldots, -k_p). \qquad (2.89)$$

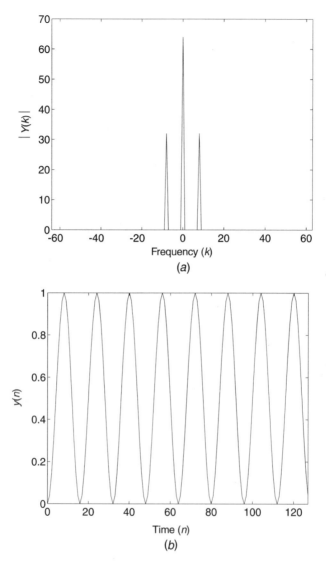

Figure 2.21 Response of the quadratic system $y(n) = x^2(n)$ to a sampled input sinusoid with $M = 128$, $k_a = 4$: (a) spectrum of the output signal; (b) output signal.

Since $H_p(k_1, k_2, \ldots, k_p)$ is periodic with period M in each dimension, this condition is identical to

$$H_p(k_1, k_2, \ldots, k_p) = H_p^*(M - k_1, M - k_2, \ldots, M - k_p). \qquad (2.90)$$

This condition is somewhat more useful since we usually specify the DFT in the range $0 \leq k_1, k_2, \ldots, k_p \leq M - 1$. The shaded rectangle in Figure 2.25 indicates the

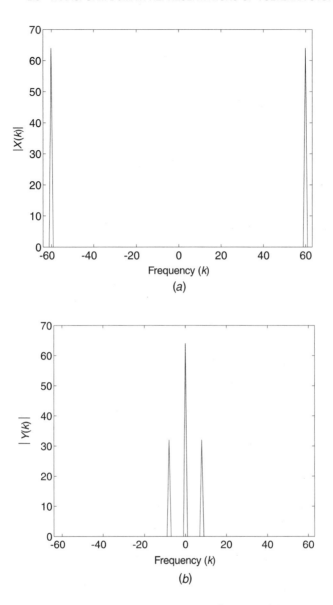

Figure 2.22 Response of the quadratic system $y(n) = x^2(n)$ to a sampled input sinusoid with $M = 128, k_a = 60$—the aliasing effect is evident: (a) input signal spectrum; (b) output signal spectrum.

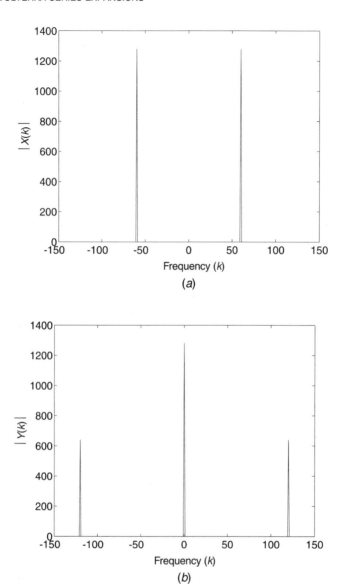

Figure 2.23   Response of the quadratic system $y(n) = x^2(n)$ to a sampled input sinusoid with $M = 2560$, $k_a = 60$: (a) input signal spectrum; (b) output signal spectrum.

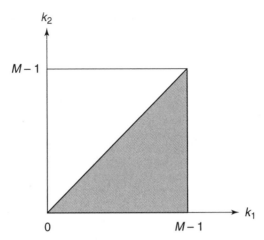

Figure 2.24 The shaded region indicates the range of (k_1, k_2) that will completely specify $H_2(k_1, k_2)$ based on the symmetry conditions.

portion of $H_2(k_1, k_2)$ that we must know to completely specify the DFT of a real quadratic kernel using the conjugate symmetry property. We get the following set of symmetries for the DFT of $h_p(m_1, m_2, \ldots, m_p)$ by combining the symmetry conditions in (2.87) and (2.88):

$$H_p(k_1, k_2, \ldots, k_p) = H_p(\pi_i(k_1, k_2, \ldots, k_p))$$
$$= H_p^*[\pi_l(M - k_1, M - k_2, \ldots, M - k_p)], \qquad (2.91)$$

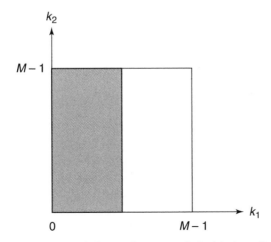

Figure 2.25 The shaded region indicates the range of (k_1, k_2) that will completely specify $H_2(k_1, k_2)$ for a real quadratic kernel based on the conjugate symmetry property.

where $\pi_i(\cdot)$ and $\pi_l(\cdot)$ denote possibly different permutations of the indices within the parentheses. For a real and symmetric quadratic kernel, these conditions become

$$H_2(k_1, k_2) = H_2(k_2, k_1)$$
$$= H_2^*(M - k_1, M - k_2)$$
$$= H_2^*(M - k_2, M - k_1). \qquad (2.92)$$

Figure 2.26 shows the range of the (k_1, k_2) plane that must be known for completely specifying the two-dimensional frequency response function using the symmetries in (2.92). The symmetries bring about significant reduction in the computational and memory requirements of the DFTs of Volterra kernels. The symmetries of the DFT of a real and symmetric cubic kernel have recently been described in detail in [192], and then exploited, together with the symmetries of the linear and quadratic kernels, to identify a Volterra system of order $p = 3$ in the frequency domain.

2.5.6.3 Simplifications in the DFT-Based Representations The symmetries described above indicate that the DFT-based representations can be significantly simplified. In addition, if the input sequence is real, it satisfies the conjugate symmetry property

$$X(k) = X^*(M - k), \qquad (2.93)$$

and therefore the product signal $X(k_1)X(k_2)\cdots X(k_p)$ satisfies all the symmetry properties satisfied by $H_p(k_1, \ldots, k_p)$. Furthermore, when the Volterra kernels and

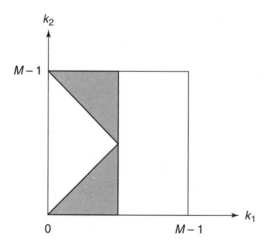

Figure 2.26 The shaded region indicates the range of (k_1, k_2) that will completely specify $H_2(k_1, k_2)$ for a real and symmetric quadratic kernel.

the input signal are real, the output sequence is also real and therefore, $Y(k)$ also satisfies the conjugate symmetry property

$$Y(k) = Y^*(M - k). \tag{2.94}$$

Consequently, it is sufficient to evaluate $Y(k)$ for approximately half of the frequency components, and thus the number of terms used in the computation of the output sequence using (2.82) is considerably reduced.

2.6 EXERCISES

2.1 Show that the nonlinear system

$$\overline{h}_p[x(n)] = \sum_{m_1=-\infty}^{\infty} \cdots \sum_{m_p=-\infty}^{\infty} h_p(m_1, \ldots, m_p)x(n - m_1) \cdots x(n - m_p)$$

is time-invariant.

2.2 Show that the filter of Figure 2.27 is not time-invariant. In that figure, $h(n)$ is a linear filter and the last block in the cascade indicates a decimation by a factor M. The decimator subsamples its input signal once every M samples to produce the output signal as $y(n) = x(nM)$.

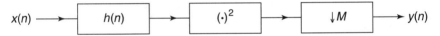

$x(n) \longrightarrow$ | $h(n)$ | $\longrightarrow$ | $(\cdot)^2$ | $\longrightarrow$ | $\downarrow M$ | $\longrightarrow y(n)$

Figure 2.27 Decimation of the output of a quadratic filter.

2.3 Consider the linear filter

$$w(n) = a_0x(n) + a_1x(n - 1) + a_2x(n - 2) + a_3x(n - 3)$$

and a static quadratic nonlinearity of the form $q(n) = b_0 + b_1v(n) + b_2v^2(n)$, where $v(n)$ and $q(n)$ represent the input and output signals, respectively, of the quadratic nonlinearity. Find the elements of the Volterra kernels formed by the interconnection of the two blocks to form **(a)** a Wiener model and **(b)** a Hammerstein model.

2.4 Consider the multiplicative interconnection of the linear system

$$w(n) = a_0x(n) + a_1x(n - 1) + a_2x(n - 2) + a_3x(n - 3)$$

and a homogeneous quadratic Wiener model as in Example 2.1 with $N=4$. Find the elements of the cubic Volterra kernel that results.

2.5 Find the kernel coefficients of the interconnected system shown in Figure 2.28 as a function of the coefficients of the linear filters $h_a(n)$ and $h_b(n)$.

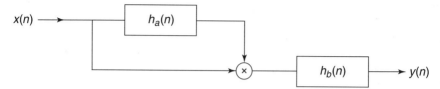

Figure 2.28 Interconnection of two linear filters.

2.6 Determine the kernel coefficients of the interconnected system formed with two linear filters $h_a(n)$ and $h_c(n)$, and a quadratic filter $h_b(n_1, n_2)$ as shown in Figure 2.29.

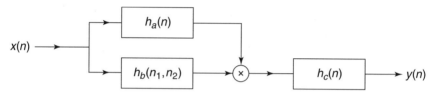

Figure 2.29 Interconnection of one quadratic and two linear filters.

2.7 Find the kernel coefficients of the interconnected system of two linear filters $h_a(n)$ and $h_b(n)$, and a quadratic filter $h_c(n_1, n_2)$ as shown in Figure 2.30.

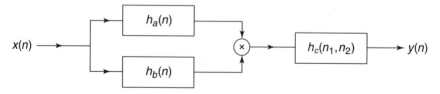

Figure 2.30 Interconnection of one quadratic and two linear filters.

2.8 Find the elements of the Volterra kernel of the nonlinear feedback system shown in Figure 2.31, where the elements of the quadratic kernel are given by

$$h_2(0, 0) = a_0 \qquad h_2(0, 1) = a_1$$
$$h_2(1, 0) = 0 \qquad h_2(1, 1) = 0.$$

2.9 Find the Volterra kernels of the system

$$y(n) = ax(n)x(n - 1) + bx(n)y(n - 1).$$

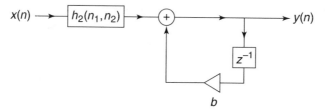

Figure 2.31 A nonlinear feedback system.

2.10 Determine whether it is possible to identify, from the impulse response of the nonlinear systems, the linear systems embedded in the following structures:
(a) The second-order Wiener model.
(b) The second-order cascade system in Figure 2.32.

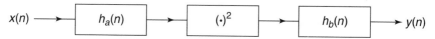

Figure 2.32 A second-order system.

2.11 Determine whether the constraint given in (2.34) is a necessary and sufficient condition for the BIBO stability of the following systems:
(a) The Wiener model.
(b) The Hammerstein model.
(c) A multiplicative system involving two linear systems $h_a(n)$ and $h_b(n)$.

2.12 Show that the kernel

$$h_2(m_1, m_2) = \left[\sum_{n=1}^{\infty} \frac{1}{(2n)^2} u_0(m_1 - 2n) \right] \left[- \sum_{k=0}^{n-1} u_0(m_2 - k) + \sum_{k=n}^{2n-1} u_0(m_2 - k) \right]$$

corresponds to a stable second-order Volterra filter, but does not satisfy the BIBO stability condition of (2.34).

2.13 Consider a homogeneous quadratic Volterra filter with six-sample memory represented by the triangular matrix

$$\mathbf{H}_{2,tri} = \begin{bmatrix} 0.60 & 1.00 & 0.70 & 1.50 & 0.20 & 0.10 \\ 0.00 & 1.00 & 0.80 & 0.90 & 1.10 & 0.40 \\ 0.00 & 0.00 & 1.30 & 0.70 & 0.20 & 0.40 \\ 0.00 & 0.00 & 0.00 & 1.00 & 0.80 & 0.50 \\ 0.00 & 0.00 & 0.00 & 0.00 & 0.80 & 1.00 \\ 0.00 & 0.00 & 0.00 & 0.00 & 0.00 & 0.50 \end{bmatrix}$$

Find the corresponding symmetric and regular kernels.

2.14 Consider a homogeneous cubic kernel of the form

$$h_3(m_1, m_2, m_3) = h(m_1)h(m_2)h(m_3)g(m_1 + m_2).$$

Find its symmetric form $h_{3,\text{sym}}(m_1, m_2, m_3)$.

2.15 Derive the responses to a complete set of pairs of impulses necessary to identify the quadratic filter described by the matrix

$$\mathbf{H}_2 = \begin{bmatrix} 0.60 & 0.00 & 0.00 & 0.80 & 0.20 & 0.00 \\ 0.00 & 1.00 & 0.00 & 0.00 & 0.00 & 0.40 \\ 0.00 & 0.00 & 1.00 & 0.00 & 0.00 & 0.20 \\ 0.80 & 0.00 & 0.00 & 1.00 & 0.00 & 0.00 \\ 0.20 & 0.00 & 0.00 & 0.00 & 0.50 & 0.00 \\ 0.00 & 0.40 & 0.20 & 0.00 & 0.00 & 0.50 \end{bmatrix}.$$

2.16 Consider the second-order system

$$y(n) = 0.8x(n) + 0.4x(n-1) - 0.5x(n-2) + x(n-3)$$
$$+ 0.5x^2(n) + x(n)x(n-1) + 0.5x(n)x(n-2) + 0.2x(n)x(n-3)$$
$$+ 0.6x^2(n-1) + 0.7x(n-1)x(n-2) + 0.5x(n-1)x(n-3)$$
$$+ 0.4x^2(n-2) + 0.2x(n-2)x(n-3) + 0.1x^2(n-3).$$

Find the output when the input is a sinusoidal signal of the form

(a) $x(n) = e^{(j\pi/4n)}$
(b) $x(n) = e^{(j\pi/2n)}$
(c) $x(n) = e^{(j3\pi/4n)}$

and determine if there is aliasing in the output signal.

2.17 Consider the discrete-time Teager energy model described by the equation

$$y(n) = x^2(n) - x(n+1)x(n-1).$$

Derive an expression for its output when the input signal is $x(n) = A\cos\omega_a n$.

2.18 Derive the intermodulation components at the output of the cubic system

$$y(n) = x^3(n) + x^2(n)x(n-1) + x(n)x^2(n-2)$$

for the input signal

$$x(n) = Ae^{j\omega_a n} + Be^{j\omega_b n}.$$

Determine if the intermodulation components produce aliasing effects for the following choices of w_a and w_b:

(a) $\omega_a = \pi/6$, $\omega_b = \pi/8$.

(b) $\omega_a = \pi/3$, $\omega_b = \pi/6$.

(c) $\omega_a = \pi/2$, $\omega_b = \pi/4$.

(d) $\omega_a = \pi/2$, $\omega_b = -\pi/2$.

2.19 Compute the response of the Volterra system $y(n) = x^2(n)$ to a signal with a discrete-time Fourier transform given by

$$X(e^{j\omega}) = \begin{cases} 1; & 0 \le \omega \le |\pi/2| \\ 0; & |\pi/2| < \omega \le |\pi| \end{cases}$$

2.20 Consider the second-order system

$$
\begin{aligned}
y(n) = {} & 0.8x(n) + 0.4x(n-1) - 0.5x(n-2) + x(n-3) \\
& + 0.5x^2(n) + x(n)x(n-1) + 0.5x(n)x(n-2) + 0.2x(n)x(n-3) \\
& + 0.6x^2(n-1) + 0.7x(n-1)x(n-2) + 0.5x(n-1)x(n-3) \\
& + 0.4x^2(n-2) + 0.2x(n-2)x(n-3) + 0.1x^2(n-3).
\end{aligned}
$$

Find a z-transform representation for this system.

2.21 Derive the input–output nonlinear relationship from the associated two-dimensional z-transform

$$
\begin{aligned}
W(z_1, z_2) = {} & (0.5 + z_2^{-1} + 0.5z_2^{-2} + 0.6z_1^{-1}z_2^{-1} + 0.7z_1^{-1}z_2^{-2} \\
& + 0.4z_1^{-2}z_2^{-2})X(z_1)X(z_2).
\end{aligned}
$$

2.22 Consider the quadratic filter

$$
\begin{aligned}
y(n) = {} & 0.5x^2(n) + x(n)x(n-1) + 0.5x(n)x(n-2) + 0.2x(n)x(n-3) \\
& + 0.1x(n)x(n-4) - 0.2x(n)x(n-5).
\end{aligned}
$$

Compute the DFT of the corresponding kernel on $M = 8$ and $M = 4$ points and explicitly show any time-domain aliasing that might exist at the output by evaluating the inverse DFT of $H_2(k_1, k_2)$ for both cases.

2.23 Consider the cubic system

$$y(n) = x^3(n) + 0.5x^3(n-1) + 0.8x^3(n-2) + 0.4x^3(n-3) + 0.2x^3(n-4).$$

Compute the DFT of the corresponding kernel using $M = 8$ and $M = 4$, and determine if any time-domain aliasing occurs due to inadequate size of the inverse DFT.

2.24   *Computing Assignment:* Compute the output of the Volterra filter

$$y(n) = 0.5x^2(n) + x(n)x(n-1) + 0.5x(n)x(n-2) + 0.2x(n)x(n-3)$$

employing the DFT of the corresponding kernel. Use 60 samples of a real random sequence as the input signal and employ $M = 64$ for the DFT size. Explain how this computation can be simplified using the symmetries of the DFT.

3

REALIZATION OF TRUNCATED VOLTERRA FILTERS

This chapter is concerned with various methods for realizing truncated Volterra filters. Different realizations such as cascade, parallel, and lattice are well-known for linear filters. Various realizations of nonlinear filters are not as well-known. We start our discussions with realizations of quadratic filters because of the pedagogical simplicity such filters provide as well as the significant number of applications in which they find use. The concepts that are developed for quadratic filters are extended to higher-order filters later in the chapter.

3.1 ALGEBRAIC REPRESENTATION OF QUADRATIC FILTERS

A causal, discrete, and time-invariant quadratic filter with finite support for its kernels is described by means of the first three terms of the Volterra series expansion as

$$y(n) = h_0 + \sum_{m_1=0}^{N-1} h_1(m_1)x(n - m_1) + \sum_{m_1=0}^{N-1}\sum_{m_2=0}^{N-1} h_2(m_1, m_2)x(n - m_1)x(n - m_2),$$

$$(3.1)$$

where h_0 represents a constant offset term, $\{h_1(m_1); 0 \leq m_1 \leq N - 1\}$ denotes the coefficients of a linear, time-invariant FIR filter and $\{h_2(m_1, m_2); 0 \leq m_1, m_2 \leq N - 1\}$ represents the coefficients of a homogeneous quadratic filter. The quadratic filter can be realized as a parallel combination of these three components as shown in Figure 3.1. As was shown in Chapter 2, the homogeneous quadratic term in (3.1) may be equivalently represented using a triangular kernel as

$$\bar{h}_2[x(n)] = \sum_{m_1=0}^{N-1}\sum_{m_2=m_1}^{N-1} h_{2,tri}(m_1, m_2)x(n - m_1)x(n - m_2), \qquad (3.2)$$

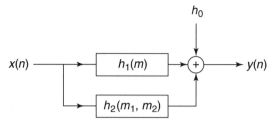

Figure 3.1 A quadratic filter realized as a parallel combination of three components.

where

$$h_{2,tri}(m_1, m_2) = \begin{cases} h_2(m_1, m_2); & m_1 = m_2 \\ h_2(m_1, m_2) + h_2(m_2, m_1); & m_1 < m_2 \\ 0; & \text{otherwise.} \end{cases} \quad (3.3)$$

These elements can also be computed as the sum of the corresponding terms of the symmetric kernel as

$$h_{2,tri}(m_1, m_2) = \begin{cases} |\pi(m_1, m_2)| h_{2,sym}(m_1, m_2); & m_1 \leq m_2 \\ 0; & \text{otherwise,} \end{cases} \quad (3.4)$$

where

$$|\pi(m_1, m_2)| = \begin{cases} 1; & m_1 = m_2 \\ 2; & m_1 < m_2. \end{cases} \quad (3.5)$$

We now introduce two different, but mathematically equivalent representations of homogeneous quadratic filters. These mathematical descriptions are well-suited for deriving efficient realizations of such filters.

3.1.1 Matrix–Vector Representation

It is usual to arrange the input values and the filter coefficients of one-dimensional linear FIR filters into two vectors $\mathbf{X}_1(n)$ and $\vec{\mathbf{H}}_1$ defined as

$$\mathbf{X}_1(n) = [x(n) \quad x(n-1) \quad \cdots \quad x(n-N+1)]^T \quad (3.6)$$

and

$$\vec{\mathbf{H}}_1 = [h_1(0) \quad h_1(1) \quad \cdots \quad h_1(N-1)]^T, \quad (3.7)$$

respectively, so that its output is given by the expression

$$\bar{h}_1[x(n)] = \mathbf{X}_1^T(n)\vec{\mathbf{H}}_1 = \vec{\mathbf{H}}_1^T\mathbf{X}_1(n). \tag{3.8}$$

It is easy to show that the input–output relation of a homogeneous quadratic filter can be written as

$$\bar{h}_2[x(n)] = \mathbf{X}_1^T(n)\mathbf{H}_2\mathbf{X}_1(n), \tag{3.9}$$

where $\mathbf{H}_2$ is an $N \times N$ matrix in which the coefficients of the quadratic kernel $h_2(m_1, m_2)$ are arranged as

$$\mathbf{H}_2 = \begin{bmatrix} h_2(0,0) & h_2(0,1) & \cdots & h_2(0,N-1) \\ h_2(1,0) & h_2(1,1) & \cdots & h_2(1,N-1) \\ \vdots & \vdots & \ddots & \vdots \\ h_2(N-1,0) & h_2(N-1,1) & \cdots & h_2(N-1,N-1) \end{bmatrix}. \tag{3.10}$$

The matrix $\mathbf{H}_2$ takes on different structures depending on the type of system realization. For example, $\mathbf{H}_2$ is symmetric for the symmetric realization of the quadratic kernel. When the triangular form is utilized, $\mathbf{H}_2$ is upper triangular. The regular form realization also results in an upper triangular matrix, but the triangularity is with respect to the antidiagonal. Examples of the quadratic coefficient matrices for symmetric, triangular and regular forms are given in Example 2.4. We now consider two examples in which the coefficient matrix assumes particular structures.

Example 3.1: The Hammerstein Model   Consider a Hammerstein model similar to the one described in Example 2.2. The output of the quadratic filter is obtained by processing the squared values of the input signal with a linear, time-invariant filter as

$$y(n) = \sum_{m=0}^{N-1} h_1(m)x^2(n-m),$$

where $h_1(m)$ represents the unit impulse response of the FIR filter. A block diagram for implementing this system is shown in Figure 3.2. It is straightforward to see that $\mathbf{H}_2$ for this coefficient matrix is given by

$$\mathbf{H}_2 = \text{diag}\,[\,h_1(0) \quad h_1(1) \quad \cdots \quad h_1(N-1)\,].$$

Example 3.2: The Wiener Model   We consider a quadratic Wiener system model similar to the one considered in Example 2.1. The system is depicted in Figure 3.3.

$x(n) \longrightarrow \boxed{(\cdot)^2} \longrightarrow \boxed{\text{FIR filter}} \longrightarrow y(n)$

Figure 3.2 Hammerstein model of Example 3.1.

Figure 3.3 Wiener model of Example 3.2.

The output of this system is obtained as the squared value of the output of a linear FIR filter with impulse response function $\{h_1(m); m = 0, 1, \ldots, N - 1\}$. Simple algebraic manipulations will show that $\mathbf{H}_2$ is a dyadic matrix, *i.e.*, $h_2(m_1, m_2) = h_1(m_1)h_1(m_2)$.

Remark 3.1 Suppose now that one of the system requirements is that the output of the homogeneous quadratic component be positive for all nonzero input signals. This is the case when the output of the quadratic filter gives an estimate of the input signal power [225]. Then, the coefficient matrix $\mathbf{H}_2$ must be positive definite. The matrix $\mathbf{H}_2$ can assume other forms in practical applications. For example, it has been shown that in some applications involving adaptive detection [225] $\mathbf{H}_2$ should be a Toeplitz matrix, i.e., $h_2(m_1, m_2) = h_2(|m_1 - m_2|)$.

3.1.2 Vector Representation

Vector representation is an alternative description for representing Volterra kernels with finite support. The output of a homogeneous quadratic filter is expressed using the vector representation as

$$\bar{h}_2[x(n)] = \mathbf{X}_2^T(n)\vec{\mathbf{H}}_2 = \vec{\mathbf{H}}_2^T\mathbf{X}_2(n), \tag{3.11}$$

where the vector $\mathbf{X}_2(n)$ contains the products of pairs of input samples in each of its entries. The input vector $\mathbf{X}_2(n)$ may be expressed compactly as

$$\mathbf{X}_2(n) = \mathbf{X}_1(n) \otimes \mathbf{X}_1(n), \tag{3.12}$$

where the symbol $\otimes$ indicates the Kronecker product operation. The Kronecker product of an $L_1 \times M_1$ element matrix $\mathbf{A}$ with an $L_2 \times M_2$-element matrix $\mathbf{B}$ results in an $L_1L_2 \times M_1M_2$ element matrix $\mathbf{A} \otimes \mathbf{B}$ expressed as

$$\mathbf{A} \otimes \mathbf{B} = \begin{bmatrix} a_{0,0}\mathbf{B} & a_{0,1}\mathbf{B} & \cdots & a_{0,M_1-1}\mathbf{B} \\ a_{1,0}\mathbf{B} & a_{1,1}\mathbf{B} & \cdots & a_{1,M_1-1}\mathbf{B} \\ \vdots & \vdots & \vdots & \vdots \\ a_{L_1-1,0}\mathbf{B} & a_{L_1-1,1}\mathbf{B} & \cdots & a_{L_1-1,M_1-1}\mathbf{B} \end{bmatrix}. \tag{3.13}$$

Some properties of Kronecker products are reviewed in Appendix A.

The Kronecker product operation implies a well-defined ordering of the elements in $\mathbf{X}_2(n)$. The coefficients of the quadratic kernel $h_2(m_1, m_2)$ are arranged in the vector $\vec{\mathbf{H}}_2$ such that the location of the coefficient $h_2(m_1, m_2)$ in $\vec{\mathbf{H}}_2$ is identical to the location of the input sample product $x(n - m_1)x(n - m_2)$ in $\mathbf{X}_2(n)$. The number of the elements in $\mathbf{X}_2(n)$ and $\vec{\mathbf{H}}_2$ is N^2. When the symmetry of the Volterra kernels is exploited, the number of independent coefficients reduces to

$$N_2 = \binom{N+1}{2},$$
(3.14)

as was shown by (2.43). Input and coefficient vectors of shorter length may be obtained by eliminating the redundant terms from $\mathbf{X}_2(n)$ and $\vec{\mathbf{H}}_2$. Such a representation is the basis for the realizations employing the so-called distributed arithmetic technique [221] described in Section 3.2.4.

3.2 REALIZATION OF QUADRATIC FILTERS

3.2.1 Direct-Form Realization

The homogeneous quadratic component of the Volterra series expansion given by (3.2) can be implemented using the *direct-form* realization [18] by means of a nonlinear combiner, a number of multipliers, and a summing bus as shown in Figure 3.4. The nonlinear combiner computes all the necessary products of the input samples. The realization shown in the figure uses the least number of multipliers needed to implement the system. A different direct-form realization is shown in

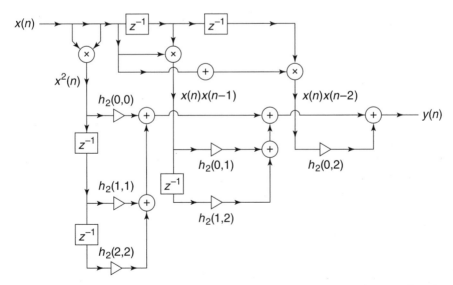

Figure 3.4 The direct-form realization of a quadratic operator as shown here requires the minimum number of multipliers.

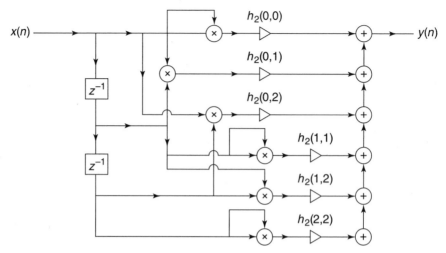

Figure 3.5 The direct-form realization of a quadratic operator as shown here requires the minimum number of delay elements.

Figure 3.5. This realization uses the minimum number of delay elements at the expense of extra multipliers. Direct-form realizations are most commonly employed in applications involving truncated Volterra systems. However, such realizations do not lend themselves easily to simplifications using approximate representations. Consequently, this chapter emphasizes alternate realizations that may have advantages over the direct form realization.

3.2.2 Realization Based on the Convolution Property

Recall from Section 2.3.2 that the output of a homogeneous quadratic filter can be obtained using a two-dimensional convolution. In this approach, we first define a two-dimensional signal from the one-dimensional input signal $x(n)$ as

$$v(n_1, n_2) = x(n_1)x(n_2). \tag{3.15}$$

The two-dimensional convolution of the coefficient matrix $\mathbf{H_2}$ and $v(n_1, n_2)$ is performed and the output of the quadratic filter is obtained as the output of the convolution operation on the principal diagonal of the (n_1, n_2) plane. However, the usefulness of a Volterra filter realization using convolution in the time domain is mostly in the analysis of Volterra filters. An efficient realization based on the ideas described above, but using fast Fourier transform techniques, is discussed in Section 3.2.5.

3.2.3 Structures Based on Matrix Decompositions

We now describe a method for implementing a quadratic filter based on matrix decompositions. This approach was developed in Chang et al. [46,47] and Venetsanopoulos et al. [322]. We first consider a realization based on an arbitrary decomposition of a coefficient matrix into a sum of rank one matrices. We then show that a judicious choice of the matrix decomposition may provide means for reduced complexity realizations of the quadratic filter. The possible choices include the LU decomposition involving lower and upper triangular matrices, the singular value (SV) decomposition, and many others [47,75].

3.2.3.1 General Matrix Decomposition

Any symmetric $(N \times N)$-element matrix $\mathbf{H}_2$ of rank r can be decomposed into a finite sum of r rank one matrices as [75]

$$\mathbf{H}_2 = \sum_{i=1}^{r} q_i \mathbf{R}_i \mathbf{R}_i^T \qquad (3.16)$$

where $\{q_i; i = 1, 2, \ldots, r\}$ are scalar numbers and $\{\mathbf{R}_i; i = 1, 2, \ldots, r\}$ are N-element vectors. Using this decomposition for the coefficient matrix $\mathbf{H}_2$ in (3.9) results in the following expression for the output of a homogeneous quadratic filter:

$$\bar{h}_2[x(n)] = \sum_{i=1}^{r} q_i [\mathbf{X}_1^T(n)\mathbf{R}_i][\mathbf{R}_i^T \mathbf{X}_1(n)]. \qquad (3.17)$$

This equation can be rewritten as

$$y(n) = \bar{h}_2[x(n)] = \sum_{i=1}^{r} q_i y_i^2(n), \qquad (3.18)$$

where

$$y_i(n) = \mathbf{X}_1^T(n)\mathbf{R}_i = \mathbf{R}_i^T \mathbf{X}_1(n). \qquad (3.19)$$

is the output of a linear filter with impulse response function described by the coefficient vector $\mathbf{R}_i$. Consequently, the overall output $y(n)$ of the quadratic filter can be computed by squaring the outputs of r linear FIR filters and then summing all the squared values after weighting them with appropriate constants. Such a realization is shown in Figure 3.6.

Remark 3.2: Finite-Memory Quadratic Systems Can Be Realized Exactly Using Parallel-Cascade Structures   Recall from Section 2.4.1 that a parallel-cascade realization involving linear filters and memoryless nonlinearities can provide arbitrarily close approximations for a large class of nonlinear systems. The realization of Figure 3.6 shows that a quadratic filter with finite memory can be exactly represented using a parallel-cascade structure.

3.2.3.2 Complexity of Parallel-Cascade Realizations

Even though the parallel-cascade structure of Figure 3.6 is conceptually simple, the complexity of its implementation is similar to that of direct-form realizations when $\mathbf{H}_2$ has full rank. However, when the rank r of the coefficient matrix is smaller than the memory

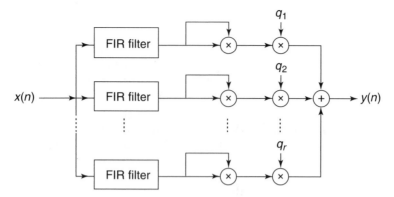

Figure 3.6   Quadratic filter realization by general matrix decomposition.

length N, the computational complexity may be much lower since we require only r parallel paths in the realization of the filter.

We now describe several special cases of the realization based on matrix decompositions.

3.2.3.3 LU Decomposition
A slightly more efficient realization of the quadratic filters arises from the application of the LU decomposition of the symmetric coefficient matrix $\mathbf{H_2}$. Under some mild conditions described in Appendix B, we can express $\mathbf{H_2}$ as

$$\mathbf{H_2} = \sum_{i=1}^{r} d_i \mathbf{L}_i \mathbf{L}_i^T, \tag{3.20}$$

where the d_i terms are scalar multipliers and the vectors $\mathbf{L}_i$ have $i-1$ leading zeros. Consequently, the ith FIR filter in the parallel-cascade realization of the quadratic filter of Figure 3.6 has $i-1$ coefficients that are zero. Furthermore, the squaring and adding operations in the ith branch can be implemented at time $n-i$ since the latest data sample used in this branch is $x(n-i)$. Thus, these operations can be performed sequentially for each branch and the complete system can be implemented with minimum throughput delay [47]. A quadratic filter realization employing the LU matrix decomposition is shown in Figure 3.7. As was the case for the realization based on the general matrix decomposition, the complexity of the implementation of Figure 3.7 is clearly related to the rank of the matrix $\mathbf{H_2}$.

Example 3.3   A homogeneous quadratic system described by a 49×49 coefficient kernel is displayed in a three-dimensional form in Figure 3.8. Even though this system has a 49-sample memory, the rank of the coefficient matrix $\mathbf{H_2}$ is only 22. Therefore, a compact realization can be obtained by applying the LU decomposition to $\mathbf{H_2}$. The direct realization of the triangular form of the quadratic kernel requires 1225 coefficients, whereas the LU decomposition-based realization requires only

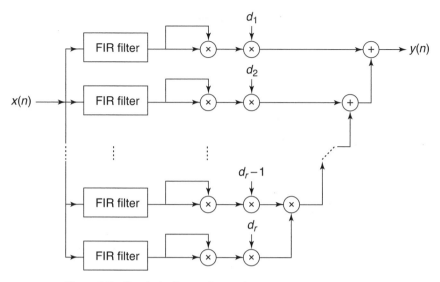

Figure 3.7 Quadratic filter realization by LU matrix decomposition.

847 coefficients for the 22 FIR filters and 22 additional multipliers to represent the scaling factor d_i. Note that we may incorporate the magnitude of d_i into the filters by scaling each coefficient in the ith filter by $\sqrt{|d_i|}$.

This would eliminate the need for additional scaling after the squaring operation. However, care must be taken to change the sign of the squared output $y_i^2(n)$ if d_i is negative.

3.2.3.4 *Singular Value Decomposition*

We have so far discussed methods for obtaining exact realizations of the quadratic filters using parallel-cascade structures. Simpler realizations arise when the rank of the coefficient matrix is smaller than the memory span of the filter. Singular value decompositions [75] can be used to derive approximate realizations with significantly reduced complexity, often with a minimal loss of performance. The SV decomposition of a symmetric matrix **H** of rank r decomposes it as

$$\mathbf{H} = \sum_{i=1}^{r} w_i \mathbf{S}_i \mathbf{S}_i^T, \tag{3.21}$$

where the w_i terms are the singular values of **H**, usually arranged in a decreasing order of magnitude, and the $\mathbf{S}_i$ terms are the corresponding singular vectors. The vectors $\mathbf{S}_i$ and $\mathbf{S}_j$ are orthogonal to each other in the sense that $\mathbf{S}_i^T \mathbf{S}_j = 0$ whenever $i \neq j$ and have unit length, i.e., $\|\mathbf{S}_i\|^2 = 1$.

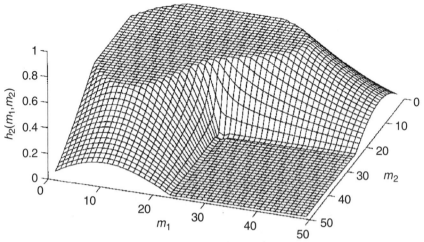

Figure 3.8 Three-dimensional representation of the elements of a 49 × 49 quadratic kernel in Example 3.3. (Courtesy of S. Marsi.)

The usefulness of singular value decompositions in quadratic filter realization comes from the fact that the best rank k approximation of $\mathbf{H}$ in the sense of minimizing the Frobenius norm of the error matrix is

$$\tilde{\mathbf{H}} = \sum_{i=1}^{k} w_i \mathbf{S}_i \mathbf{S}_i^T . \tag{3.22}$$

where $\{w_i; i = 1, 2, \ldots, k\}$ are the largest k singular values of $\mathbf{H}$. The Frobenius norm of an $(N \times N)$-element matrix $\mathbf{A}$ is defined as

$$\|\mathbf{A}\|_F = \sqrt{\sum_{i=1}^{N} \sum_{j=1}^{N} a_{i,j}^2}, \tag{3.23}$$

where $a_{i,j}$ is the (i,j)th element of $\mathbf{A}$. It is easy to show, using the orthogonality of the vectors in the set $\{\mathbf{S}_i: i = 1, 2, \ldots, r\}$, that

$$\|\mathbf{H} - \tilde{\mathbf{H}}\|_F = \sqrt{\sum_{i=k+1}^{r} w_i^2}. \tag{3.24}$$

In many cases, an approximate expansion of the coefficient matrix $\mathbf{H_2}$ using a small number of terms of the singular value decomposition could result in substantial savings in computation and memory requirements. We can also apply the LU decomposition to the approximate coefficient matrix $\tilde{\mathbf{H}}_2$ to achieve even more reduction in the system complexity. We now present an example of an approximate realization of a quadratic filter obtained in this manner.

Example 3.4 Figure 3.9 displays the three-dimensional representation of the approximation of the quadratic kernel of Example 3.3 using only the four highest singular values. The squared value of the norm of the coefficient error matrix, as defined in (3.23), is only 0.3567. Comparing this value with the Frobenius norm of the coefficient matrix itself, which is 270, we conclude that an approximate parallel-cascade realization using only four channels is efficient and accurate for this system. If we use only the two highest singular values in the matrix approximation, the squared norm of the coefficient error matrix becomes 1.9084. Figure 3.10 displays the corresponding kernel. Note that we can notice some of the distortions in this case.

3.2.3.5 Some Additional Advantages of the Parallel-Cascade Structure

As was previously shown, the quadratic filter realizations derived using matrix decompositions consist of a set of parallel FIR filters in cascade with a set of squaring and adding operations. Each stage can operate simultaneously on the same input array, resulting in a substantial decrease in the data throughput delay over completely sequential realizations. These structures also offer modularity and regularity, and therefore are well suited for very large scale integrated (VLSI) circuit implementation.

3.2.4 Structures Based on Distributed Arithmetic

Filter realizations using distributed arithmetic *distribute* the multiplication and addition operations at the level of single bits in a finite-wordlength environment. Such realizations were first conceived for linear systems in Peled and Liu [221]. In this section, we describe efficient realizations of quadratic filters obtained by extending the basic approach of those authors [221] to the nonlinear system models. Two different distributed arithmetic realizations are presented in this section [284,286]. The first one uses combinatorial structures based on logical operations at

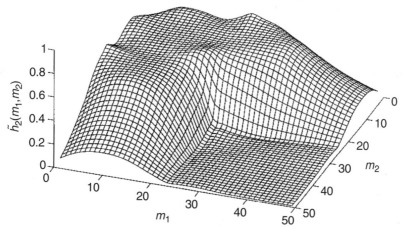

Figure 3.9 Three-dimensional representation of the elements of the 49×49-element quadratic kernel of Figure 3.8 approximated with the first four eigenvectors of the SV decomposition. (Courtesy of S. Marsi.)

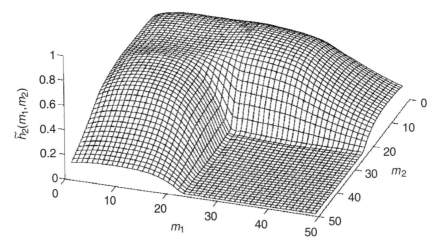

Figure 3.10 Three-dimensional representation of the elements of a 49 × 49-element quadratic kernel of Figure 3.8 approximated with the first two eigenvectors of the SV decomposition. (Courtesy of S. Marsi.)

the bit level. The second approach involves memory-oriented realizations that use certain functions of binary arguments stored in read-only or random-access memory locations.

3.2.4.1 *The Principle of Distributed Arithmetic Realizations*   Assume
that the input signal is normalized so that $|x(n)| \leq 1$ for all n. In order to perform the multiplication and addition operations at the bit level, each input sample is coded by means of logical variables $x_b(n)$ that assume binary values such that

$$x(n) = \sum_{b=1}^{B} p(b)x_b(n), \tag{3.25}$$

where B is the wordlength used and the weights $p(b)$ depend on the binary code employed. For the two's-complement code,

$$p(1) = -1 \quad \text{and} \quad p(b) = 2^{-b+1}; \qquad b = 2, \ldots, B \tag{3.26}$$

and $x_b(n)$ is either 1 or 0. For the offset-binary code, $x_b(n) = \pm 1$ and

$$p(b) = 2^{-b}, \qquad b = 1, \ldots, B. \tag{3.27}$$

Remark 3.3   Note that $+1$ is not represented in the two's-complement code. Similarly, we cannot represent 0 exactly in the offset-binary code.

By extending the definition in (3.25) to the past $N - 1$ input samples, the input vector of (3.6) can be expressed in the form

$$\mathbf{X}_1(n) = \mathbf{Q}_1(n)\mathbf{P}_1, \tag{3.28}$$

where $\mathbf{P}_1$ is a B-element vector of the binary weights given by

$$\mathbf{P}_1 = [p(1) \quad p(2) \quad \cdots \quad p(B)]^T \tag{3.29}$$

and $\mathbf{Q}_1(n)$ is an $N \times B$-element matrix obtained by expanding each component of the vector $\mathbf{X}_1(n)$ in its B elementary bits as

$$\mathbf{Q}_1(n) = \begin{bmatrix} x_1(n) & x_2(n) & \cdots & x_B(n) \\ x_1(n-1) & x_2(n-1) & \cdots & x_B(n-1) \\ \vdots & \vdots & \ddots & \vdots \\ x_1(n-N+1) & x_2(n-N+1) & \cdots & x_B(n-N+1) \end{bmatrix}. \tag{3.30}$$

Consequently, the output of a linear FIR filter can be expressed using the vector notation of (3.8), in the form

$$\bar{h}_1[x(n)] = \mathbf{X}^T(n)\vec{\mathbf{H}}_1 = \mathbf{P}_1^T\mathbf{Q}_1^T(n)\vec{\mathbf{H}}_1 \tag{3.31}$$

The key to implementing a linear FIR filter using distributed arithmetic is to first perform the computations involved in the matrix multiplication $\mathbf{Q}_1^T(n)\vec{\mathbf{H}}_1$. Since all the elements of $\mathbf{Q}_1(n)$ are either ones or zeroes, or $+1$ or -1, depending on the type of code employed, the above matrix multiplication can be accomplished using only additions and subtractions. Calculation of the product $\mathbf{P}_1^T(\mathbf{Q}_1^T(n)\vec{\mathbf{H}}_1)$ involves only bit-shifting, additions, and subtractions since the elements of the vector $\mathbf{P}_1$ are -1 or a negative integer power of two.

3.2.4.2 *Distributed Arithmetic Representation of Quadratic Filters* To derive a distributed arithmetic realization of a homogeneous quadratic filter, we start by substituting (3.28) in (3.12), which gives

$$\mathbf{X}_2(n) = \mathbf{X}_1(n) \otimes \mathbf{X}_1(n)$$

$$= \mathbf{Q}_1(n)\mathbf{P}_1 \otimes \mathbf{Q}_1(n)\mathbf{P}_1$$

$$= [\mathbf{Q}_1(n) \otimes \mathbf{Q}_1(n)][\mathbf{P}_1 \otimes \mathbf{P}_1]$$

$$= \mathbf{Q}_2(n)\mathbf{P}_2, \tag{3.32}$$

where

$$\mathbf{Q}_2(n) = \mathbf{Q}_1(n) \otimes \mathbf{Q}_1(n) \tag{3.33}$$

is an $N^2 \times B^2$-element matrix, and

$$P_2 = P_1 \otimes P_1. \tag{3.34}$$

is a B^2-element vector. We made use of the mixed product rule of Kronecker products described in Appendix A for deriving (3.32). The elements of $Q_2(n)$ still take values from a binary set, and the elements of P_2 are all still integer powers of two with a possible negative sign. The output of the homogeneous quadratic filter is given by

$$\bar{h}_2[x(n)] = X_2^T(n)\vec{H}_2 = P_2^T Q_2^T(n)\vec{H}_2. \tag{3.35}$$

The use of the symmetry property of the Volterra kernels permits a reduction of the implementation complexity. We can define a smaller vector $\vec{H}_{2r}$ containing only $N_2 = N(N+1)/2$ independent coefficients in the quadratic kernel, and a corresponding reduced-size input vector $X_{2r}(n)$ obtained by deleting all redundant elements from $X_2(n)$. The output of the homogeneous quadratic filter is then given by

$$\bar{h}_2[x(n)] = X_{2r}^T(n)\vec{H}_{2r} = P_2^T Q_{2r}^T(n)\vec{H}_{2r}, \tag{3.36}$$

where $Q_{2r}(n)$ is an $N_2 \times B^2$-element matrix obtained by expanding the vector $X_{2r}(n)$ in its B^2 elementary bits.

3.2.4.3 Relationship between $Q_2(n)$ and $Q_{2r}(n)$ The vectors $\vec{H}_2$ and $\vec{H}_{2r}$ are related by means of an $N_2 \times N^2$ matrix transformation of the form

$$\vec{H}_{2r} = M_1 \vec{H}_2. \tag{3.37}$$

The matrix M_1 can be defined so that all its elements are either 1 or 0. Deletion of the redundant elements from $X_2(n)$ to get $X_{2r}(n)$ can be described using matrix notation as

$$X_{2r}(n) = M_2 X_2(n), \tag{3.38}$$

where M_2 is an $N_2 \times N^2$-element matrix whose elements are all either 1 or 0. Furthermore, each row of M_2 contains only a single 1, so that the matrix multiplication clearly corresponds to picking N_2 elements from the vector $X_2(n)$. The bit-level decomposition of $X_{2r}(n)$ can be achieved easily since

$$X_{2r}(n) = M_2 X_2(n) = M_2 Q_2(n)P_2, \tag{3.39}$$

and thus

$$Q_{2r}(n) = M_2 Q_2(n). \tag{3.40}$$

Example 3.5 Consider a generic kernel with $N = 2$ and

$$\vec{\mathbf{H}}_2 = [\, h_2(0, 0) \quad h_2(0, 1) \quad h_2(1, 0) \quad h_2(1, 1) \,]^T.$$

The matrix $\mathbf{M}_1$ for this coefficient vector is given by

$$\mathbf{M}_1 = \begin{bmatrix} 1 & 0 & 0 & 0 \\ 0 & 1 & 1 & 0 \\ 0 & 0 & 0 & 1 \end{bmatrix},$$

and the matrix $\mathbf{M}_2$ is given by

$$\mathbf{M}_2 = \begin{bmatrix} 1 & 0 & 0 & 0 \\ 0 & 1 & 0 & 0 \\ 0 & 0 & 0 & 1 \end{bmatrix}.$$

Obviously, the matrix multiplication $\mathbf{M}_2\mathbf{X}_2(n)$ corresponds to picking $N_2 = 3$ elements from the vector $\mathbf{X}_2(n)$ to form $\mathbf{X}_{2r}(n)$.

Figure 3.11 shows the block diagram of the structure derived from (3.36). First, the elements of the vector $\mathbf{Q}_{2r}^T(n)\vec{\mathbf{H}}_{2r}$ are computed using delayed samples of the input signal $x(n)$. Subsequently, the shifts and the additions implied by the multiplication of $\mathbf{Q}_{2r}^T(n)\vec{\mathbf{H}}_{2r}$ with $\mathbf{P}_2^T$ are implemented. The elements of the vector $\mathbf{P}_2$ can take at most $2B - 1$ different values. Consequently, the multiplication of $\mathbf{P}_2^T$ with $\mathbf{Q}_{2r}^T(n)\vec{\mathbf{H}}_{2r}$ can be performed by first adding together all the entries of $\mathbf{Q}_{2r}^T\vec{\mathbf{H}}_{2r}$ that are

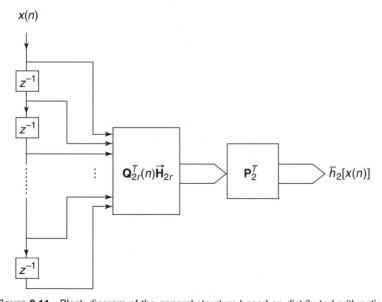

Figure 3.11   Block diagram of the general structure based on distributed arithmetic.

multiplied by the same value, and then multiplying the sum by the appropriate value. Since this multiplication involves an integer power of two, it can be accomplished by a bit-shift operation. Furthermore, by performing the operations as described above, no more-than $2B - 1$ shifts need to be executed. The key to an efficient realization of the quadratic filter is the computation of the vector $\mathbf{Q}_{2r}^T(n)\vec{\mathbf{H}}_{2r}$. We describe two methods for efficiently implementing the matrix product next. Both methods are useful for realizations using general-purpose DSPs or dedicated integrated circuits.

3.2.4.4 Combinatorial Structures A block diagram for the distributed arithmetic realization of a homogeneous quadratic filter that uses a combinatorial structure and exploits the symmetry property of the Volterra kernels is shown in Figure 3.12. This method first generates the entries of the matrix $\mathbf{Q}_{2r}(n)$ from the bit-level description of the input signal $x(n)$ and its past samples as shown in the top part of Figure 3.13 for an example involving $N = 2$ and $B = 2$. The structure shown in

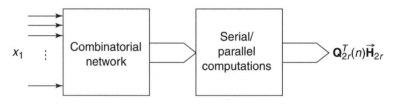

Figure 3.12   Combinatorial implementation.

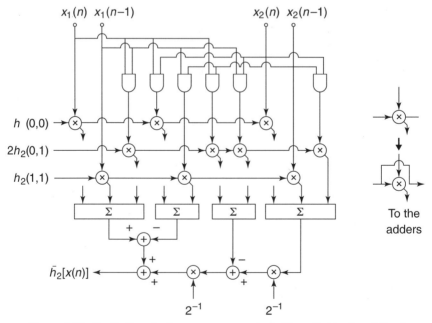

Figure 3.13   Realization of a homogeneous quadratic filter with $N = 2$ and $B = 2$.

the figure uses AND gates to generate the entries of $\mathbf{Q}_{2r}(n)$ when two's-complement code is employed. The AND gates should be replaced with exclusive NOR gates for realization using offset-binary code.

The next step in the realization is the computation of the vector $\mathbf{Q}_{2r}^T(n)\vec{\mathbf{H}}_{2r}$ which is implemented by using a serial/parallel arithmetic. The outputs of the AND gates that generate the entries of $\mathbf{Q}_{2r}(n)$ are transferred to the multipliers as shown in Figure 3.13. The coefficients $h_2(m_1, m_2)$ enter serially along the horizontal lines, and the multiplications are executed in parallel at the bit level. The bit-level multiplication can be accomplished using an AND gate and a shift register that contains the bit-level description of the coefficients. The multiplier structure is shown in Figure 3.14.

The complexity of the combinatorial structure depends on the memory of the quadratic filter and the number of bits used. The matrix $\mathbf{Q}_{2r}(n)$ has $N_2 \times B^2$ entries. Consequently, we require $N_2 \times B^2$ AND gates for generating $\mathbf{Q}_{2r}(n)$. However, NB of these AND gates corresponding to the squared values of the same input bits degenerate to direct connections. In addition, $N_2 \times B^2$ AND gates are necessary to implement at the bit level the serial multiplications with the elements of $\vec{\mathbf{H}}_{2r}$. Thus, the number of AND gates in the combinatorial network is given by

$$N_c = 2N_2B^2 - NB. \tag{3.41}$$

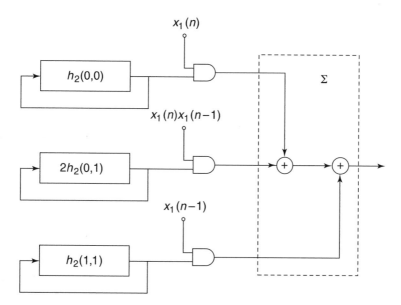

Figure 3.14 Hardware implementation of three multipliers and an accumulator in a combinatorial implementation.

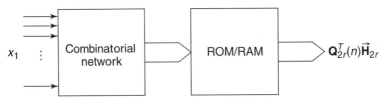

Figure 3.15   Memory-oriented implementation.

3.2.4.5 Memory-Oriented Structures
A block diagram for the memory-oriented realization of a homogeneous quadratic filter that uses distributed arithmetic and exploits the symmetry property of the Volterra kernels is shown in Figure 3.15. This structure also uses a combinatorial network to compute the entries of the matrix $\mathbf{Q}_{2r}(n)$ from the bit-level description of the input samples. Now, each row of the matrix $\mathbf{Q}_{2r}^T(n)$ can be considered as an address to a location in a ROM or RAM (read-only or random-access memory) that stores the product of the binary row vector with the column vector $\vec{\mathbf{H}}_{2r}$. The ROM or RAM outputs are then shifted and added according to (3.36).

This solution is quite demanding in terms of hardware even though the symmetry of the quadratic kernels has been exploited. In addition to the AND gates used in the combinatorial network, this technique requires a large amount of memory to store the products of all possible values of the row vectors of $\mathbf{Q}_{2r}^T(n)$ with the column vector $\vec{\mathbf{H}}_{2r}$. The storage requirement corresponds to

$$N_m = 2^{N_2} B_m \text{ bits,} \qquad (3.42)$$

where B_m denotes the wordlength in the hardware memory. Consequently, the complexity of the structure increases exponentially with the length of the filter.

3.2.4.6 Address Compression in Memory-Oriented Realizations
The calculation of the storage requirement in (3.42) was based on the fact that each row of the matrix $\mathbf{Q}_{2r}^T(n)$ has N_2 elements. However, recall that $\mathbf{Q}_{2r}^T(n)$ is obtained from $\mathbf{Q}_2^T(n)$, which is formed as a Kronecker product of $\mathbf{Q}_1^T(n)$ with itself. Consequently, the number of independent rows of $\mathbf{Q}_2^T(n)$ is significantly smaller than N^2. To see this, recall that any row of the Kronecker product of two $M \times N$-element matrices $\mathbf{A}$ and $\mathbf{B}$ is completely specified by a row of $\mathbf{A}$ and a row of $\mathbf{B}$ so that only $2N$ elements are necessary to define that row. Since the elements of $\mathbf{Q}_1^T(n)$ take only one of two values, and since each row of $\mathbf{Q}_1^T(n)$ has N elements, the preceding statement implies that there are at most 2^{2N} independent rows in $\mathbf{Q}_2^T(n)$. Consequently, we require no more than 2^{2N} memory locations, each of which corresponds to an independent row of $\mathbf{Q}_2^T(n)$, to store all possible values of the binary sums of the entries of the vector $\vec{\mathbf{H}}_2$. For single-bit realizations of the quadratic filter, the matrix $\mathbf{Q}_1^T(n)$ has a single row and, therefore, the memory requirements are even lower.

Additional reductions in the number of distinct rows occur because of certain constraints and symmetries that exist in the binary number representation, as shown in the next subsection.

3.2.4.7 Additional Redundancy in the Realization Using Two's-Complement Code
In the case of two's-complement code, additional redundancies occur because one of the possible 2^N values that a row of $\mathbf{Q}_1^T(n)$ can take is that of an all-zero row. As a result, $2(2^N) - 1$ of the 2^{2N} possible values each row of $\mathbf{Q}_2^T(n)$ can take are zero. Thus the number of distinct rows are further reduced to $N_{2,m} = 2^{2N} - 2^{N+1} + 2$. This calculation counts one of the $2^{N+1} - 1$ rows of all zero values as a distinct row.

3.2.4.8 Additional Redundancy in the Realization Using Offset Binary Code
Consider all possible values of the rows of the matrix $\mathbf{Q}_1^T(n)$. It is not difficult to see that for every possible value of a row, there is another row that is its negative. Consequently, the row corresponding to the Kronecker product of any row $\mathbf{q}_i$ with another row $\mathbf{q}_j$ is identical to the row corresponding to the Kronecker product of $-\mathbf{q}_i$ with $-\mathbf{q}_j$. Thus, the number of independent rows of $\mathbf{Q}_2^T(n)$ is only $N_{2,m} = 2^{2N-1}$. This analysis results in a significant saving over the 2^{N^2} memory locations that would have been required if the address compression that naturally occurs in the structure of $\mathbf{Q}_2^T(n)$ were not utilized. The hardware memory requirement associated with offset binary codes is approximately half of that associated with two's-complement code for large values of N.

Example 3.6 Suppose that we wish to implement a homogeneous quadratic filter. Table 3.1 shows the number of memory locations required to implement the system using memory-oriented distributed arithmetic architecture for different choices of memory length N and $B \geq 2$.

3.2.4.9 Implementation of the Address Compression Technique
In realizations using distributed arithmetic each row of the matrix $\mathbf{Q}_2^T(n)$ is used as an address to a ROM or RAM location where the result of the scalar product of that row

TABLE 3.1 Hardware Memory Requirement of Memory-Oriented Realizations of Quadratic Filters

	Memory Locations			
N	Offset-Binary	Two's-Complement	2^{N_2}	2^{N^2}
1	2	2	2	2
2	8	10	8	16
4	128	226	1024	65,536
8	32,768	65,026	6.8719×10^{10}	1.845×10^{19}
16	2.1475×10^9	4.2948×10^9	8.7112×10^{40}	1.1579×10^{77}
32	9.2234×10^{18}	1.8447×10^{19}	2^{528}	2^{1024}

and the vector $\vec{\mathbf{H}}_2$ is stored. In order to store the addresses in a reduced-size memory when the address compression is applied, it is necessary to map each string of N_2 bits to an address formed with Int $[\log_2 N_{2,m}]$ bits, where Int $[\alpha]$ represents the smallest integer greater than or equal to α. In software realizations on a DSP, it is easy to define such a mapping, while in hardware realizations some additional circuitry is required. When the offset binary code is used, the address compression can be achieved by eliminating the redundant bits. Each row of $\mathbf{Q}_2^T(n)$, composed of N subrows of N elements each, is completely determined by the first subrow, together with the first bit of the remaining $N - 1$ subrows. The $2N - 1$ bits obtained as above also allows the reconstruction of the complete row of $\mathbf{Q}_2^T(n)$ starting from the compressed address. It is left as an exercise for the reader to derive the reconstruction procedure of the complete addresses from its compressed representation.

It is obvious from this discussion that the address compression technique reduces the complexity of implementing the system. However, it should also be clear that only systems with moderate memory lengths can be implemented in this manner. We now discuss a method for reducing the hardware requirements of systems with long memory spans using a split-address technique.

3.2.4.10 The Split-Address Technique for Memory-Oriented Realizations

In the split-address method, the ROM or RAM is divided in K submemories, each one addressed by a shorter group of bits. Each subdivision of the memory contains results of multiplying a subset of the coefficients of $\vec{\mathbf{H}}_2$ with the corresponding bits in $\mathbf{Q}_2^T(n)$ and adding them together. In one method of implementing the split-address technique, we first divide $\mathbf{Q}_2^T(n)$ into K submatrices $\mathbf{Q}_{2,1}^T(n)$, $\mathbf{Q}_{2,2}^T(n), \ldots, \mathbf{Q}_{2,K}^T(n)$. Let $\mathbf{Q}_{2,i}^T$ contain L_i columns of $\mathbf{Q}_2^T$, and let

$$L_1 + L_2 + \cdots + L_K = N^2, \tag{3.43}$$

where N^2 is the total number of columns of $\mathbf{Q}_2(n)$. Recall that N^2 also represents the number of coefficients in $\vec{\mathbf{H}}_2$. Now, divide the coefficient vector $\vec{\mathbf{H}}_2$ into K smaller vectors $\vec{\mathbf{H}}_{2,1}, \vec{\mathbf{H}}_{2,2} \ldots \vec{\mathbf{H}}_{2,K}$ such that the elements of $\vec{\mathbf{H}}_{2,i}$ correspond to the columns in $\mathbf{Q}_{2,i}^T(n)$. It can be shown by direct manipulation of the matrices involved that

$$\mathbf{Q}_2^T(n)\vec{\mathbf{H}}_2 = \sum_{i=1}^{K} \mathbf{Q}_{2,i}^T\vec{\mathbf{H}}_{2,i} \tag{3.44}$$

The ith matrix multiplication on the right-hand side of this equation can be implemented in memory-oriented distributed arithmetic using 2^{L_i} address locations. In most practical situations

$$2^{L_1} + 2^{L_2} + \cdots + 2^{L_K} << 2^{N^2} \tag{3.45}$$

implying a reduction in the memory requirements. The same splitting procedure can be conveniently applied to the matrix $\mathbf{Q}_{2r}^T$ by choosing $L_1, L_2, \ldots, L_K$ such that

$$L_1 + L_2 + \cdots + L_K = N_2. \tag{3.46}$$

As long as

$$2^{L_1} + 2^{L_2} + \cdots + 2^{L_K} << 2^{N_2} << 2^{N^2}, \tag{3.47}$$

we can obtain substantial reduction in the memory requirements at the expense of needing $K - 1$ additional adders to complete the multiplications.

3.2.4.11 Combining Split-Address and Address Compression Techniques

Significant additional reduction in the hardware requirements can be accomplished by combining the address compression technique with the split-address technique. One way to combine the two techniques is to decompose $\mathbf{Q}_1^T(n)$ into K smaller matrices in a manner similar to the decomposition of $\mathbf{Q}_2^T(n)$ described earlier. For simplicity, assume that all the submatrices have identical dimensions of L columns so that

$$LK = N \tag{3.48}$$

It can be shown that each column of $\mathbf{Q}_2^T(n)$ is a column in one of the matrices of the form $\mathbf{Q}_{1,i}^T(n) \otimes \mathbf{Q}_{1,j}^T(n)$, where $\mathbf{Q}_{1,j}^T(n)$ is a submatrix in the decomposition of $\mathbf{Q}_1^T(n)$. Thus, the columns of $\mathbf{Q}_2^T(n)$ are divided into K^2 smaller matrices. We can now divide the coefficient vector $\vec{\mathbf{H}}_2$ into K^2 smaller vectors, each of which corresponds to one submatrix of the form $\mathbf{Q}_{1,i}^T(n) \otimes \mathbf{Q}_{1,j}^T(n)$. As a result, the matrix product $\mathbf{Q}_2^T(n)\vec{\mathbf{H}}_2$ can be realized using K^2 lookup tables and $K^2 - 1$ additions. Each lookup table has 2^{2L-1} address locations if the offset binary code is employed. For two's-complement code, the hardware requirement corresponds to $2^{2L} - 2 \cdot 2^L + 2$ memory locations per lookup table.

A different method of combining the split-address technique with address compression when the offset-binary code is used is to directly subdivide the $N_{2,m}$ memory cells that result from the application of the address compression into two submemories of 2^N and 2^{N-1} cells, respectively, and then proceed as in the general case described by Sicuranza and Ramponi [288]. Exercise 3.9 illustrates this approach.

Example 3.7 The number of memory cells in a memory-oriented realization of a quadratic filter with $N = 8$ is 2^{36} when the symmetry property of the Volterra kernels is exploited. If the address compression is applied without using the split-address technique, the number of memory cells decreases to $2^{15} = 32{,}768$ for the offset-binary code and 65,026 for the two's-complement code. If we now apply a split-address technique with $K = 2$, we require four lookup tables. Each set of addresses require only 128 memory cells for offset-binary code and 226 memory cells for the two's-complement code. In addition, we require three adders to complete the matrix multiplication. For the offset-binary code it is also possible to obtain a split-address compression by dividing the memory locations into two subdivisions containing 256 and 128 memory cells, respectively, as described in [288].

It can be seen from Example 3.7 that it is possible with the split address technique to implement quadratic filters with relatively long memory spans in an efficient manner. In addition, they enjoy the regularity and modularity properties of all the memory-based structures, and thus they are well suited for implementation using modern VLSI technology.

Remark 3.4   Each branch of the parallel-cascade realization of a quadratic filter contains a linear FIR filter. Different realizations of such filters result in different parallel structures for the quadratic filter. Several such structures have been compared [47] in the context of VLSI implementation. Using various figures of merit such as data throughput delay, chip area, and the variance of the roundoff error at the filter output, the authors of [47] concluded that the singular-value decomposition-based realizations implemented using systolic arrays offer the best compromise among the different conflicting figures of merit. The distributed arithmetic implementations of such parallel-cascade structures also offer a similar compromise, but such realizations are not as modular as realizations using systolic arrays.

3.2.5 FFT-Based Realization of Quadratic Filters

We know that efficient realizations of discrete-time linear filters are possible using fast Fourier transform (FFT) techniques. We explore the usefulness of FFT-based realization of quadratic filters in this section.

Consider a signal $x(n)$ of length L samples that is processed by a homogeneous quadratic system with N-sample memory. We saw in Section 2.5.5 that the quadratic kernel $h_2(m_1, m_2)$ of this system can be represented by its two-dimensional discrete Fourier transform given by

$$H_2(k_1, k_2) = \sum_{m_1=0}^{M-1} \sum_{m_2=0}^{M-1} h_2(m_1, m_2)e^{-j(2\pi/M)(m_1 k_1 + m_2 k_2)}. \tag{3.49}$$

Furthermore, the system output can be evaluated as

$$y(n) = \text{IDFT}\{H_2(k_1, k_2)X(k_1)X(k_2)\}|_{n_1=n_2=n}$$

$$= \sum_{k_1=0}^{M-1} \sum_{k_2=0}^{M-1} H_2(k_1, k_2)X(k_1)X(k_2)e^{j(2\pi/M)(k_1+k_2)n}. \tag{3.50}$$

In this expression, $X(k)$ represents the one-dimensional, M-point DFT of $x(n)$, which is given by

$$X(k) = \sum_{n=0}^{M-1} x(n)e^{-j(2\pi/M)nk}. \tag{3.51}$$

The reader should verify that the expression for $y(n)$ in (3.50) is accurate only if $M \geq L + N - 1$. If the DFT size is too small, the output of the system for times larger than $M - 1$ will wrap around and add to the output values in the range 0 to $L + N - M - 2$. The preceding discussion assumes that M is at least as large as both N and L.

The phenomenon of time-domain aliasing described above is illustrated in Figure 3.16. The basic property that enables a DFT-based realization of truncated Volterra filters is the fact that the output calculated using the DFT operations corresponds to the actual output of the system for a large number of samples. Specifically, the actual

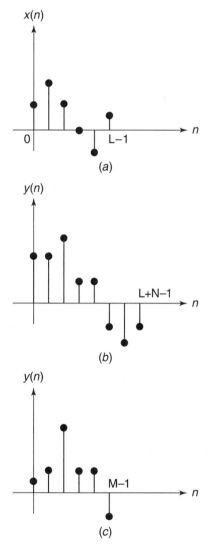

Figure 3.16 Illustration of time-domain aliasing due to inadequate resolution of the DFT: (a) input signal of length L samples; (b) output signal of length $L + N - 1$ samples; (c) output calculated by M-point DFT.

output and the DFT-based calculations coincide in the range $N + L - M - 1$ to $M - 1$. The realization that we describe next is known as the *overlap-save* technique.

3.2.5.1 The Overlap-Save Technique

We assume that the system to be realized is a homogeneous quadratic system. Realization of inhomogeneous systems is a straightforward extension of the method described here. Figure 3.17 illustrates the basic ideas behind the overlap-save technique. The input signal is first partitioned into M-point segments. The first segment is defined as

$$x_0(n) = \begin{cases} 0; & n = 0, 1, 2, \ldots, N - 2 \\ x(n - N + 1); & n = N - 1, N, N + 1, \ldots, M - 1 \end{cases} \tag{3.52}$$

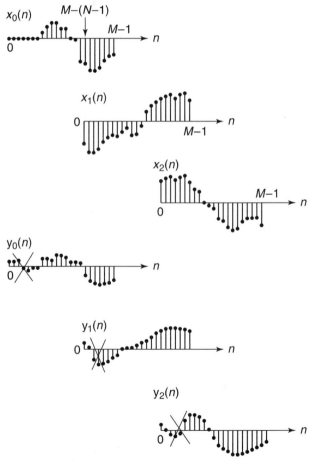

Figure 3.17 Overlap-save realization of truncated Volterra filters.

and the lth segment is represented mathematically as

$$x_l(n) = x(n + lR - N + 1); n = 0, 1, 2, \ldots, M - 1, \qquad (3.53)$$

where $R = M - N + 1$ is the number of output samples that can be accurately calculated in each segment using M-point DFTs. Let $X_l(k)$ denote the M-point one-dimensional DFT of the lth segment of the input signal. The DFT-based realization first computes

$$y_l(n) = IDFT\{H_2(k_1, k_2)X_l(k_1)X_l(k_2)\}|_{n_1=n_2=n}. \qquad (3.54)$$

Recall that the inverse DFT is defined to be zero outside the range $[0, M - 1]$.

Now, the output $y(n)$ can be obtained by retaining the portion of $y_l(n)$ that matches the system's actual output. This operation may be described using the following equation:

$$y(n) = y_l(n - lR + N - 1); lR \leq n \leq (l+1)R - 1. \qquad (3.55)$$

3.2.5.2 Computational Complexity

It is typical to choose the FFT size to be an integer power of two. If we assume, for simplicity, that $N = 2^i$ for some integer value of i and that $M = 2N$, each segment results in the calculation of $N + 1$ samples of the output. We assume that $H_2(k_1, k_2)$ has already been calculated and is available for additional processing. Table 3.2 provides the number of complex multiplications required for calculating $N + 1$ samples of the output. The calculations assume that computation of an M-point DFT requires $(M/2) \log_2 M$ complex multiplications. While counting the number of multiplications, we made use of the symmetry in the calculation of $H_2(k_1, k_2)X_l(k_1)X_l(k_2)$. In addition, we also utilized the fact that the inverse DFT operation in (3.54) can be performed by first adding together all the values of $H_2(k_1, k_2)X_l(k_1)X_l(k_2)$ for which $k_1 + k_2$ is a constant and then computing the one-dimensional IDFT. For this sequence of operations, we first compute

$$Y_l(k) = \sum_{(k_1+k_2)_M=k} H_2(k_1, k_2)X_l(k_1)X_l(k_2) \qquad (3.56)$$

TABLE 3.2 Computational Complexity of FFT-Based Realization of a Quadratic Filter

Operation	Number of Complex Multiplications
DFT of input segment	$N \log_2(2N)$
$H_2(k_1, k_2)X_l(k_1)X_l(k_2)$	$N(N + 1)$
IDFT of output segment	$N \log_2(2N)$
Total	$N(N + 1) + 2N \log_2(2N)$

and then evaluate $Y_l(n)$ as

$$y_l(n) = \frac{1}{M} \sum_{k=0}^{M-1} Y_l(k) e^{j(2\pi/M)kn}. \tag{3.57}$$

In (3.56), $(k_1 + k_2)_M$ represents modulo-M addition of k_1 and k_2. It can be seen from Table 3.2 that the computational complexity of implementing the filter corresponds to $O(N + 2\log_2 2N)$ complex multiplications per sample, which is an order of magnitude lower than the complexity associated with the direct form realization.

3.3 CONSTRAINTS IMPOSED ON QUADRATIC FILTERS

Computational simplifications beyond those provided by the symmetry conditions can be achieved in many applications by imposing additional constraints on the filter coefficients. Three types of constraints are often employed in the design of truncated Volterra filters.

1. Equalities among coefficients that guarantee some specific shape of the filter mask. Examples of this type of constraint involve additional symmetry conditions such as isotropy conditions.
2. Conditions on the filter coefficients that guarantee that the dynamic range of the filter output possesses some specific features. Examples of this approach include preservation of a constant input level at the output and conditions on the response to biased input signals.
3. Constraints imposed on the Volterra kernels when it is known that the input signal is band-limited.

In addition to reducing the computational complexity of the filters, such constraints also facilitate easier filter designs since they reduce the number of degrees of freedom in the filter design procedures. We now describe how the constraints listed above can be applied in the design of quadratic filters in more detail. The first two constraints have found heavy use in many applications of quadratic filters in image processing, where the computational complexity of such filters is otherwise very high. Examples of several such applications are discussed in Chapters 4 and 10.

3.3.1 Isotropy

The isotropy constraint in one dimension reflects the invariance of the filter output with respect to a 180° rotation in the input data, i.e., the output of the filter is identical when the input vector is

$$\mathbf{X}_1(n) = [x(n) \quad x(n-1) \quad \cdots \quad x(n-N+1)]^T \tag{3.58}$$

or

$$\mathbf{X}'_1(n) = [x(n - N + 1) \quad x(n - N + 2) \quad \cdots \quad x(n)]^T. \qquad (3.59)$$

For linear FIR filters the isotropy constraint is satisfied by some linear phase filters. To see this, note that the output of a linear FIR filter that exhibits the isotropy property is given by

$$\overline{h}_1[x(n)] = \vec{\mathbf{H}}_1^T \mathbf{X}_1(n) = \vec{\mathbf{H}}_1^T \mathbf{X}'_1(n). \qquad (3.60)$$

The last equality can be guaranteed by selecting the coefficients such that

$$h_1(m_1) = h_1(N - 1 - m_1); \qquad m_1 = 0, 1, \ldots, N - 1, \qquad (3.61)$$

which results in a linear phase filter. We can extend this result to homogeneous quadratic filters by considering two input vectors having N^2 elements, defined as

$$\mathbf{X}_2(n) = \mathbf{X}_1(n) \otimes \mathbf{X}_1(n) \qquad (3.62)$$

and

$$\mathbf{X}'_2(n) = \mathbf{X}'_1(n) \otimes \mathbf{X}'_1(n), \qquad (3.63)$$

along with a coefficient vector $\vec{\mathbf{H}}_2$ of the same dimension. The isotropy condition in this case implies that the quadratic filter outputs for $\mathbf{X}_2(n)$ and $\mathbf{X}'_2(n)$ are identical, i.e.,

$$\overline{h}_2[x(n)] = \vec{\mathbf{H}}_2^T \mathbf{X}_2(n) = \vec{\mathbf{H}}_2^T \mathbf{X}'_2(n). \qquad (3.64)$$

This condition can be achieved if

$$h_2(m_1, m_2) = h_2(N - 1 - m_1, N - 1 - m_2); \qquad m_1, m_2 = 0, 1, \ldots, N - 1. \qquad (3.65)$$

The two sets of constraints (3.61) and (3.65) can be used to reduce the number of independent coefficients during the design of a quadratic filter [245]. The number of independent coefficients of the linear component is reduced from N to $N/2$ when N is even or to $(N + 1)/2$ when N is odd. For the homogeneous quadratic component, the symmetry of the kernel, combined with the isotropy condition, implies that the $N \times N$-element coefficient matrix $\mathbf{H}_2$ is symmetric with respect to both its principal diagonals. The number of independent coefficients for the homogeneous quadratic part is reduced from $N(N + 1)/2$ to $[(N/2) + 1]N/2$ when N is even or to $[(N + 1)/2]^2$ when N is odd.

Remark 3.5 As stated earlier, the isotropy constraints for a linear FIR filter result in one of the conditions that guarantee the linearity of the phase of its frequency response. The linear phase characteristic is a desirable property in many applications of linear filters. We can consider the isotropy condition on the quadratic filters as an extension of the linear phase condition to the nonlinear system.

3.3.2 Preservation of a Constant Input Value

One useful condition that is often imposed on many types of filters is that they should preserve the value of a constant input signal at the output also [240,241]. A quadratic filter can satisfy this condition if

$$h_0 = 0, \tag{3.66}$$

$$\sum_{m_1=0}^{N-1} h_1(m_1) = 1, \tag{3.67}$$

and

$$\sum_{m_1=0}^{N-1} \sum_{m_2=0}^{N-1} h_2(m_1, m_2) = 0. \tag{3.68}$$

3.3.3 Preservation of Bias

A condition somewhat stronger than that of preserving constant input values is that a uniform shift in the input level is also preserved at the output [240], that is, if $y(n)$ is the system output when $x(n)$ is the input signal, then the system response to $x(n) + \mu$ at the input should be $y(n) + \mu$, where μ is a constant value. A quadratic filter can meet these requirements if it satisfies the following conditions:

$$\sum_{m_1=0}^{N-1} h_1(m_1) = 1, \tag{3.69}$$

$$\sum_{m_1=0}^{N-1} h_2(m_1, m_2) = 0 \quad \text{for} \quad m_2 = 0, 1, \ldots, N-1 \tag{3.70}$$

and

$$\sum_{m_2=0}^{N-1} h_2(m_1, m_2) = 0 \quad \text{for} \quad m_1 = 0, 1, \ldots, N-1. \tag{3.71}$$

Note that conditions (3.70) and (3.71) imply (3.68).

3.3.4 Spectral Constraints

As already shown, the huge number of coefficients associated with Volterra kernels is one of the main drawbacks of such filtering structures. In order to reduce their complexity, it is possible to consider suitable approximations based on the projection onto a coefficient subspace of reduced dimensionality. As a consequence, the approximating filter has fewer degrees of freedom. Approximations of this form have been used to model bandpass systems and filter bandpass signals in [199].

3.4 REALIZATION OF HIGHER-ORDER VOLTERRA FILTERS

All the realizations that were discussed in the previous section for quadratic filters can be extended to higher-order filters. Because of the similarity in the derivations of the various realizations to those for quadratic systems, our discussion is relatively brief.

3.4.1 Parallel-Cascade Realization of Higher-Order Volterra Filters

The parallel-cascade realization of quadratic filters can be extended to the higher-order filters in several different ways [202,214,215]. In this section, we consider the problem of realizing a homogeneous pth-order system with N-sample memory using a parallel-cascade structure involving lth-order and $(p - l)$th-order Volterra systems as shown in Figure 3.18. The results are applicable to any $l < p$, and each component in the structure shown can be further decomposed using smaller-order components.

Recall that the output of a homogeneous pth-order Volterra system with N-sample memory is given by

$$y(n) = \sum_{m_1=0}^{N-1} \sum_{m_2=m_1}^{N-1} \cdots \sum_{m_p=m_{p-1}}^{N-1} h_p(m_1, m_2 \ldots, m_p)x(n - m_1)x(n - m_2) \cdots x(n - m_p).$$

$$(3.72)$$

This relationship can be compactly written using matrix notation as

$$y(n) = \mathbf{X}_l^T(n)\mathbf{H}_{l,p-l}\mathbf{X}_{p-l}(n), \tag{3.73}$$

where $\mathbf{H}_{p_1,p_2}$ is a matrix of dimension $\begin{pmatrix} N + p_1 - 1 \\ p_1 \end{pmatrix} \times \begin{pmatrix} N + p_2 - 1 \\ p_2 \end{pmatrix}$ elements in which the coefficients in (3.72) are arranged in some orderly manner. The vector $\mathbf{X}_{N,p_1}(n)$ has $\begin{pmatrix} N + p_1 - 1 \\ p_1 \end{pmatrix}$ elements and contains all possible p_1th-order product signals of the form $x(n - k_1)x(n - k_2) \cdots x(n - k_{p_1})$, where $0 \leq k_1, k_2, \ldots,$ $k_{p_1} \leq N - 1$. Since $\mathbf{H}_{p_1,p_2}$ contains more elements than there are independent

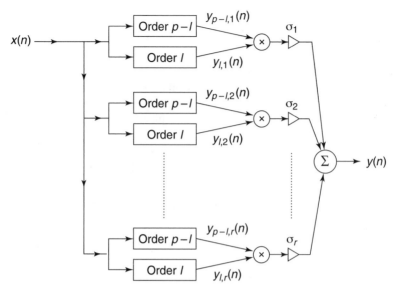

Figure 3.18 A parallel-cascade realization of a pth-order Volterra kernel. Each block represents a Volterra system of the order shown within.

coefficients, several entries of the coefficient matrix may be zero. Let the rank of $\mathbf{H}_{l,p-l}$ be r. Then we can express $\mathbf{H}_{l,p-l}$ as

$$\mathbf{H}_{l,p-l} = \sum_{i=1}^{r} \sigma_i \mathbf{U}_i \mathbf{V}_i^T, \tag{3.74}$$

using singular value decomposition. In the equations above, the σ_i terms are the nonzero singular values of $\mathbf{H}_{l,p-l}$ and the $\mathbf{U}_i$ and $\mathbf{V}_i$ terms are the left and right singular vectors, respectively, of the matrix. The sets $\{\mathbf{U}_i; i = 1, 2, \ldots, r\}$ and $\{\mathbf{V}_i; i = 1, 2, \ldots, r\}$ consists of orthonormal vectors. Substituting (3.74) in (3.73), we get

$$y(n) = \sum_{i=1}^{r} \sigma_i [\mathbf{X}_l^T(n)\mathbf{U}_i][\mathbf{V}_i^T \mathbf{X}_{p-l}(n)]$$

$$= \sum_{i=1}^{r} \sigma_i y_{l,i}(n) y_{p-l,i}(n), \tag{3.75}$$

where we have defined $y_{l,i}(n)$ as the output of a homogeneous lth-order Volterra system given by

$$y_{l,i}(n) = \mathbf{X}_l^T(n)\mathbf{U}_i. \tag{3.76}$$

The signal $y_{p-l,i}(n)$ is also defined in a similar manner. From the preceding analysis, it should be clear that the left and right singular vectors define the coefficients of the lower-order components used in the decomposition shown in Figure 3.18. Furthermore, the output of the system is the sum of products of outputs of lth- and $(p - l)$th-order systems weighted by the singular values.

3.4.1.1   *Approximation of Higher-Order Volterra Systems*

The decomposition in (3.74) immediately results in a systematic method for approximating Volterra systems using simpler structures in a manner similar to the approximation of quadratic filters. As was done for the quadratic filters, we can obtain the best k-rank approximation for $\mathbf{H}_{l,p-l}$ as

$$\mathbf{H}_{l,p-l} = \sum_{i=1}^{k} \sigma_i \mathbf{U}_i \mathbf{V}_i^T, \tag{3.77}$$

where we assumed that $\sigma_1, \sigma_2, \ldots, \sigma_k$ are the k largest singular values of $\mathbf{H}_{l,p-l}$. Furthermore, we can characterize the error in the approximation using

$$\|\mathbf{H}_{l,p-l} - \tilde{\mathbf{H}}_{l,p-l}\|_F = \sqrt{\sum_{i=k+1}^{r} \sigma_i^2}. \tag{3.78}$$

Consequently, we can approximate the system using a lower-complexity realization by removing the branches corresponding to low singular values.

Example 3.8   We consider the realization of a homogeneous fourth-order system with twelve-sample memory using second-order components. The coefficients of the system are given by

$$h_4(m_1, m_2, m_3, m_4) =$$

$$\begin{cases} \dfrac{100}{2\pi(1.5^4 + a_1^4 + a_2^4 + a_3^4 + a_4^4)^{3/4}} + u(m_1, m_2, m_3, m_4); \\ \qquad\qquad\qquad\qquad\qquad\qquad\qquad 0 \le m_1 \le m_2 \le m_3 \le m_4 \le 11 \\ 0; \qquad\qquad\qquad\qquad\qquad\qquad\quad \text{otherwise,} \end{cases}$$

where $a_i = m_i - 5$ and $u(m_1, m_2, m_3, m_4)$ is a random variable that is uniformly distributed between -0.1 and 0.1. The maximum number of branches for this particular case is 78. The coefficients of the two filters in each branch are the same because of the symmetry of the matrix $\mathbf{H}_{p1,p2}$ resulting in further simplification of the structure. Figure 3.19 displays the normalized squared norm of the coefficient error matrix as a function of the number of branches used in the approximate realization. It can be seen that significant errors are introduced only after more than half the branches are discarded.

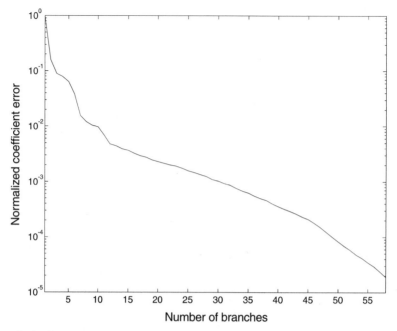

Figure 3.19 Squared norm of the coefficient error. (Courtesy of T. M. Panicker; Copyright ©
1998 IEEE.)

3.4.2 Distributed Arithmetic Realizations of Higher-Order Operators

The vector representation discussed in Section 3.1.2 can be extended to higher-order
Volterra operators. Consequently, such filters can be realized using distributed
arithmetic.

In order to derive the vector representation, we define the input vector $\mathbf{X}_p(n)$ of a
homogeneous pth order Volterra filter recursively as

$$\mathbf{X}_p(n) = \mathbf{X}_1(n) \otimes \mathbf{X}_{p-1}(n). \tag{3.79}$$

The output of the operator of order p is given by the equation

$$\overline{h}_p[x(n)] = \mathbf{X}_p^T(n)\vec{\mathbf{H}}_p, \tag{3.80}$$

where $\vec{\mathbf{H}}_p$ contains all the coefficients of the filter ordered in an appropriate manner.
The derivation of the distributed arithmetic realization of higher-order Volterra filters
is similar to the derivation for quadratic filters. Recall that $\mathbf{X}_1(n)$ can be expressed in
the form

$$\mathbf{X}_1(n) = \mathbf{Q}_1(n)\mathbf{P}_1, \tag{3.81}$$

where $\mathbf{Q}_1(n)$ and $\mathbf{P}_1$ are as defined in Section 3.2. It immediately follows that

$$\mathbf{X}_p(n) = \mathbf{Q}_p(n)\mathbf{P}_p, \tag{3.82}$$

where $\mathbf{Q}_p(n)$ and $\mathbf{P}_p$ are defined recursively as

$$\mathbf{Q}_p(n) = \mathbf{Q}_1(n) \otimes \mathbf{Q}_{p-1}(n) \tag{3.83}$$

and

$$\mathbf{P}_p = \mathbf{P}_1 \otimes \mathbf{P}_{p-1}, \tag{3.84}$$

respectively. When two's-complement code is employed, $\mathbf{Q}_p(n)$ is an $N^p \times B^p$-element matrix containing only 1s or 0s. Similarly, all the elements of $\mathbf{Q}_p(n)$ are $+1$ or -1 when the offset-binary code is employed. All the entries of the B^p-element vector $\mathbf{P}_p(n)$ are positive or negative values of integer powers of two. The output of the homogeneous pth-order Volterra filter can now be written as

$$y(n) = \mathbf{X}_p^T(n)\vec{\mathbf{H}}_p = \mathbf{P}_p^T\mathbf{Q}_p^T(n)\vec{\mathbf{H}}_p. \tag{3.85}$$

As was the case for quadratic filters, more efficient realizations can be obtained by utilizing the symmetry properties of the Volterra kernels. For this purpose, we define the reduced coefficient vector $\vec{\mathbf{H}}_{pr}$ containing the N_p independent coefficients of the pth-order Volterra kernel and the corresponding reduced input vector $\mathbf{X}_{pr}(n)$. The input–output relationship of the homogeneous filter becomes

$$\bar{h}_p[x(n)] = \mathbf{X}_{pr}^T(n)\vec{\mathbf{H}}_{pr} = \mathbf{P}_p^T\mathbf{Q}_{pr}^T(n)\vec{\mathbf{H}}_{pr}, \tag{3.86}$$

where $\mathbf{Q}_{pr}(n)$ is an $N_p \times B^p$-element matrix obtained by expanding the vector $\mathbf{X}_{pr}(n)$ in its B^p elementary bits. As in the case of quadratic filters, the reduced coefficient vector $\vec{\mathbf{H}}_{pr}$ and the reduced input vector $\mathbf{X}_{pr}(n)$ can be expressed as

$$\vec{\mathbf{H}}_{pr} = \mathbf{M}_{p1}\vec{\mathbf{H}}_p \tag{3.87}$$

and

$$\mathbf{X}_{pr}(n) = \mathbf{M}_{p2}\mathbf{X}_p(n), \tag{3.88}$$

respectively, where $\mathbf{M}_{p1}$ and $\mathbf{M}_{p2}$ are matrices of size $N_p \times N^p$ elements that contain only 1s and 0s. Consequently, the matrix $\mathbf{Q}_{pr}(n)$ can be also expressed as

$$\mathbf{Q}_{pr}(n) = \mathbf{M}_{p2}\mathbf{Q}_p(n). \tag{3.89}$$

3.4.2.1 *Combinatorial Structures* The similarity between distributed arithmetic representations of the pth-order Volterra filter given by (3.86) and the corresponding representation for quadratic filters given by (3.36), indicates that the realizations based on the distributed arithmetic can be directly extended to higher-order kernels. The main difference is that the combinatorial network must compute all possible products of p input bits that define the entries of the matrix $\mathbf{Q}_{pr}(n)$. It is left as an exercise for the reader to show that a combinatorial circuit realization of the homogeneous pth-order Volterra filter requires

$$N_c = 2N_p \times B^p - NB \tag{3.90}$$

AND gates when two's-complement code is employed. Realization using the offset binary code requires the same number of exclusive NOR gates.

3.4.2.2 *Memory-Oriented Structures* In the memory-oriented structures, the rows of $\mathbf{Q}_{pr}^T(n)$ are used as addresses to a ROM or RAM in which all the possible binary sums of the elements of $\vec{\mathbf{H}}_{pr}$ are stored. The memory requirement for the realization of such filters is given by

$$N_m = 2^{N_p} B_m \tag{3.91}$$

bits of hardware memory, where B_m is the wordlength used in the memory.

It is even more imperative for higher-order filters to obtain hardware complexity reductions using techniques such as address compression and split address. It is left as an exercise for the reader to show that there are, in principle, no more than $N_{p,m} = 2^{pN}$ distinct rows in $\mathbf{Q}_p^T(n)$ for $B \geq p$, as a consequence of its definition as a Kronecker product of matrices. Additional reduction in the memory requirement can be achieved because of the structure of the two's-complement code and the offset binary code. Finally, we can also combine the split-address technique with the address compression technique, as described in Section 3.2.4. Therefore, significant reductions in the hardware complexity can be obtained.

3.4.2.3 *Additional Redundancy in Realization Using the Two's-Complement Code* Using the same notation as before, we can obtain a recursive expression for $N_{p,m}$, the number of distinct values that any row of $\mathbf{Q}_p^T(n)$ can take as

$$N_{p,m} = N_{p-1,m} N_{1,m} - (N_{p-1,m} + N_{1,m} - 1) + 1. \tag{3.92}$$

Since $N_{1,m} = 2^N$, we can obtain the following closed form expression for $N_{p,m}$:

$$N_{p,m} = (2^N - 1)^{p-1} 2^N - (2^N - 2) \frac{[1 - (2^N - 1)^{p-1}]}{1 - (2^N - 1)}. \tag{3.93}$$

3.4.2.4 Additional Redundancy in Realization Using the Offset Binary Code

Let the number of possible distinct values each row of $\mathbf{Q}_p^T(n)$ can take be $N_{p,m}$. Recall that

$$\mathbf{Q}_p^T(n) = \mathbf{Q}_{p-1}^T(n) \otimes \mathbf{Q}_1^T(n). \tag{3.94}$$

Furthermore, one can show that for every possible value the rows of $\mathbf{Q}_{p-1}^T(n)$ and $\mathbf{Q}_1^T(n)$ can take, they can also take the corresponding negative values. Consequently, the number of distinct values that each row of $\mathbf{Q}_p^T(n)$ can assume is given by

$$N_{p,m} = \frac{N_{p-1,m}N_{1,m}}{2}. \tag{3.95}$$

Since $N_{1,m} = 2^N$, we can see that

$$N_{p,m} = \frac{2^{pN}}{2^{p-1}} = 2^{p(N-1)+1}. \tag{3.96}$$

The following example illustrates the usefulness of address compression and the split-address technique in reducing the complexity of memory-oriented structures for higher-order Volterra filters.

Example 3.9 Consider a homogeneous cubic symmetric operator with a memory span $N = 8$ samples. The number of coefficients N_p is 120 in this case. Implementation of the memory-oriented structure without address compression requires 2^{120} words, each of length B_m bits, to be stored. Obviously, this represents an impossible requirement for the foreseeable future. Applying address compression to this problem suggests that there are no more than 2^{22} different address locations necessary for this system when binary offset code is employed and approximately 2^{24} address locations when two's-complement code is employed. Finally, we can use the split-address technique by dividing $\mathbf{Q}_1^T(n)$ into two submatrices of four columns each. This procedure requires eight ROMs, each containing 1024 address locations for the offset binary code. Split-address technique using the two's-complement code and the same parameters require eight ROMs with 3376 address locations. Both implementations also need seven adders to sum the ROM outputs to implement the system using the split-address technique. Therefore, this technique makes it possible to implement high-order Volterra systems with moderate memory spans in an efficient manner.

3.4.3 FFT-Based Realization of Higher-Order Volterra Filters

We saw in Section 2.5.5 that the pth-order kernel $h_p(m_1, m_2, \ldots, m_p)$ of a truncated Volterra system can be represented by its p-dimensional discrete Fourier transform given by

$$H_p(k_1, k_2, \ldots, k_p) = \sum_{m_1=0}^{M-1} \sum_{m_2=0}^{M-1} \cdots \sum_{m_p=0}^{M-1} h_p(m_1, m_2, \ldots, m_p)e^{-j(2\pi/M)m_1 k_1} \cdots e^{-j(2\pi/M)m_p k_p}. \tag{3.97}$$

Furthermore, the system output can be evaluated as

$$y(n) = \text{IDFT}\{H_p(k_1, k_2, \ldots, k_p)X(k_1)X(k_2)\cdots X(k_p)\}|_{n_1=n_2=\cdots=n_p=n}$$

$$= \sum_{k_1=0}^{M-1}\sum_{k_2=0}^{M-1}\cdots\sum_{k_p=0}^{M-1} H_p(k_1, k_2, \ldots, k_p)X(k_1)X(k_2)\cdots X(k_p)e^{j(2\pi/M)(k_1+k_2+\cdots+k_p)n}.$$

$$(3.98)$$

As was the case for quadratic systems, the expression for $y(n)$ in (3.98) is accurate only if $M \geq L + N - 1$. If the DFT size is too small, the output of the system for time indices larger than $M - 1$ will wrap around and add to the output values in the range 0 to $L + N - M - 2$. We can derive the overlap-save technique for implementing a homogeneous pth-order Volterra filter with N-sample memory using the preceding result. Realization of inhomogeneous Volterra systems is a straightforward extension of the method described here. The input signal is first partitioned into M-point segments in an identical manner to the partition for implementing the quadratic filter. The first segment is defined as

$$x_1(n) = \begin{cases} 0; & n = 0, 1, 2, \ldots, N - 2 \\ x(n - N + 1); & n = N - 1, N, N + 1, \ldots, M - 1. \end{cases} \quad (3.99)$$

The lth segment is represented mathematically as

$$x_l(n) = x(n + lR - N + 1); \quad n = 0, 1, 2, \ldots, M - 1, \quad (3.100)$$

where $R = M - N + 1$ is the number of output samples that can be accurately calculated in each segment using M-point DFTs. Let $X_l(k)$ denote the M-point one-dimensional DFT of the lth segment of the input signal. The DFT-based realization first computes

$$y_l(n) = \text{IDFT}\{H_p(k_1, k_2, \ldots, k_p)X_l(k_1)X_l(k_2)\cdots X_l(k_p)\}|_{n_1=n_2=\cdots=n_p=n}. \quad (3.101)$$

The inverse DFT is defined to be zero outside the range $[0, M - 1]$.

Now, the output $y(n)$ can be obtained by retaining the portion of $y_l(n)$ that matches the system's actual output. This operation may be described using the following equation:

$$y(n) = y_l(n - lR + N - 1); \quad lR \leq n \leq (l + 1)R - 1. \quad (3.102)$$

It can be shown that the computational complexity of this method is $O(N^{p-1})$ complex multiplications per input sample. Even though this complexity is an order of magnitude smaller than that of direct realization, the relative reduction in the computational complexity is not as significant for higher-order Volterra filters as the complexity reduction for quadratic filters.

3.5 EXERCISES

3.1 Consider the Wiener model involving the linear filter

$$w(n) = a_0 x(n) + a_1 x(n-1) + a_2 x(n-2) + a_3 x(n-3)$$

in cascade with a squarer. Compare the complexity of this realization with the direct realization of the corresponding quadratic kernel represented in a triangular form.

3.2 *Computing Assignment:* Assume that the elements of the matrix $\mathbf{H}_2$ defining a quadratic kernel are given by

$$h_2(m_1, m_2) = \frac{1.5}{2\pi(2.25 + (m_1 - 10)^2 + (m_2 - 10)^2)^{3/2}},$$
$$m_1, m_2 = 0, 1, \ldots, 20.$$

Derive the parallel-cascade realizations based on
(a) The LU matrix decomposition.
(b) The singular value decomposition.
(c) The singular value decomposition truncated to the largest four singular values of $\mathbf{H}_2$ implemented via LU decomposition.
Evaluate the approximation error when the filter is realized using the singular value decomposition and four branches by computing the Frobenius norm of the error matrix.

3.3 Show that the number of distinct bit shifts in a combinatorial realization of a homogeneous quadratic filter employing the two's-complement code is given by $2(B-1)$:
(a) Modify the matrix notation in (3.35) and (3.36) in order to obtain a realization employing the distinct bit shifts obtained above only once each.
(b) Extend this result to any $p > 2$.

3.4 Show that a combinatorial circuit realization of the homogeneous pth-order Volterra filter requires $N_c = 2N_p \times B^p - NB$ AND gates when the two's-complement code is employed.

3.5 Show that if $B < p$, the number of the independent addresses to the ROM in a memory-oriented realization cannot exceed 2^{BN}. Verify this result for a quadratic filter.

3.6 Enumerate all the independent rows that may form the matrix $\mathbf{Q}_2^T(n)$ for a homogeneous quadratic filter with $N = 3$ and $B = 1$ employing the offset binary and the two's-complement codes.

3.7 Determine the memory requirements of a memory-oriented realization of the cubic filter

$$y(n) = w^3(n),$$

where

$$w(n) = ax(n) + bx(n - 1)$$

using the offset-binary code with $B = 2$.

3.8 Find a procedure for defining in a memory-oriented realization the mapping from the set of the complete addresses given as rows of the matrix $\mathbf{Q}_p^T(n)$ to the set of the compressed addresses when the offset-binary code is employed. Verify the result for $p = 2, N = 2, B = 2$, and $p = 3, N = 2, B = 3$.

3.9 Given a homogeneous quadratic filter with $N = 8$, apply the split-address compression by dividing the memory locations into two subdivisions containing 256 and 128 memory cells employing the offset binary code.

3.10 Suppose that we know the values stored in the memory of a memory-oriented realization of a homogeneous quadratic filter employing the offset binary code.

(**a**) Develop a method to compute the coefficients of the quadratic kernel.

(**b**) Extend this result to any $p > 2$.

3.11 Show that all the filter coefficients cannot be recovered from a memory-oriented realization as in Exercise 3.10 if $B < p$. Verify this result with an example for $p = 2$.

3.12 *Computing Assignment:* Implement the overlap-save technique and apply it to the homogeneous quadratic filter described by the input–output relationship

$$
\begin{aligned}
y(n) = {} & 0.5x^2(n) + 0.2x(n)x(n-1) + 0.1x(n)x(n-2) + 0.05x(n)x(n-3) \\
& + 0.3x^2(n-1) - 0.5x(n-1)x(n-2) - 0.2x(n-1)x(n-3) \\
& + 0.2x^2(n-2) + 0.1x(n-2)x(n-3) - 0.1x^2(n-3),
\end{aligned}
$$

using 1024 samples of an arbitrary input signal partitioned into 64-point segments.

3.13 *Computing Assignment:* Consider the parallel-cascade realization of a fourth-order system with 12 sample memory using first- and third-order components.

(**a**) Determine the maximum number of branches in the exact realization.

(**b**) Compute the squared norm of the coefficient error matrix as a function of the number of branches used in the approximate realization.

(**c**) Compute the mean-square difference of the signal at the output of the exact and approximate realizations. Assume as input signal a zero-mean pseudo-Gaussian sequence with unit variance and compute the error variance as the average over 1000 samples and 10 independent experiments, normalized by the variance of the output of the exact realization. You may select a kernel whose multidimensional frequency response corresponds to that of a lowpass filter.

(**d**) Comment on the results obtained.

3.14 *Computing Assignment:* Repeat the computations of the Exercise 3.13 using second-order components and compare the results.

3.15 Show that if the input signal to a homogeneous pth-order Volterra filter is amplified by a factor K, the output should be divided by the value $K^{(p-1)}$ in order to guarantee the same amplification of the output signal.

4

MULTIDIMENSIONAL
VOLTERRA FILTERS

This chapter extends the results of the previous two chapters to the multidimensional case. Many applications of truncated two-dimensional Volterra filters in image processing are also described in this chapter. We start our discussion with multidimensional Volterra series expansions and realizations of multidimensional truncated Volterra systems. We will see that the greatest difficulty in using multidimensional Volterra system theory in practical applications is the large complexity associated with such system models. Consequently, a large portion of this chapter is devoted to the derivation of efficient realizations as well as the design of multidimensional Volterra filters through the application of judicious constraints imposed on their kernels.

For the sake of simplicity, we consider only systems with finite planes of support in this chapter. In addition, most of the results and examples involve two-dimensional quadratic systems. However, extensions and generalizations of the topics discussed in this chapter to higher dimensions are conceptually straightforward.

4.1 MULTIDIMENSIONAL VOLTERRA SERIES EXPANSION

The discrete-time Volterra series expansions of one-dimensional systems can be easily extended to characterize multidimensional nonlinear systems. In the following, we consider the representation of multidimensional nonlinear filters using truncated Volterra series expansions. As in the one-dimensional case, much of our development depends on vector and matrix representation of the systems.

4.1.1 Vector Representation of Linear Multidimensional Filters with Finite Memory

Consider a linear, M-dimensional finite impulse response filter with input–output relationship given by

$$\overline{h}_1[x(n_1, n_2, \ldots, n_M)]$$
$$= \sum_{m_{11}=0}^{N_1-1} \sum_{m_{12}=0}^{N_2-1} \cdots \sum_{m_{1M}=0}^{N_M-1} h_1(m_{11}, m_{12}, \ldots, m_{1M})$$
$$\times x(n_1 - m_{11}, n_2 - m_{12}, \ldots, n_M - m_{1M}). \tag{4.1}$$

We can express this relationship more compactly using vector notation for the independent variables as

$$\overline{h}_1[x(\mathbf{n})] = \sum_{\mathbf{m}_1}^{\mathcal{N}} h_1(\mathbf{m}_1)x(\mathbf{n} - \mathbf{m}_1), \tag{4.2}$$

where the vectors $\mathbf{n}$ and $\mathbf{m}_1$ are defined as

$$\mathbf{n} = [n_1 \quad n_2 \cdots n_M] \tag{4.3}$$

and

$$\mathbf{m}_1 = [m_{11} \quad m_{12} \cdots m_{1M}], \tag{4.4}$$

respectively, and the symbol $\sum_{\mathbf{m}_1}^{\mathcal{N}}$ indicates summation over all the components of the vector $\mathbf{m}_1$ in the set $\mathcal{N}$. We assume throughout that the set $\mathcal{N}$ given by

$$\mathcal{N} = \{(m_{11}, m_{12}, \ldots, m_{1M}) : 0 \leq m_{11} \leq N_1 - 1, 0 \leq m_{12}$$
$$\leq N_2 - 1, \ldots, 0 \leq m_{1M} \leq N_M - 1\} \tag{4.5}$$

defines the rectangular plane of support of the system of (4.2).

Let $\mathbf{X}_1(\mathbf{n})$ represent an ordered arrangement of the elements of the set $x(\mathbf{n}_1 - \mathbf{m}_1) : \mathbf{m}_1 \in \mathcal{N}\}$. Similarly, let $\vec{\mathbf{H}}_1$ denote the coefficient vector in which the coefficients $h_1(\mathbf{m}_1)$ are arranged in the same order as the arrangement of $x(\mathbf{n}_1 - \mathbf{m}_1)$ in $\mathbf{X}_1(\mathbf{n})$. The input–output relation of the multidimensional linear filter can now be expressed as

$$\overline{h}_1[x(\mathbf{n})] = \mathbf{X}_1^T(\mathbf{n})\vec{\mathbf{H}}_1. \tag{4.6}$$

This representation is identical to the vector representation of one-dimensional, linear, time-invariant systems. The similarities that arise between vector and matrix representations of one- and multidimensional systems will prove to be useful in the analysis, design, and realization of multidimensional Volterra systems.

4.1.2 Vector Representation of Finite-Support Multidimensional Volterra Filters

Using the notation introduced in the previous subsection, we can express the input–output relationship of a homogeneous pth-order truncated M-dimensional Volterra system as

$$\overline{h}_p[x(\mathbf{n})] = \sum_{m_1}^{\mathcal{N}} \sum_{m_2}^{\mathcal{N}} \cdots \sum_{m_p}^{\mathcal{N}} h_p(\mathbf{m}_1, \mathbf{m}_2, \ldots, \mathbf{m}_p)x(\mathbf{n} - \mathbf{m}_1)x(\mathbf{n} - \mathbf{m}_2)\cdots x(\mathbf{n} - \mathbf{m}_p),$$

(4.7)

where

$$\mathbf{m}_j = [m_{j1} \, m_{j2} \, \cdots \, m_{jM}]; \qquad j = 1, 2, \ldots, p. \tag{4.8}$$

and

$$0 \leq m_{j1} \leq N_1 - 1, 0 \leq m_{j2} \leq N_2 - 1, \ldots, 0 \leq m_{jM} \leq N_M - 1 \tag{4.9}$$

By ordering the elements of the kernel $h_p(\mathbf{m}_1, \mathbf{m}_2, \ldots, \mathbf{m}_p)$ in some predetermined manner, it is possible to derive a vector representation for the elements of the pth-order Volterra kernel and for the products of p input samples. Let $\vec{\mathbf{H}}_p$ and $\mathbf{X}_p(\mathbf{n})$ represent the coefficient vector and input vector, respectively, that result from the preceding process. Then, we can rewrite (4.7) as

$$\overline{h}_p[x(\mathbf{n})] = \mathbf{X}_p^T(\mathbf{n})\vec{\mathbf{H}}_p. \tag{4.10}$$

Both the vectors are formed by N_T^p elements, with $N_T = \prod_{i=1}^M N_i$ if the symmetry of the Volterra kernels is not exploited. If such a symmetry is considered, the independent coefficients of the pth kernel can be derived on the basis of the invariance of the products of p input signals with respect to any permutation of the elements in the vectors $\mathbf{m}_1, \mathbf{m}_2, \ldots, \mathbf{m}_p$. Using compact vector notation, (4.7) can be rewritten as

$$\overline{h}_p[x(\mathbf{n})] = \sum_{m_1}^{S} \sum_{m_2}^{S} \cdots \sum_{m_p}^{S} |\pi(\mathbf{m}_1, \mathbf{m}_2, \ldots, \mathbf{m}_p)|h_p(\mathbf{m}_1, \mathbf{m}_2, \ldots, \mathbf{m}_p)$$
$$\times x(\mathbf{n} - \mathbf{m}_1)x(\mathbf{n} - \mathbf{m}_2)\cdots x(\mathbf{n} - \mathbf{m}_p), \tag{4.11}$$

where S is an appropriate sub-region of $\mathcal{N}$ and $|\pi(\mathbf{m}_1, \mathbf{m}_2, \ldots, \mathbf{m}_p)|$ denotes the number of permutations of $\mathbf{m}_1, \mathbf{m}_2, \ldots, \mathbf{m}_p$ that leaves the product $x(\mathbf{n} - \mathbf{m}_1)x(\mathbf{n} - \mathbf{m}_2)\cdots x(\mathbf{n} - \mathbf{m}_p)$ unchanged. The representation in (4.11) corresponds to that of the triangular kernel. The constant values $|\pi(\mathbf{m}_1, \mathbf{m}_2, \ldots, \mathbf{m}_p)|$ can be derived using a combinatorial approach [287]. Let us denote the number of distinct vectors in a specific set of $(\mathbf{m}_1, \mathbf{m}_2, \ldots, \mathbf{m}_p)$ as r. Let $k_1, k_2, \ldots, k_r$ denote the number of times these vectors appear in $(\mathbf{m}_1, \mathbf{m}_2, \ldots, \mathbf{m}_p)$. Then

$$|\pi(\mathbf{m}_1, \mathbf{m}_2, \ldots, \mathbf{m}_p)| = \frac{p!}{k_1! k_2! \cdots k_r!}. \tag{4.12}$$

The number of the independent coefficients in the representation of the system can be found using similar techniques as in Chapter 2 to be

$$N_{pM} = \binom{N_T + p - 1}{p}. \tag{4.13}$$

Comparing N_{pM} with $(N_T)^p$, we find that significant complexity reduction is possible using the symmetry properties of Volterra kernels.

Example 4.1   The lexicographic rule is typically employed to order the elements of the coefficient and input arrays.

Consider an M-dimensional system such that the vector $\mathbf{m}_1 = [m_{11} \cdots m_{1,M-1} \quad m_{1M}]$. The elements of the kernel and the input arrays of the M-dimensional linear filter in (4.2) are ordered in the lexicographic arrangement in accordance with the arrangement of the indices shown below:

$(0, \ldots, 0, 0)$	$(0, \ldots, 0, 1)$	$\cdots$	$(0, \ldots, 0, N_M - 1)$
$(0, \ldots, 1, 0)$	$(0, \ldots, 1, 1)$	$\cdots$	$(0, \ldots, 1, N_M - 1)$
$\cdots$	$\cdots$	$\cdots$	$\cdots$
$(0, \ldots, N_{M-1} - 1, 0)$	$(0, \ldots, N_{M-1} - 1, 1)$	$\cdots$	$(0, \ldots, N_{M-1} - 1, N_M - 1)$
$\cdots$	$\cdots$	$\cdots$	$\cdots$
$\cdots$	$\cdots$	$\cdots$	$\cdots$
$(N_1 - 1, \ldots, N_{M-1} - 1, 0)$	$(N_1 - 1, \ldots, N_{M-1} - 1, 1)$	$\cdots$	$(N_1 - 1, \ldots, N_{M-1} - 1, N_M - 1)$.

In this arrangement, we have let the variable m_{1M} vary first, $m_{1,M-1}$ vary next and so on till we let m_{11} vary in the last several rows of indices.

It is possible to derive the value of a single index $I_1(m_{11}, m_{12}, \ldots, m_{1M})$ that indicates the positions of the corresponding elements in the coefficient and input vectors of (4.6) by analyzing the arrangement of the indices in the above sequence. This index can be shown to be

$$I_1(m_{11}, m_{12}, \ldots, m_{1M}) = 1 + \sum_{i=1}^{M} m_{1i} \prod_{k=i+1}^{M} N_k,$$

where $\prod_{k=i+1}^{M} N_k$ is assumed to be equal to 1 for $i = M$. Since a homogeneous pth-order M-dimensional Volterra filter involves p vectors $\mathbf{m}_j$, we need to first define the indices

$$I_j(m_{j1}, m_{j2}, \ldots, m_{pM}) = 1 + \sum_{i=1}^{M} m_{ji} \prod_{k=i+1}^{M} N_k$$

with $j = 1, 2, \ldots, p$, and then to nest the p ordering sequences identified by such indexes I_j. If, again, the index I_p is allowed to vary before I_{p-1}, I_{p-1} before I_{p-2}, and so on, the total lexicographic order is represented by the index

$$I(\mathbf{m}_1, \mathbf{m}_2, \ldots, \mathbf{m}_p) = 1 + \sum_{j=1}^{p} N_T^{p-j}(I_j - 1),$$

which specifies the ordering sequence for the products of input samples and for the corresponding elements in the kernel $h_p(\mathbf{m}_1, \mathbf{m}_2, \ldots, \mathbf{m}_p)$.

It is also possible to show that, if the symmetry of the Volterra kernels is considered, the subregion S in (4.11) is defined according to the constraints

$$I_1(m_{11}, m_{12}, \ldots, m_{1M}) \leq I_2(m_{21}, m_{22}, \ldots, m_{2M}) \leq \cdots \leq I_p(m_{p1}, m_{p2}, \ldots, m_{pM}).$$

4.2 REALIZATIONS OF MULTIDIMENSIONAL VOLTERRA FILTERS

In this section, we describe straightforward extensions of the one-dimensional truncated Volterra system realizations to multidimensional systems. For simplicity of presentation, we concentrate primarily on two-dimensional quadratic systems.

4.2.1 Direct-Form Realization

Direct-form realizations are based on the direct implementation of (4.7) or, preferably, of (4.11). A nonlinear combiner is required to form all possible products of the input samples corresponding to the samples in the multidimensional plane of support. The structures so obtained are straightforward extensions of direct form realizations of one-dimensional Volterra filters.

4.2.2 Realization of Quadratic Filters through Matrix Decomposition

A homogeneous M-dimensional quadratic filter with finite region of support is described by the input–output relationship

$$\overline{h}_2[x(\mathbf{n})] = \sum_{\mathbf{m}_1}^{N} \sum_{\mathbf{m}_2}^{N} h_2(\mathbf{m}_1, \mathbf{m}_2)x(\mathbf{n} - \mathbf{m}_1)x(\mathbf{n} - \mathbf{m}_2), \tag{4.14}$$

where the variables $\mathbf{n}$, $\mathbf{m}_1$, and $\mathbf{m}_2$ represent M-dimensional vectors. Let us form a vector $\mathbf{X}_1(\mathbf{n})$ in which the samples of the input signal $x(\mathbf{n})$ employed in the representation of the system in (4.14) are arranged in a predetermined manner. Then, it is relatively easy to show that the quadratic filter output can be expressed as

$$\overline{h}_2[x(\mathbf{n})] = \mathbf{X}_1^T(\mathbf{n})\mathbf{H}_2\mathbf{X}_1(\mathbf{n}), \qquad (4.15)$$

where $\mathbf{H}_2$ is an $N_T \times N_T$-element block symmetric matrix in which the coefficients of the Volterra kernel are arranged in an appropriate manner. Let $x(\mathbf{n} - \mathbf{m}_1)$ and $x(\mathbf{n} - \mathbf{m}_2)$ represent the ith and jth elements of $\mathbf{X}_1(\mathbf{n})$. Then the (i, j)th element of $\mathbf{H}_2$ is $h_2(\mathbf{m}_1, \mathbf{m}_2)$. It immediately follows from the representation in (4.15) that the matrix decomposition techniques presented in Chapter 3 can also be applied to the multidimensional systems to obtain parallel-cascade realizations. The filter output in such realizations is computed as shown in Figure 4.1 for the general case by squaring the outputs of r, M-dimensional linear FIR filters having the same region of support as the quadratic filter, weighting the results with appropriate coefficients, and then adding together all the weighted partial results. In the preceding description, r is the rank of the matrix $\mathbf{H}_2$. We illustrate the ideas described above for two-dimensional quadratic filters in the following subsections.

4.2.2.1 Parallel-Cascade Realizations of Two-Dimensional Quadratic Filters
The parallel-cascade realizations of two-dimensional quadratic filters are of particular interest because of their many applications in image processing. The

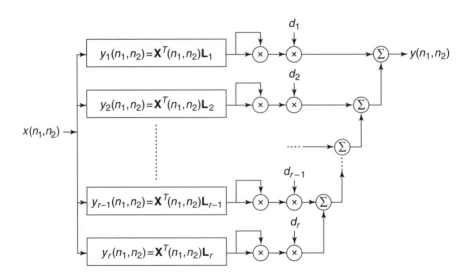

Figure 4.1 Parallel-cascade realization of a homogeneous two-dimensional quadratic filter.

quadratic two-dimensional operator can be expressed explicitly as[1]

$$\bar{h}_2[x(n_1, n_2)] = \sum_{m_{11}=0}^{N_1-1} \sum_{m_{12}=0}^{N_2-1} \sum_{m_{21}=0}^{N_1-1} \sum_{m_{22}=0}^{N_2-1} h_2(m_{11}, m_{12}, m_{21}, m_{22})$$

$$\times x(n_1 - m_{11}, n_2 - m_{12}) x(n_1 - m_{21}, n_2 - m_{22}). \quad (4.16)$$

We now illustrate how to evaluate the coefficient matrix $\mathbf{H_2}$ for this case [177]. We assume for our derivation that the kernel $h_2(m_{11}, m_{12}, m_{21}, m_{22})$ is symmetric with respect to the pairs of indices (m_{11}, m_{12}) and (m_{21}, m_{22}) so that $h_2(m_{11}, m_{12}, m_{21}, m_{22}) = h_2(m_{21}, m_{22}, m_{11}, m_{12})$. The filter output can be written in the form

$$y(\mathbf{n}) = y(n_1, n_2) = \mathbf{X}^T(n_1, n_2) \mathbf{H_2} \mathbf{X}(n_1, n_2), \quad (4.17)$$

where

$$\mathbf{X}(n_1, n_2) = \begin{bmatrix} x(n_1, n_2) \\ x(n_1 - 1, n_2) \\ \vdots \\ x(n_1 - N_1 + 1, n_2) \\ \hline \vdots \\ \hline x(n_1, n_2 - N_2 + 1) \\ x(n_1 - 1, n_2 - N_2 + 1) \\ \vdots \\ x(n_1 - N_1 + 1, n_2 - N_2 + 1) \end{bmatrix}. \quad (4.18)$$

The coefficient matrix $\mathbf{H_2}$ has $N_1 N_2 \times N_1 N_2$ elements, and it consists of N_2^2 submatrices $\mathbf{H}(i, j)$ containing $N_1 \times N_1$ elements each, arranged as follows:

$$\mathbf{H_2} = \begin{bmatrix} \mathbf{H}(0, 0) & \mathbf{H}(0, 1) & \cdots & \mathbf{H}(0, N_2 - 1) \\ \mathbf{H}(1, 0) & \mathbf{H}(1, 1) & \cdots & \mathbf{H}(1, N_2 - 1) \\ \vdots & \vdots & \cdots & \vdots \\ \mathbf{H}(N_2 - 1, 0) & \mathbf{H}(N_2 - 1, 1) & \cdots & \mathbf{H}(N_2 - 1, N_2 - 1) \end{bmatrix}, \quad (4.19)$$

---

[1] We have assumed in much of the derivations that the filter under consideration is causal. This is often not the case for finite-support multidimensional systems, but the analysis of noncausal, finite-support multidimensional systems is straightforward.

where each submatrix $\mathbf{H}(i, j)$ has the form

$$\mathbf{H}(i, j) = \begin{bmatrix} h(0, i, 0, j) & h(0, i, 1, j) & \cdots & h(0, i, N_1 - 1, j) \\ h(1, i, 0, j) & h(1, i, 1, j) & \cdots & h(1, i, N_1 - 1, j) \\ \vdots & \vdots & \cdots & \vdots \\ h(N_1 - 1, i, 0, j) & h(N_1 - 1, i, 1, j) & \cdots & h(N_1 - 1, i, N_1 - 1, j) \end{bmatrix}. \tag{4.20}$$

Since the quadratic kernel is symmetric, we can show using the definitions in (4.19) and (4.20) that the matrix $\mathbf{H_2}$ is also symmetric. The matrix decomposition of $\mathbf{H_2}$ can be performed in a number of different ways, as was the case for one-dimensional systems. One of the most convenient matrix decompositions is the LU decomposition discussed in Chapter 3. We assume that the rows and columns of $\mathbf{H_2}$ have been permuted so that the first r successive principal minors are different from zero. We further assume that the entries of $\mathbf{X}(n_1, n_2)$ have also been appropriately rearranged if the columns of $\mathbf{H_2}$ have been rearranged. Then, the matrix can be decomposed in the form

$$\mathbf{H_2} = \sum_{i=1}^{r} d_i \mathbf{L}_i \mathbf{L}_i^T, \tag{4.21}$$

where $r = \text{rank} \, [\mathbf{H_2}] \leq N_1 N_2$ and the scalar multipliers d_i values are real numbers. By replacing the matrix $\mathbf{H_2}$ in (4.17) with (4.21), it is possible to express the output of the two-dimensional quadratic filter in the form

$$y(n_1, n_2) = \sum_{i=1}^{r} d_i [\mathbf{X}^T(n_1, n_2) \mathbf{L}_i][\mathbf{L}_i^T \mathbf{X}(n_1, n_2)]$$

$$= \sum_{i=1}^{r} d_i y_i^2(n_1, n_2), \tag{4.22}$$

where

$$y_i(n_1, n_2) = \mathbf{X}^T(n_1, n_2) \mathbf{L}_i \tag{4.23}$$

Recall from Appendix B that the first $i - 1$ elements of $\mathbf{L}_i$ are zero and that its ith element is one. Consequently, the output of the linear filters can be explicitly written as

$$y_i(n_1, n_2) = x_i(n_1, n_2) + \sum_{j=i+1}^{N_1 N_2} l_i(j) x_j(n_1, n_2). \tag{4.24}$$

In this equation, $x_i(n_1, n_2)$ denotes the ith element of the vector $\mathbf{X}(n_1, n_2)$ and $l_i(j)$ is the jth element of the vector $\mathbf{L}_i$. The parallel-cascade realization that results from the representation of the system in (4.22) is similar to the one depicted in Figure 4.1.

Example 4.2 The homogeneous, two-dimensional quadratic filter with a 3×3-sample plane of support, and described by the coefficient matrix

$$
\mathbf{H_2} = \begin{bmatrix}
1 & \frac{1}{2} & 0 & \frac{1}{2} & 0 & -\frac{1}{2} & 0 & -\frac{1}{2} & -1 \\
\frac{1}{2} & 0 & \frac{1}{2} & 0 & 0 & 0 & -\frac{1}{2} & 0 & -\frac{1}{2} \\
0 & \frac{1}{2} & 1 & -\frac{1}{2} & 0 & \frac{1}{2} & -1 & -\frac{1}{2} & 0 \\
\frac{1}{2} & 0 & -\frac{1}{2} & 0 & 0 & 0 & \frac{1}{2} & 0 & -\frac{1}{2} \\
0 & 0 & 0 & 0 & 0 & 0 & 0 & 0 & 0 \\
-\frac{1}{2} & 0 & \frac{1}{2} & 0 & 0 & 0 & -\frac{1}{2} & 0 & \frac{1}{2} \\
0 & -\frac{1}{2} & -1 & \frac{1}{2} & 0 & -\frac{1}{2} & 1 & \frac{1}{2} & 0 \\
-\frac{1}{2} & 0 & -\frac{1}{2} & 0 & 0 & 0 & \frac{1}{2} & 0 & \frac{1}{2} \\
-1 & -\frac{1}{2} & 0 & -\frac{1}{2} & 0 & \frac{1}{2} & 0 & \frac{1}{2} & 1
\end{bmatrix}
$$

was employed in [234] to perform noise reduction and edge extraction simultaneously on input images. Since there are only four independent columns or rows in this matrix, it is a rank 4 matrix, and we can express it using LU decomposition as

$$\mathbf{H_2} = d_1 \mathbf{L}_1 \mathbf{L}_1^T + d_2 \mathbf{L}_2 \mathbf{L}_2^T + d_3 \mathbf{L}_3 \mathbf{L}_3^T + d_4 \mathbf{L}_4 \mathbf{L}_4^T,$$

where we calculate the parameters of this decomposition using the expressions given in Appendix B as

$$d_1 = 1, \quad d_2 = -\tfrac{1}{4}, \quad d_3 = 2, \quad d_4 = \tfrac{1}{2},$$

$$\mathbf{L}_1 = [1 \quad \tfrac{1}{2} \quad 0 \quad \tfrac{1}{2} \quad 0 \quad -\tfrac{1}{2} \quad 0 \quad -\tfrac{1}{2} \quad -1]^T,$$

$$\mathbf{L}_2 = [0 \quad 1 \quad -2 \quad 1 \quad 0 \quad -1 \quad 2 \quad -1 \quad 0]^T,$$

$$\mathbf{L}_3 = [0 \quad 0 \quad 1 \quad -\tfrac{1}{2} \quad 0 \quad \tfrac{1}{2} \quad -1 \quad 0 \quad 0]^T$$

and

$$\mathbf{L}_4 = [0 \quad 0 \quad 0 \quad 1 \quad 0 \quad -1 \quad 0 \quad 0 \quad 0]^T.$$

The realization resulting from this decomposition is shown in Figure 4.2. Each of the four linear filters in the realization has a 3×3-pixel plane of support.

It is useful to compare the computational complexity of the direct-form and parallel-cascade realizations in this example. Since the coefficients of the direct-form realization are all integer powers of two or zero, the computation of the output can be performed using only bit shifts and additions. However, before these operations can be performed, we need to evaluate the product of input samples of the form $x(n_1 - m_{1i}, n_2 - m_{2j})x(n_1 - m_{1k}, n_2 - m_{2l})$. Forming the product samples for the 22 nonzero coefficients of the triangular kernel derived from the symmetric matrix

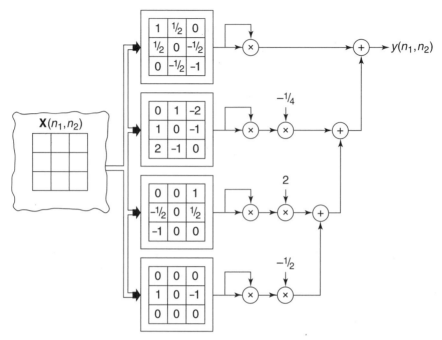

Figure 4.2   The LU decomposition-based parallel-cascade realization of the filter in Example 4.2.

H_2 requires 22 multiplications. Thus, implementing the direct form realization of the quadratic filter requires 22 multiplications to compute the products of pairs of input samples, one shift for applying the filter coefficient -2, and 21 additions per output sample. For the parallel-cascade realization also, the coefficients of the linear filters as well as the scaling factor are integer powers of two. As a result, this realization requires four squaring operations, six bit shifts, and 17 additions for computing each output sample.

4.2.2.2 Approximation Using Singular Value Decomposition

As was the case in the one-dimensional case, it is possible to decompose the coefficient matrix using the singular value decomposition and then approximate the system using a lower-rank matrix to obtain a parallel-cascade realization using fewer branches than the rank of H_2. Further computational complexity reduction can be achieved by applying the LU decomposition to the approximate coefficient matrix that results from the singular value decomposition. The next example illustrates the combined use of singular value decomposition and LU decomposition to obtain efficient realizations of multidimensional truncated Volterra systems.

Example 4.3   Singular value decomposition of the coefficient matrix H_2 of Example 4.2 results in the singular values and the singular vectors displayed in

TABLE 4.1 Singular Values and Singular Vectors of the Coefficient Matrix H_2 of Example 4.2

	$i = 1$	$i = 2$	$i = 3$	$i = 4$
Singular values	2.7321	2.7321	−0.7321	−0.7321
Singular vectors	0.6280	0.0000	0.2965	0.1332
	0.2299	−0.2299	−0.5870	0.2231
	0.0000	−0.6280	0.1332	−0.2965
	0.2299	0.2299	−0.2231	−0.5870
	0.0000	0.0000	0.0000	0.0000
	−0.2299	−0.2299	0.2231	0.5870
	0.0000	0.6280	−0.1332	0.2965
	−0.2299	0.2299	0.5870	−0.2231
	−0.6280	0.0000	−0.2965	−0.1332

Table 4.1. An approximate representation $\tilde{H}_2$ may be obtained for H_2 by retaining only the two most significant terms of the singular value decomposition as

$$\tilde{H}_2 = w_1 S_1 S_1^T + w_2 S_2 S_2^T,$$

where $w_1 = w_2 = 2.7321$ and the singular vectors S_1 and S_2 are given by the nine-element vectors defined by the first two columns of Table 4.1.

We can obtain an additional reduction in the implementation complexity by applying an LU decomposition to $\tilde{H}_2$ and then obtaining the realization on the basis of this decomposition. The LU decomposition of $\tilde{H}_2$ gives

$$\tilde{H}_2 = d_1 L_1 L_1^T + d_2 L_2 L_2^T,$$

where $d_1 = 1.0775$, $d_2 = 0.1445$,

$$L_1 = [1 \quad 0.3661 \quad 0 \quad 0.3661 \quad 0 \quad -0.3661 \quad 0 \quad -0.3661 \quad -1]^T$$

and

$$L_2 = [0 \quad 1 \quad 2.7301 \quad -1 \quad 0 \quad 1 \quad -2.7301 \quad -1 \quad 0]^T.$$

Thus we can obtain a realization based on two (two-dimensional) FIR filters with 3×3-sample planes of support. The computational complexity corresponds to four multiplications, two squaring operations, and 11 additions for computing each output sample. This measure of the computational complexity is slightly larger than the complexity measure for the exact parallel-cascade realization that utilizes all four branches. This is a consequence of the fact that the coefficients in this example lent themselves to implementational simplicity because their values were all integer powers of two. In applications where such simplifications are not available, the approximate realizations will in general provide significant reduction in the computational complexity over exact implementations.

The quality of the approximate realization was evaluated in two different ways. First, the Frobenius of the error matrix normalized by the Frobenius norm of $\mathbf{H_2}$ was evaluated to be

$$\frac{\|\mathbf{H}_2 - \tilde{\mathbf{H}}_2\|_F}{\|\mathbf{H}_2\|_F} = 0.2588.$$

Even though this value is relatively large and may not be appropriate in many applications, we will see that the performance of our filter is acceptable for applications in edge extraction. The second method for evaluating the quality of the approximation involved comparing the performance of the exact filter and its approximation in this application. Figure 4.3c shows the result of processing the image displayed in Figure 4.3b with the original filter. The input image was obtained by corrupting the image shown in Figure 4.3a with a multiplicative noise that is uniformly distributed with a mean value of one and variance equal to 0.05. The images contain 512×512 pixels with a gray-scale resolution of eight bits per pixel

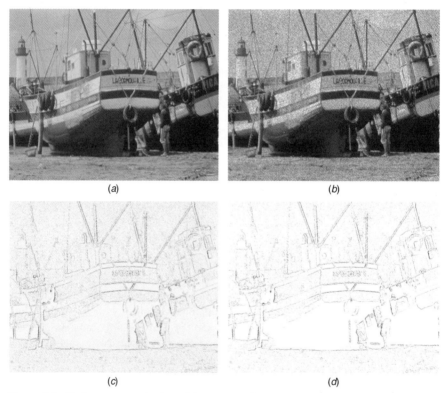

Figure 4.3 Performance evaluation of the exact and approximate realization of the quadratic filter of Example 4.3 obtained in an edge detection task: (a) original image; (b) corrupted image; (c) output of the exact realization of the quadratic filter; (d) output of the approximate realization of the quadratic filter.

in the range 0 to 255. Figure 4.3d shows the results of applying the approximate filter to the noisy image. We can see that the quadratic filter is able to provide a reasonably good estimate of the edges in spite of the noise present in the input image. The two edge maps obtained appear to have similar visual characteristics, indicating that the approximate filter works almost as good as the original filter in this application.

4.2.3 Realization of Multidimensional Volterra Filters Using Distributed arithmetic

The vector representation of multidimensional truncated Volterra systems as in (4.10) allows us to decompose the coefficient and input vectors using their elementary bits with distributed arithmetic. The derivation of the distributed arithmetic realization as well as issues related to complexity reduction remain substantially the same as in the case of one-dimensional Volterra systems. Consequently, they are not repeated here.

4.3 CONSTRAINTS IMPOSED ON TWO-DIMENSIONAL QUADRATIC FILTERS

A generic M-dimensional, pth-order Volterra filter with a memory of N samples along each dimension has $O(N^{Mp})$ coefficients. As shown in (4.13), this number reduces to $O(N_{pM})$ when the kernel is symmetric. It is obvious that the number of coefficients of the system can become extremely large even for moderately large values of the dimension M and system order p. Consequently, it is important to develop design techniques that result in systems that can significantly reduce the implementation complexity of multidimensional Volterra systems. One way of achieving this objective is to impose certain constraints on the Volterra kernels of the system during its design so that a reduction in the number of independent coefficients is obtained. Many of the conditions discussed in this section are quite natural in applications involving images, and such conditions have been successfully applied in image processing environments [245]. In the following subsections, we present several constraints imposed in the design of two-dimensional quadratic filters. These conditions are based on the concept of isotropy and other properties related to the desired response of the systems for specific gray-scale levels in the input image.

4.3.1 Isotropy of Two-Dimensional Quadratic systems

Recall from Section 3.3.1 that the isotropy was defined for one-dimensional systems as the invariance of the filter output with respect to a 180° rotation of the input data. The isotropy of multidimensional systems denotes the invariance of the filter output to each of the exact rotations of the data allowed by the sampling lattice. The exact data rotations possible in the case of two-dimensional images sampled on a rectangular lattice involve integer multiples of 90 degrees: 90°, 180°, and 270°. Another constraint that is imposed on the coefficients of isotropic two-dimensional systems is the invariance of their outputs to the reflection of the input pixels around their central horizontal or vertical axis. Isotropy conditions can be derived for the coefficients by imposing equality of the coefficients that weight the same input samples in the case of linear operators and the same products of input samples in the case of quadratic operators before and after each rotation or reflection.

4.3.1.1 The Linear Operator

Let us arrange the input samples contained in the plane of support of the two-dimensional system in a matrix $X_{1M}(n_1, n_2)$ as

$$
X_{1M}(n_1, n_2) = \begin{bmatrix} x(n_1, n_2) & x(n_1, n_2 - 1) & \cdots & x(n_1, n_2 - N + 1) \\ x(n_1 - 1, n_2) & x(n_1 - 1, n_2 - 1) & \cdots & x(n_1 - 1, n_2 - N + 1) \\ \vdots & \vdots & \ddots & \vdots \\ x(n_1 - N + 1, n_2) & x(n_1 - N + 1, n_2 - 1) & \cdots & x(n_1 - N + 1, n_2 - N + 1) \end{bmatrix},
$$
(4.25)

where we have assumed that $N_1 = N_2 = N$ in order to correctly define all the mentioned isotropy conditions. Let us also define a coefficient matrix H_{1M} in which the coefficients of the linear kernel are arranged in the same order as the samples are arranged in the data matrix in (4.25). The input–output relationship of a linear operator can be expressed in the following alternative form:

$$
y(n_1, n_2) = \sum_{m_1=0}^{N-1} \sum_{m_2=0}^{N-1} h_1(m_1, m_2) x(n_1 - m_1, n_2 - m_2)
$$

$$
= \mathrm{Tr}[H_{1M} X_{1M}^T(n_1, n_2)],
$$
(4.26)

where $\mathrm{Tr}[\cdot]$ indicates the trace of a matrix. To develop the isotropy constraints related to a clockwise $90°$ rotation, consider the matrix $X'_{1M}(n_1, n_2, 90)$ resulting from such a rotation of $X_{1M}(n_1, n_2)$ and given by

$$
X'_{1M}(n_1, n_2, 90)
$$
$$
= \begin{bmatrix} x(n_1 - N + 1, n_2) & x(n_1 - N + 2, n_2) & \cdots & x(n_1, n_2) \\ x(n_1 - N + 1, n_2 - 1) & x(n_1 - N + 2, n_2 - 1) & \cdots & x(n_1, n_2 - 1) \\ \vdots & \vdots & \ddots & \vdots \\ x(n_1 - N + 1, n_2 - N + 1) & x(n_1 - N + 2, n_2 - N + 1) & \cdots & x(n_1, n_2 - N + 1) \end{bmatrix}.
$$
(4.27)

Then, the condition

$$
y(n_1, n_2) = \mathrm{Tr}[H_{1M} X_{1M}^T(n_1, n_2)] = \mathrm{Tr}[H_{1M} X'^T_{1M}(n_1, n_2, 90)]
$$
(4.28)

can be met using the following set of constraints on the filter coefficients:

$$h_1(m_1, m_2) = h_1(N - 1 - m_2, m_1); \qquad m_1, m_2 = 0, 1, \ldots, N - 1. \qquad (4.29)$$

By applying similar considerations, it can be shown that the rotations of 180° and 270° imply the following sets of constraints:

$$h_1(m_1, m_2) = h_1(N - 1 - m_1, N - 1 - m_2); \qquad m_1, m_2 = 0, 1, \ldots, N - 1 \qquad (4.30)$$

and

$$h_1(m_1, m_2) = h_1(m_2, N - 1 - m_1); \qquad m_1, m_2 = 0, 1, \ldots, N - 1, \qquad (4.31)$$

respectively. Finally, to meet the isotropy condition related to reflecting $\mathbf{X}_{1M}(n_1, n_2)$ about its central horizontal axis, the coefficients must satisfy the conditions

$$h_1(m_1, m_2) = h_1(N - 1 - m_1, m_2); \qquad m_1, m_2 = 0, 1, \ldots, N - 1. \qquad (4.32)$$

The reflection about the vertical axis can be obtained using a horizontal reflection and a rotation by 180°. Consequently, the constraints imposed by the conditions in (4.32) and (4.30) together imply the condition related to a vertical reflection given by

$$h_1(m_1, m_2) = h_1(m_1, N - 1 - m_2); \qquad m_1, m_2 = 0, 1, \ldots, N - 1. \qquad (4.33)$$

The constraints imposed by (4.29)–(4.32) correspond to the octagonal symmetry of the linear filter both in the data and in the frequency domains. As a result, the linear filter can be completely specified by knowing the coefficients in the shaded region of Figure 4.4. The derivation of this result is left as an exercise for the reader.

4.3.1.2 *The Quadratic Operator* The derivation of the constraints resulting from the isotropy of the homogeneous quadratic operator can be performed as in the previous section. Let us define the array $\mathbf{X}_{2M}(n_1, n_2)$ containing products of pairs of input samples belonging to the plane of support of the filter and obtained as the Kronecker product of $\mathbf{X}_{1M}(n_1, n_2)$ with itself:

$$\mathbf{X}_{2M}(n_1, n_2) = \mathbf{X}_{1M}(n_1, n_2) \otimes \mathbf{X}_{1M}(n_1, n_2). \qquad (4.34)$$

Let us also arrange the quadratic coefficients $h_2(m_{11}, m_{12}, m_{21}, m_{22})$ into the array $\mathbf{H}_{2M}$, so that the coefficients in a specific location of this matrix scales the entry in the same location of $\mathbf{X}_{2M}(n_1, n_2)$. The arrangements of the coefficients can be best illustrated by partitioning the coefficient matrix into N^2 submatrices, each of which contains $N \times N$ elements as described below [245]. We use a lexicographic ordering

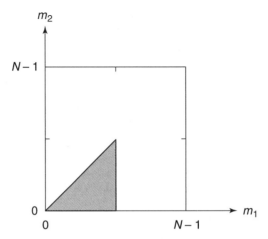

Figure 4.4 The shaded portion of the plane of support of a two-dimensional, isotropic linear filter completely specifies such a system.

for the elements of $\mathbf{X}_{1M}(n_1, n_2)$ so that each entry of this matrix is characterized by a single index as

$$
\mathbf{X}_{1M}(n_1, n_2) = \begin{bmatrix} x_1 & x_2 & \cdots & x_N \\ x_{N+1} & x_{N+2} & \cdots & x_{2N} \\ \vdots & \vdots & \ddots & \vdots \\ x_{N^2-N+1} & x_{N^2-N+2} & \cdots & x_{N^2} \end{bmatrix}. \tag{4.35}
$$

Let us define an $N \times N$-element subset of the coefficient matrix $\mathbf{h}_{2L}(i)$ as

$$
\mathbf{h}_{2L}(i) = \begin{bmatrix} h_{2L}(i, 1) & h_{2L}(i, 2) & \cdots & h_{2L}(i, N) \\ h_{2L}(i, N+1) & h_{2L}(i, N+2) & \cdots & h_{2L}(i, 2N) \\ \vdots & \vdots & \ddots & \vdots \\ h_{2L}(i, N^2-N+1) & h_{2L}(i, N^2-N+2) & \cdots & h_{2L}(i, N^2) \end{bmatrix}, \tag{4.36}
$$

where the coefficient denoted by $h_{2L}(i, j)$ scales the product signal represented by $x_i x_j$ in the input–output relationship. Then, the matrix $\mathbf{H}_{2M}$ is given by

$$
\mathbf{H}_{2M} = \begin{bmatrix} \mathbf{h}_{2L}(1) & \mathbf{h}_{2L}(2) & \cdots & \mathbf{h}_{2L}(N) \\ \mathbf{h}_{2L}(N+1) & \mathbf{h}_{2L}(N+2) & \cdots & \mathbf{h}_{2L}(2N) \\ \vdots & \vdots & \ddots & \vdots \\ \mathbf{h}_{2L}(N^2-N+1) & \mathbf{h}_{2L}(N^2-N+2) & \cdots & \mathbf{h}_{2L}(N^2) \end{bmatrix}. \tag{4.37}
$$

The input–output relationship of the quadratic operator can be rewritten with the help of these definitions as

$$y(n_1, n_2) = \mathrm{Tr}\ [\mathbf{H}_{2M}\mathbf{X}_{2M}^T(n_1, n_2)].\tag{4.38}$$

To proceed further, we first evaluate the effects of rotating $\mathbf{X}_{1M}(n_1, n_2)$ by multiples of $90°$ and of reflection about its horizontal axis on the matrix $\mathbf{X}_{2M}(n_1, n_2)$. The $90°$ clockwise rotation of $\mathbf{X}_{1M}(n_1, n_2)$ gives

$$\mathbf{X}_{1M}'(n_1, n_2, 90) = \begin{bmatrix} x_{N^2-N+1} & x_{N^2-2N+1} & \cdots & x_1 \\ x_{N^2-N+2} & x_{N^2-2N+2} & \cdots & x_2 \\ \vdots & \vdots & \ddots & \vdots \\ x_{N^2} & x_{N^2-N} & \cdots & x_N \end{bmatrix}.\tag{4.39}$$

To find the isotropy constraints related to a $90°$ rotation of the data matrix, we seek conditions on the coefficients that force the filter output to be identical for input sequences corresponding to $\mathbf{X}_{2M}(n_1, n_2)$ and $\mathbf{X}_{2M}'(n_1, n_2, 90)$, i.e.,

$$y(n_1, n_2) = \mathrm{Tr}[\mathbf{H}_{2M}\mathbf{X}_{2M}^T(n_1, n_2)] = \mathrm{Tr}[\mathbf{H}_{2M}\mathbf{X}_{2M}'^T(n_1, n_2, 90)],\tag{4.40}$$

where the matrix $\mathbf{X}_{2M}'(n_1, n_2, 90)$ is obtained by applying a $90°$ clockwise rotation to $\mathbf{X}_{2M}(n_1, n_2)$. For this purpose, it is convenient to first derive a relationship between the indices of a lexicographic-ordered matrix before and after the rotation. Let us consider an entry in an $N \times N$-element matrix with row and column indices r and c, respectively. This element of the matrix goes to the location defined by

$$(r', c') = (N + 1 - c, r)\tag{4.41}$$

after a clockwise $90°$ rotation. The index corresponding to the (r, c)th element in the lexicographic ordering of the matrix is

$$k = N(r - 1) + c.\tag{4.42}$$

After the $90°$ rotation, the index becomes

$$k'(90) = N(r' - 1) + c' = N^2 - cN + r.\tag{4.43}$$

Conversely, the kth entry in the lexicographical ordering of the matrix elements corresponds to r and c defined by

$$c = [k - 1]_N + 1\tag{4.44}$$

and

$$r = \frac{k - [k - 1]_N - 1}{N} + 1,\tag{4.45}$$

respectively, where $[(\cdot)]_N$ indicates the modulo-N representation of $(\cdot)$, i.e., the remainder of the integer division of $(\cdot)$ by N. By substituting (4.44) and (4.45) in (4.43), we get

$$k'(90) = N^2 - N[k-1]_N - N + \frac{k - [k-1]_N - 1}{N} + 1. \qquad (4.46)$$

A homogeneous quadratic filter satisfies (4.40) if the coefficients of the form $h_{2L}(i,j)$ and $h_{2L}(i',j')$ are equal to each other. The relationships between i and i' and between j and j' are identical to the relationship between k and k' given by (4.46).

For the 180° rotation, the following relationship can be derived for the lexicographic indices before and after the rotation:

$$k'(180) = N^2 - k + 1. \qquad (4.47)$$

The corresponding results for the 270° rotation is

$$k'(270) = N[k-1]_N + N - \frac{k - [k-1]_N - 1}{N}. \qquad (4.48)$$

Finally, for reflection about the horizontal axis, we get

$$k'(rh) = N^2 - N - k + 2[k-1]_N + 2. \qquad (4.49)$$

The details of these derivations are left as exercises for the reader.

Remark 4.1 Let us express the relationship in (4.46) as $k'(90) = u(k)$. Then, (4.47) can be derived by applying the function $u(\cdot)$ twice on k. Specifically, $k'(180) = u(u(k))$ and $k'(270) = u(u(u(k)))$.

4.3.1.3 Complexity Reduction
The isotropy constraints result in a significant reduction of the complexity by reducing the number of independent coefficients in the filter. When the symmetry and the isotropy constraints given by (4.29)–(4.32) are applied to the linear component of the filter, only a half-quadrant of the coefficient matrix $\mathbf{H}_{1M}$ contains independent coefficients. Thus the number of independent coefficients in the linear component decreases from N^2 to

$$\frac{N}{4}\frac{N+2}{2} \qquad (4.50)$$

TABLE 4.2   The Number of Independent Quadratic Coefficients for Various Filter Supports

N	$N^2 \times N^2$	$N^2(N^2+1)/2$	C_q
2	16	10	3
3	81	45	11
4	256	136	24
5	625	325	55

for even values of N and

$$\frac{N+1}{4}\frac{N+3}{2} \tag{4.51}$$

for odd values of N.

For the quadratic coefficients, arranged in the matrix $\mathbf{H}_{2M}$, it is convenient to apply (4.46)–(4.48) first, taking simultaneously into account the symmetry condition of the kernel expressed by the equality $h_{2L}(i,j) = h_{2L}(j,i)$. Then, (4.49) can be used to determine the additional constraints that are independent of those due to the rotations. A formal procedure to derive the number of independent quadratic coefficients is described in Ramponi [245]. The results are summarized in Table 4.2 for small filter supports, where $N^2(N^2+1)/2$ is the number of independent coefficients of a quadratic symmetric kernel and C_q is the number of the independent coefficients after application of the isotropy constraints.

It can be clearly seen from the results of Table 4.2 that the isotropy constraints permit a significant reduction in the number of independent coefficients.

Example 4.4   Consider a homogeneous 3×3 quadratic filter described by a symmetric and isotropic kernel. Let the pixels on the filter support be numbered according to the lexicographic order as in (4.35) as

$$\mathbf{X}_{1M}(n_1, n_2) = \begin{bmatrix} x_1 & x_2 & x_3 \\ x_4 & x_5 & x_6 \\ x_7 & x_8 & x_9 \end{bmatrix}.$$

The application of the kernel symmetry reduces the $N^2 = 81$ filter coefficients $h_{2L}(i,j)$ to 45. The indices of these coefficients may be specified by the following set: $1 \le i \le j \le 9$. The application of the isotropy constraints (4.46–4.49) further reduces the independent coefficients to the following 11 coefficients:

$$\begin{array}{llllll}
h_{2L}(1,1) & h_{2L}(1,2) & h_{2L}(1,3) & h_{2L}(1,5) & h_{2L}(1,6) & h_{2L}(1,9) \\
h_{2L}(2,2) & h_{2L}(2,4) & h_{2L}(2,5) & h_{2L}(2,8) \\
h_{2L}(5,5)
\end{array}$$

The other 34 coefficients are consequently determined according to the following equalities:

$$h_{2L}(1,4) = h_{2L}(2,3) = h_{2L}(3,6) = h_{2L}(4,7) = h_{2L}(6,9) = h_{2L}(7,8)$$
$$= h_{2L}(8,9) = h_{2L}(1,2)$$
$$h_{2L}(1,7) = h_{2L}(3,9) = h_{2L}(7,9) = h_{2L}(1,3)$$
$$h_{2L}(1,8) = h_{2L}(2,7) = h_{2L}(2,9) = h_{2L}(3,4) = h_{2L}(3,8) = h_{2L}(4,9)$$
$$= h_{2L}(6,7) = h_{2L}(1,6)$$
$$h_{2L}(2,6) = h_{2L}(4,8) = h_{2L}(6,8) = h_{2L}(2,4)$$
$$h_{2L}(3,3) = h_{2L}(7,7) = h_{2L}(9,9) = h_{2L}(1,1)$$
$$h_{2L}(3,5) = h_{2L}(5,7) = h_{2L}(5,9) = h_{2L}(1,5)$$
$$h_{2L}(3,7) = h_{2L}(1,9)$$
$$h_{2L}(4,4) = h_{2L}(6,6) = h_{2L}(8,8) = h_{2L}(2,2)$$
$$h_{2L}(4,5) = h_{2L}(5,6) = h_{2L}(5,8) = h_{2L}(2,5)$$
$$h_{2L}(4,6) = h_{2L}(2,8)$$

4.3.2 Typical Gray-Scale Constraints Imposed on Two-Dimensional Filters in Image Processing Applications

It is typical to impose additional constraints on the output of filters in image processing applications for certain types of input signals. These constraints result in additional simplifications of such filters. We describe some of the most commonly used constraints below.

4.3.2.1 *Preservation of a Constant Input Value*   In many applications, we specify that the output of a quadratic filter is identical to the value of its input signal whenever its input is a constant gray level. A set of conditions that satisfy this constraint is

$$h_0 = 0, \tag{4.52}$$

$$\sum_{m_1=0}^{N-1} \sum_{m_2=0}^{N-1} h_1(m_1, m_2) = 1, \tag{4.53}$$

and

$$\sum_{m_{11}=0}^{N-1} \sum_{m_{12}=0}^{N-1} \sum_{m_{21}=0}^{N-1} \sum_{m_{22}=0}^{N-1} h_2(m_{11}, m_{12}, m_{21}, m_{22}) = 0. \tag{4.54}$$

4.3.2.2 Scaling Conditions Often, a stronger condition than the preservation of a constant bias is imposed during the design of two-dimensional filters. This condition requires that the quadratic filter also preserves a uniform shift in its input at the output, that is, if $y(n_1, n_2)$ is the output of the system when its input is $x(n_1, n_2)$, its output must be $y(n_1, n_2) + \mu$ whenever its input signal is $x(n_1, n_2) + \mu$. It is not difficult to show that a quadratic filter whose coefficients satisfy the conditions

$$\sum_{m_1=0}^{N-1} \sum_{m_2=0}^{N-1} h_1(m_1, m_2) = 1, \qquad (4.55)$$

$$\sum_{m_{11}=0}^{N-1} \sum_{m_{12}=0}^{N-1} h_2(m_{11}, m_{12}, m_{21}, m_{22}) = 0 \qquad (4.56)$$

for any $m_{21}, m_{22} = 0, \ldots, N - 1$, and

$$\sum_{m_{21}=0}^{N-1} \sum_{m_{22}=0}^{N-1} h_2(m_{11}, m_{12}, m_{21}, m_{22}) = 0 \qquad (4.57)$$

for any $m_{11}, m_{12} = 0, \ldots, N - 1$ meets this requirement. Clearly, conditions (4.56) and (4.57) imply (4.54).

4.3.2.3 Amplitude Dependence One of the relatively well-known facts about human perception of visual stimuli is Weber's law [99]. In a simplified form, Weber's law states that larger differences in the gray levels are necessary in the bright zones of an image than in the dark ones to produce the same visual effect. Consequently, in many applications of two-dimensional quadratic filters in which the final judge of quality is a human observer, it is necessary to design filters whose response to a constant shift in the gray level over the filter's plane of support is dependent on the average gray level in the area surrounding the pixel of interest in addition to the value of the shift. In the following example, we demonstrate the ability of a homogeneous quadratic filter to provide responses that are tuned to the human visual system in the manner described above.

Example 4.5 Suppose that the input image is formed with two contiguous regions having constant gray levels a and b, respectively, such that a step increase in the gray level with a magnitude $a - b$ is encountered when the image is scanned in the horizontal direction. Let us apply a homogeneous quadratic filter defined on a 3 × 3-sample plane of support to this synthetic image. We now consider the situation in which the filter satisfies the isotropy conditions. The output of the quadratic filter in the first column of the transition region is given by

$$y_a(n_1, n_2) = C_1 a^2 + C_2 ab + C_3 b^2,$$

where the constants C_1, C_2, and C_3 depend on the values of the filter coefficients. The input image takes the value of a in this column and to its left. The output value in the second column of the transition region is given by

$$y_b(n_1, n_2) = C_3 a^2 + C_2 ab + C_1 b^2.$$

The input image takes the value b in this column and to its right. The derivation of this result makes use of the isotropy conditions. After some simple algebraic manipulations, the step change Δ in the output image that corresponds to the input step change of magnitude $b - a$ can be shown to be

$$\Delta = y_b(n_1, n_2) - y_a(n_1, n_2) = 2(C_1 - C_3)(b - a)\frac{(b + a)}{2}.$$

This fact means that, unlike linear isotropic filters that respond to input step changes with signals having proportional amplitudes, we are able to design quadratic filters that respond with outputs that are proportional to the mean values of the backgrounds as well as amplitudes of the step changes. Thus, it is possible to design quadratic filters that respond to input stimuli in a manner tuned to the Weber's law.

Remark 4.2   As pointed out above, the quadratic filter is capable of responding to changes in the input signal differently according to their location within the allowed dynamic range. Therefore, in problems such as the enhancement of noisy images, it is useful to define a threshold τ so that magnitude changes in the input image larger than τ are considered as relevant details of the image, while magnitude changes smaller than τ are considered as noise. On the other hand, it is often convenient to define the coefficients of a prototype reference filter, conventionally designed with reference, for example, to unit steps. The threshold τ is then used as a denormalization factor to scale down before processing the coefficients of the quadratic prototype filter so that gray steps below τ are smoothed since they are considered due to noise, while steps above τ are amplified since they are due to details. The value τ can be determined by a trial-and-error procedure or, more conveniently, from knowledge of the statistics of the noise present in the image. Methods for determining the threshold are described by Ramponi in [240,242].

4.4 DESIGN OF TWO-DIMENSIONAL QUADRATIC FILTERS

Formal design of multidimensional Volterra filters is a difficult problem because of the complexity of the filters for higher dimensions and higher orders, and because it is often difficult to define a mathematically tractable optimization problem for the design. As a result, almost all of the results presently available for multidimensional Volterra filter design deal only with two-dimensional quadratic filters. Higher-order filter designs available at this time involve heuristic approaches, and are usually derived for specific applications. Several such application-dependent design techniques are described in Chapter 10. In this section, we consider some formal methods for designing two-dimensional quadratic filters. Design of such filters are of importance because of their usefulness in a large number of image processing applications.

4.4.1 Optimization Methods for Quadratic Filter Design

The formulation of the design of a truncated Volterra filter as an optimization problem that minimizes a cost function is conceptually useful for designing multi-dimensional and higher-order nonlinear filters. However, the application of such a method to higher-order kernels or to higher-dimensional operators may suffer from difficulties related to the large number of parameters that must be computed using standard optimization algorithms. Consequently, the method described here has been applied so far only for designing two-dimensional quadratic filters [235,237,241].

The first step in the filter design problem is the *formulation phase,* which consists of determining the number of design variables that must be optimized, the optimization criterion, and any additional constraints the parameters resulting from the optimization process must satisfy. The design problem can be formally stated as follows.

Problem Statement. Find a vector $\mathbf{h}^0$ such that

$$f(\mathbf{h}^0) = \min_{\mathbf{h} \in \mathcal{D}} f(\mathbf{h}), \qquad (4.58)$$

where $\mathbf{h}$ is a k-dimensional design variable, $\mathcal{D}$ is the space of all vectors $\mathbf{h}$ that satisfy the constraint function $g(\mathbf{h}) = 0$ so that $\mathcal{D} = \{\mathbf{h} : g(\mathbf{h}) = 0\}$, and $f(\cdot)$ is a real and nonnegative function of the design variable and is usually called the *cost function* or the *objective function.* The constraint function $g(\cdot)$ is typically a vector function that is defined from the space of k-dimensional vectors to a space of l-dimensional vectors, where $l \leq k$.

The Formulation Phase. The determination of the filter structure, the objective function and the constraint function are dependent on the specific application that we are interested in. In many applications, it is impossible or difficult to formulate an objective function using theoretical analysis. In such situations the common practice is to design the filter with the help of a set of one or more training images. The training images consist of input images that are representative of the class of input images that are typically applied to the filter when it is used in practical situations, and the desired output images when the input images are processed by the filter. The desired output images should be such that they satisfy the properties we wish the output images to have. It is also possible to create synthetic training images that satisfy the desired properties of input and output images of the filter to be designed.

The Constraint Function. The constraint function for the filter design can be formulated using properties of quadratic filters such as symmetry and isotropy. These constraints reduce the number of the independent coefficients to be determined and thereby simplify the optimization task. Furthermore, these constraints can be built into the objective function, which will allow the use of unconstrained optimization techniques for determining the optimal filter coefficients.

The Optimization Process. Once the optimization problem has been set up, there are several numerical techniques for finding the minimum of the objective function. The commonly used techniques for this purpose include the steepest-descent method [329], the conjugate gradient technique [97], and the simulated annealing technique [113]. The steepest-descent method is computationally simple and straightforward to implement. However, this technique has slow convergence properties. The conjugate gradient method is more complex than the steepest gradient method, but provides convergence to the correct solution in far fewer iterations. Both methods may converge to a local minimum of the objective function if the objective function is not a convex function of the design variables. The simulated annealing technique is capable of finding the global minimum point of the objective function even when there are local minima. However, this method is the slowest to converge and the most computation-intensive of the three methods. In most of the filter design problems of interest to us, it is possible to formulate an objective function without local minima, and therefore, the extra expense of implementing the simulated annealing algorithm may not be useful in such situations. Ramponi et al. compared the performance of these three algorithms in the design of a quadratic filter for application in an image enhancement problem [241]. The conclusion of this study indicated a preference for the conjugate gradient method among the three techniques.

4.4.1.1   *A Design Example*   We illustrate the concepts described above and their usefulness in practice using the design of a quadratic filter for enhancing a blurred and noisy image. We assume, without loss of generality, that the filter structure has already been fixed to have a linear component and a quadratic component, and that the design variables are the filter coefficients. Therefore, an initial formulation of the problem in (4.58) involves $N^2 + N^4$ design variables. We can impose the isotropy constraints to reduce the number of design variables. Since the isotropy conditions lead to the equality of several coefficients, we can directly impose the constraints on the filter structure and avoid the need for a constraint function specifically for the isotropy conditions.

The conditions such as preservation of constant input values or the preservation of a constant bias in the input signals result in various relationships, such as those given by (4.53) and (4.54), or (4.55)–(4.57). These conditions may be incorporated into the problem formulation as constraint functions. For example, preservation of constant input values at the output of the filter results in the two constraints.

$$g_1(\mathbf{h}) = \left| 1 - \sum_{m_1} \sum_{m_2} h_1(m_1, m_2) \right| \tag{4.59}$$

and

$$g_2(\mathbf{h}) = \left| \sum_{m_{11}} \sum_{m_{12}} \sum_{m_{21}} \sum_{m_{22}} h_2(m_{11}, m_{12}, m_{21}, m_{22}) \right|. \tag{4.60}$$

Although it is possible to devise optimization procedures that search for the solution only in the space of the design variables where these constraints are met, an easier approach is to use the Lagrange multiplier method in which the constraint function is incorporated into the objective function as a penalty function. Such an approach results in the formulation of an unconstrained optimization problem that is easier to solve. Thus, the formulation of the optimization problem may involve the minimization of

$$f(\mathbf{h}) + w_1 g_1(\mathbf{h}) + w_2 g_2(\mathbf{h}), \tag{4.61}$$

where w_1 and w_2 are positive weighting factors.

A critical task that is yet to be performed is the choice of the objective function $f(\cdot)$. The objective function can be formulated using an appropriate measure of the differences between an ideal or desired filter output $y_d(n_1, n_2)$ and the actual output $y(n_1, n_2)$. The output $y(n_1, n_2)$ may be the result of processing a noisy input $x'(n_1, n_2)$ with the designed filter. In many situations, we will have reasonable knowledge of the noise characteristics. In such situations, we can simulate the noise conditions by superimposing a noise sequence with appropriate distribution on a set of reference images that are noise-free. For example, in the problem that is considered in Example 4.6, the objective is to enhance photographs shot in poor illumination conditions. Such photographs can be simulated by using an image obtained in ideal illumination, adding noise to this image and compressing the pixel values to have a smaller dynamic range. Computationally simple objective functions can be obtained with the help of simple reference inputs such as unit steps and/or impulses on windows of limited extension with respect to the filter support. The deviation measure we use in the examples is the sum of the squared differences between the actual and the ideal output images in the training sequence. Therefore, the expression of the objective function becomes

$$f(\mathbf{h}) = \sum_{n_1}\sum_{n_2}\{\bar{h}_1[x'(n_1, n_2)] + \bar{h}_2[x'(n_1, n_2)] - y_d(n_1, n_2)\}^2, \qquad (4.62)$$

where the summation is over all the samples in the reference window. We obtain the final version of the objective function after substituting (4.59), (4.60), and (4.62) in (4.61).

In addition to the ability of the filter to enhance the visual quality of distorted images, the robustness of the design procedure to different types of noise present in the input images is a major factor that determines its usefulness. Ramponi et al. [241] studied this issue and found that the objective function is often relatively flat in the space surrounding the minimum of the global objective function. Consequently, a design procedure that employs a specific type of noise often works well for a large class of noise characteristics.

Example 4.6 In this example, we consider the design of a quadratic filter that can enhance the visual quality of a photograph shot in poor lighting conditions. Photographs obtained from such situations typically exhibit compressed gray-level dynamics and also are more noisy than photographs shot in more optimal lighting conditions. Scaling the pixel values so that they span the available output range of the display system yields results in which the noise is amplified. A linear lowpass filter combined with a contrast enhancement operation can reduce the noise, but blurs small details. More satisfactory results can be obtained by using nonlinear filters. In this example, we design a quadratic filter that consists of a linear and a second-order nonlinear component defined on a 3×3-pixel plane of support. By imposing the isotropy constraints of (4.29)–(4.32), we reduce the number of independent coefficients of the linear part of the filter from nine to three, while reducing that of the quadratic part to 11 from a maximum of 81. We formulate the optimization problem in the form of the unconstrained minimization of the function (4.61). The isotropy conditions imposed on the filter coefficients allow us to define the objective function in (4.62) using a training set that consists of two desired output images in the form of two synthetic vertical edges defined by

$$y_{d,1}(n_1, n_2) = \begin{bmatrix} 1 & 2 & 2 \\ 1 & 2 & 2 \\ \cdot & \cdot & \cdot \\ 1 & 2 & 2 \end{bmatrix}$$

and

$$y_{d,2}(n_1, n_2) = \begin{bmatrix} 1 & 1 & 2 \\ 1 & 1 & 2 \\ \cdot & \cdot & \cdot \\ 1 & 1 & 2 \end{bmatrix}.$$

The corresponding input images to the filter, denoted by $x'_1(n_1, n_2)$ and $x'_2(n_1, n_2)$ are the same synthetic edges corrupted by an additive Gaussian noise sequence with zero mean value and variance as specified later. Training images with 100 rows turned out to be sufficient to design a filter that exhibits satisfactory robustness to noise present in the input images.

The conjugate gradient algorithm was employed to design different quadratic filters for different values of the variance of superimposed input noise. As the magnitude of the noise used in the training images increases, operators offering more pronounced smoothing effects, and thus reduced mean-square error, are obtained. In addition, different solutions were also obtained using different values of the weighting factors in (4.61), and thus relaxing the condition of preservation of a constant input value. Table 4.3 lists the 14 coefficients obtained by the optimization process in one of these cases. The 45 coefficients of the triangular quadratic operator can be obtained from this set using the relations in Example 4.4. The set of coefficients was obtained using a training sequence in which the input image contained additive Gaussian noise with zero mean value and variance equal to 0.05.

We now present the results of applying the quadratic filter as well as some other competing systems on a test image different from the training image. Figure 4.5 shows the original reference image. This image contains 512×512 pixels with eight bits per pixel gray-scale resolution. It was then corrupted with additive Gaussian noise with zero mean value and variance equal to 1300. Figure 4.6 is the input image to the filters and was obtained by applying a linear mapping of the dynamic range of the input signal to the range $[110, 125]$, and then quantizing each pixel of the result to its nearest integer value to simulate the effects of reduction in the dynamic range that occurs during photography in low illumination levels.

We can evaluate the effects of different operators on this image using both quantitative and qualitative criteria. The quantitative evaluations are done by computation of the mean-square error (MSE) with respect to the original image of Figure 4.5. The qualitative evaluations are performed by visual inspection of the

TABLE 4.3   Coefficients of the Quadratic Filters Designed in Example 4.6

m_1, m_2	$h_1(m_1, m_2)$	i, j	$h_{2L}(i, j)$
1, 1	−0.051	1, 1	0.0014
1, 2	0.025	1, 2	0.0001
2, 2	0.1	1, 3	−0.0001
		1, 5	0.0002
		1, 6	−0.0001
		1, 9	−0.0002
		2, 2	0.0005
		2, 4	0.0000
		2, 5	0.0001
		2, 8	−0.0002
		5, 5	0.0018

Figure 4.5   Original image of Example 4.6.

different results. Figure 4.7 shows the effect of processing the input image with a linear filter that averages the input over a 3×3-pixel plane of support at each location and then scales the averaged image to fit the desired dynamic range of the output image. The MSE between the original and the output image is 246. Even though this is the best MSE value obtained among all our comparisons, it appears to be the worst in terms of edge preservation and overall visual quality.

Figure 4.6   Test image of Example 4.6 obtained from the original image after superposition of Gaussian noise and reduction of gray-scale dynamics.

Figure 4.7 Effect of linear averaging and gray-level expansion on the test image in Example 4.6.

Median filters are employed in many applications in which noise reduction and edge preservation are desired. Figure 4.8 shows the result of a 3×3-pixel cross-shaped median filtering on the input image. The output was scaled to fit the desired range of the values of the output pixels. The MSE between the resulting image and the original is equal to 311. However, the perceptual quality of this image is spoiled by a "patchwork" effect due to the reduced number of gray levels in the compressed image.

Figure 4.8 Effect of median filtering and gray-level expansion on the test image in Example 4.6.

Figure 4.9 Effect of quadratic filter and gray-level expansion on the test image in Example 4.6.

Figure 4.9 shows the result of applying the filter of Table 4.3, designed using a noisy version of the synthetic edge, on the test image. The MSE value is equal to 435, but the detail preservation is better than in the other techniques. The best overall visual quality seems to be obtained in this last image. The MSE measurements were made on the test image, and therefore it can be higher for the quadratic filter than for the linear filter, as is the case in this example. However, since the linear filter solution is a possible solution to the quadratic filter design problem also, the MSE measured on the training image will be no greater for the quadratic filter than for the linear filter.

4.4.2 Design Using the Bi-Impulse Response Signals

Two-dimensional quadratic filters can be characterized using their bi-impulse response functions similar to the characterization of one-dimensional quadratic operators using bi-impulse response functions [245]. The bi-impulse response functions can be employed to develop a deterministic method for identifying the coefficients of a quadratic filter. It is also possible to design quadratic filters by specifying its bi-impulse response function.

4.4.2.1 Bi-Impulse Response Signals Let the output of a homogeneous and symmetric two-dimensional quadratic filter be given by

$$y(n_1, n_2) = \bar{h}_2[x(n_1, n_2)]. \tag{4.63}$$

Consider an input signal to this system that is decomposed as a sum of two components as

$$x(n_1, n_2) = x_a(n_1, n_2) + x_b(n_1, n_2). \tag{4.64}$$

The filter output for this input is given by

$$y(n_1, n_2) = y_a(n_1, n_2) + y_b(n_1, n_2) + 2\overline{h}_{ab}[x_a(n_1, n_2), x_b(n_1, n_2)], \tag{4.65}$$

where $y_a(n_1, n_2)$ and $y_b(n_1, n_2)$ are the outputs of the quadratic filter when its input is equal to $x_a(n_1, n_2)$ and $x_b(n_1, n_2)$, respectively. The term that depends on both $x_a(n_1, n_2)$ and $x_b(n_1, n_2)$, given by

$$\overline{h}_{ab}[x_a(n_1, n_2), x_b(n_1, n_2)]$$

$$= \sum_{m_{11}=0}^{N-1} \sum_{m_{12}=0}^{N-1} \sum_{m_{21}=0}^{N-1} \sum_{m_{22}=0}^{N-1} h_2(m_{11}, m_{12}, m_{21}, m_{22}) x_a(n_1 - m_{11}, n_2 - m_{12})$$
$$\times x_b(n_1 - m_{21}, n_2 - m_{22}) \tag{4.66}$$

can be used to determine the two-dimensional *bi-impulse response* of the quadratic filter. In order to develop a deterministic approach for identifying the coefficients of the quadratic filter, we consider an input signal that consists of two distinct impulses so that it can be decomposed as in (4.64) with

$$x_a(n_1, n_2) = u_0(n_1 - k_{11}, n_2 - k_{12}) \tag{4.67}$$

and

$$x_b(n_1, n_2) = u_0(n_1 - k_{21}, n_2 - k_{22}) \tag{4.68}$$

as two unit impulse signals located in different positions within the plane of support of the filter. By evaluating the bi-impulse response function using (4.66), we find that

$$\overline{h}_{ab}[u_0(n_1 - k_{11}, n_2 - k_{12}), u_0(n_1 - k_{21}, n_2 - k_{22})]$$
$$= h_2(n_1 - k_{11}, n_2 - k_{12}, n_1 - k_{21}, n_2 - k_{22}) \tag{4.69}$$

for this input signal. Consequently, the operator $\overline{h}_{ab}$ allows the estimation of the coefficients of the quadratic filter by means of appropriately located pairs of impulse functions in the input signal. The complete set of responses corresponding to the

$N^2(N^2 + 1)/2$ independent coefficients can be estimated from the following expressions:

$$\bar{h}_{ab}[u_0(n_1, n_2), u_0(n_1 - k_1, n_2 - k_2)]$$
$$= h_2(n_1, n_2, n_1 - k_1, n_2 - k_2); \quad k_1, k_2 = 0, 1, \ldots, N - 1 \quad (4.70)$$
$$\bar{h}_{ab}[u_0(n_1, n_2 - k_1), u_0(n_1 - k_2, n_2)]$$
$$= h_2(n_1, n_2 - k_1, n_1 - k_2, n_2); k_1, k_2 = 1, \ldots, N - 1. \quad (4.71)$$

These equations show that we can estimate the coefficients of the quadratic kernel by finding the responses of the filter to sets of two appropriately placed impulse functions in the input signal. We demonstrate the use of bi-impulse response functions for characterizing an isotropic quadratic filter with a 3×3-element plane of support in the next example.

Example 4.7   Very often, nonlinear filters defined on a small plane of support have proved to be effective in image processing applications. In this example, we analyze a homogeneous quadratic filter with a 3×3-sample plane of support. The purpose of this example is to illustrate the ideas discussed above using a relatively simple situation.

The quadratic filter contains 81 coefficients, of which only 45 are independent because of the symmetry of the kernel. By imposing the isotropy constraints (4.29)–(4.32), we can reduce the number of independent coefficients to $C_q = 11$, as can be seen from Table 4.2. We now explicity consider the structure of the bi-impulse response of the filter. For $N = 3$, (4.70) and (4.71) become a set of 13 independent equations corresponding to pairs of impulses located on the filter support as shown in Figure 4.10. Each of the 45 coefficients is involved in one of these 13 responses. The indices of these coefficients are listed in Table 4.4 using their lexicographic representation.

As was demonstrated in Example 4.4, the isotropy constraints in (4.46)–(4.49) force the equality between pairs of coefficients resulting in only 11 independent coefficients for the filter. These 11 coefficients can be identified using the six independent bi-impulse responses tabulated in Table 4.5. Comparison of the entries in Table 4.5 with the results in Example 4.4 permits the computation of all the 45 independent coefficients. The six bi-impulse responses are denoted α, β, δ, ϵ, θ, μ in Table 4.5. The locations of the six pairs of impulses that define the independent bi-impulse responses of this system when the isotropy constraints are imposed are shown in Figure 4.11.

It is typical to group the bi-impulse responses according to the distance between the two impulses at the input that drive the system to produce the bi-impulse response. The response denoted by α is called a *type 0* response since the impulses are coincident in this case, and therefore, the distance between the two impulses is zero. The responses denoted by β and ϵ are called *type 1* responses since the impulses that drive the system are a single sample apart from each other. Finally, the

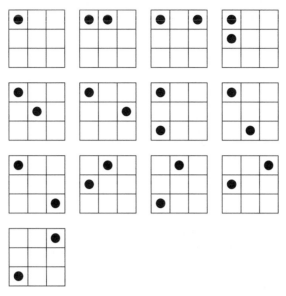

Figure 4.10 Location of the input impulses defining the response of the homogeneous quadratic operator of Example 4.7.

responses δ, θ, and μ are called *type 2* bi-impulse responses because the input impulses are two samples apart in this case. Bi-impulse responses of larger quadratic filters with an $N \times N$-sample plane of support are classified in a similar manner up to *type* $(N - 1)$.

4.4.2.2 Filter Design

The design method based on the bi-impulse response of a two-dimensional quadratic filter exploits the fact that its linear and homogeneous

TABLE 4.4 Indices of the Coefficients Determined by the 13 Independent Bi-Impulse Responses in Example 4.7

	$x_a(n_1, n_2)$	$x_b(n_1, n_2)$	Indices of Coefficients Identified								
1	$u_0(n_1, n_2)$	$u_0(n_1, n_2)$	11	22	33	44	55	66	77	88	99
2	$u_0(n_1, n_2)$	$u_0(n_1, n_2 - 1)$	12	23	45	56	78	89			
3	$u_0(n_1, n_2)$	$u_0(n_1, n_2 - 2)$	13	46	79						
4	$u_0(n_1, n_2)$	$u_0(n_1 - 1, n_2)$	14	25	36	47	58	69			
5	$u_0(n_1, n_2)$	$u_0(n_1 - 1, n_2 - 1)$	15	26	48	59					
6	$u_0(n_1, n_2)$	$u_0(n_1 - 1, n_2 - 2)$	16	49							
7	$u_0(n_1, n_2)$	$u_0(n_1 - 2, n_2)$	17	28	39						
8	$u_0(n_1, n_2)$	$u_0(n_1 - 2, n_2 - 1)$	18	29							
9	$u_0(n_1, n_2)$	$u_0(n_1 - 2, n_2 - 2)$	19								
10	$u_0(n_1, n_2 - 1)$	$u_0(n_1 - 1, n_2)$	24	35	57	68					
11	$u_0(n_1, n_2 - 1)$	$u_0(n_1 - 2, n_2)$	27	38							
12	$u_0(n_1, n_2 - 2)$	$u_0(n_1 - 1, n_2)$	34	67							
13	$u_0(n_1, n_2 - 2)$	$u_0(n_1 - 2, n_2)$	37								

TABLE 4.5   Indices of the Coefficients Determined
by the Six Independent Bi-Impulse Responses
When Isotropy Constraints Are Applied to the Filter
in Example 4.7

	Indices from Table 4.4	Indices of Coefficients Identified		
α	1	11	22	55
β	2 or 4	12	25	
δ	3 or 7	13	28	
ϵ	5 or 10	15	24	
θ	6, 8, 11 or 12	16		
μ	9 or 13	19		

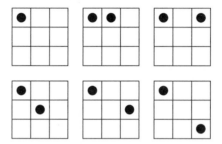

Figure 4.11   Location of the input impulses defining the six independent bi-impulse responses of the isotropic quadratic operator in Example 4.7.

quadratic components are uniquely defined by the impulse response of the linear component and the set of bi-impulse responses of the homogeneous quadratic component, respectively. The characterization of a quadratic filter by means of the components of its bi-impulse response, classified as type $0, 1, \ldots N - 1$, allows the introduction of a simple but effective design procedure. In what follows, we assume that the quadratic filter satisfies the isotropy constraints as well as the conditions for preservation of a constant input signal at its output. As we have already seen, these conditions simplify the design problem considerably. For example, in the case of a 3×3 filter, (4.53) and (4.54) become, respectively

$$4h_{1L}(1) + 4h_{1L}(2) + h_{1L}(5) = 1 \tag{4.72}$$

$$\begin{aligned}
4h_{2L}(1, 1) + 16h_{2L}(1, 2) &+ 8h_{2L}(1, 3) + 8h_{2L}(1, 5) + 16h_{2L}(1, 6) \\
&+ 4h_{2L}(1, 9) + 4h_{2L}(2, 2) + 8h_{2L}(2, 4) + 8h_{2L}(2, 5) \\
&+ 4h_{2L}(2, 8) + h_{2L}(5, 5) = 0.
\end{aligned} \tag{4.73}$$

A design strategy that employs the bi-impulse response functions can be the following:

1. Define the linear filter according to the given specifications.
2. Select the coefficients of the homogeneous quadratic components so that they
   a. Compensate for any undesirable effects introduced by the linear component.
   b. Exploit the ability of the different bi-impulse response functions to process pairs of impulses at the input in different ways based on their distance from each other.

This methodology can then be exploited to design two-dimensional quadratic filters to perform classic image processing tasks. The next example present the design of such a filter for an application in texture discrimination.

Example 4.8: Design of a Quadratic Filter for Texture Discrimination   It is well-known that it is often difficult for human observers to discriminate textures with very similar first- and second-order statistics [106]. Furthermore, linear filters are able to discriminate their input signal characteristics only on the basis of their power spectrum, and therefore, are not able to discriminate between textures with identical power spectra. A quadratic filter can be used to identify and discriminate between two such textures because of its intrinsic ability to exploit high-order moments.

Figure 4.12   The noisy texture composition of Example 4.8. (Courtesy of G. Ramponi, *Proc. IEEE,* Copyright © 1990 IEEE.)

Figure 4.12 shows a noisy version of a synthetic texture containing two different components. In a rectangular portion in the middle of the image, the texture is formed by a repeating pattern of a 4 × 4-pixel block with impulses two samples apart from each other. The outer portion of the image is formed by another repeating pattern where the impulses are adjacent. The two parts of the image have similar mean values and variances. The texture described above has been corrupted to obtain Figure 4.12 using a multiplicative noise sequence having uniform distribution with a mean value of 1 and variance equal to 0.2. We wish to design an isotropic quadratic filter, with a 3 × 3-sample plane of support, that is able to process without attenuation the texture in the inner portion of the image and eliminate the texture in the outer portion of the image at the output of the filter. We further specify that the filter contains a linear component and a quadratic component, and that it preserves constant inputs without change at the output.

The general requirements on the quadratic filter may be stated in the following manner:

1. The filter should be able to eliminate isolated or adjacent impulses in the input image at the output.
2. It should be able to process pairs of impulses that are two samples apart in the input image without attenuation.

It may be possible to meet these requirements using several different designs. One possible solution to the problem may be obtained as described below. The indices correspond to a lexicographic ordering of the input data matrix. We also assume that the input image is normalized to have a maximum value of one before being processed by the filter. The output of the filter can then be scaled appropriately to match the dynamic range of the display system.

We choose the linear component to be

$$h_{1L}(1) = h_{1L}(2) = 0 \quad h_{1L}(5) = 1$$

so that the constraint in (4.72) is satisfied, and the linear filter is able to preserve constant inputs at its output. Recall from the results of Example 4.7 that there are six independent bi-impulse responses for the quadratic filter. The response denoted by α is a type 0 filter; those denoted by β and ϵ are type 1 responses; and δ, θ, and μ are type 2 responses. Since the requirements on the filter are based on the response of the filter to impulse inputs, we can specify the three types of response of the quadratic component of the filter to meet our objectives:

1. The filter should attenuate isolated impulses. To achieve this objective, we choose all except one of the coefficients defined by the type 0 response α to be zero. Since we set $h_{1L}(5)$ to be one, we set $h_{2L}(5, 5) = -1$, so that this coefficient cancels out the component at the output due to the linear coefficient $h_{1L}(5)$. Recall that the input is normalized to have a maximum value of one. This normalization ensures that the effects of isolated impulses at the output due to $h_{1L}(5)$ and $h_{2L}(5, 5)$ approximately cancel each other. As a result of our

choice of the type 1 response of the filter, we find that

$$h_{2L}(1, 1) = h_{2L}(2, 2) = 0, h_{2L}(5, 5) = -1.$$

2. Since we want the filter to eliminate the pairs of input impulses that are adjacent to each other, we set the type 1 response β and ϵ to zero. Consequently, we have

$$h_{2L}(1, 2) = h_{2L}(2, 5) = h_{2L}(1, 5) = h_{2L}(2, 4) = 0.$$

3. Finally, the responses denoted by δ, θ, and μ control the output of the system when the input at any time contains two impulses separated by two samples. A possible design choice for the coefficients specified by these responses is the following:

$$h_{2L}(1, 3) = 0.03125 \quad h_{2L}(2, 8) = 0.0625,$$

$$h_{2L}(1, 6) = 0.015625,$$

and

$$h_{2L}(1, 9) = 0.0625.$$

The values were obtained by imposing that the coefficients related to the responses δ, θ, and μ satisfy the constraint in (4.73). Obviously, these coefficients may be selected in several other ways also to satisfy this constraint.

Figure 4.13 Output of the quadratic filter in Example 4.8. (Courtesy of G. Ramponi, *Proc. IEEE,* Copyright © 1990 IEEE.)

The result of processing the image of Figure 4.12 is shown in Figure 4.13. The locations of the two types of texture are clearly identified at the output, indicating a successful design strategy.

4.5 EXERCISES

4.1 Consider a homogeneous, two-dimensional, symmetric quadratic filter with arbitrary, but finite support. Compute the constant values $|\pi(\mathbf{m}_1, \mathbf{m}_2, \ldots, \mathbf{m}_p)|$ that scale the symmetric coefficients in its triangular kernel.

4.2 The two-dimensional quadratic filter described by the matrix

$$
\mathbf{H}(i,j) =
\begin{bmatrix}
-1 & 2 & -\frac{1}{2} & 2 & 0 & 0 & -\frac{1}{2} & 0 & -2 \\
2 & -1 & 2 & -\frac{1}{2} & 0 & -\frac{1}{2} & 0 & -2 & 0 \\
-\frac{1}{2} & 2 & -1 & 0 & 0 & 2 & -2 & 0 & -\frac{1}{2} \\
2 & -\frac{1}{2} & 0 & -1 & 0 & -2 & 2 & -\frac{1}{2} & 0 \\
0 & 0 & 0 & 0 & 0 & 0 & 0 & 0 & 0 \\
0 & -\frac{1}{2} & 2 & -2 & 0 & -1 & 0 & -\frac{1}{2} & 2 \\
-\frac{1}{2} & 0 & -2 & 2 & 0 & 0 & -1 & 2 & -\frac{1}{2} \\
0 & -2 & 0 & -\frac{1}{2} & 0 & -\frac{1}{2} & 2 & -1 & 2 \\
-2 & 0 & -\frac{1}{2} & 0 & 0 & 2 & -\frac{1}{2} & 2 & -1
\end{bmatrix}
$$

is capable of performing edge extraction in the presence of impulsive noise in the input image.

Derive the parallel-cascade realizations on the basis of

(a) The LU matrix decomposition.

(b) The singular value decomposition.

(c) The singular value decomposition truncated to the largest four singular values of $\mathbf{H}_2$ implemented via the LU decomposition of the approximate matrix.

Characterize the quality of the approximation of $\mathbf{H}_2$ when only four singular values are used in the matrix approximation by computing the Frobenius norm of the coefficient error matrix.

4.3 Derive the isotropy constraints imposed on the coefficients of a two-dimensional $N \times N$-coefficient linear filter for rotations of the input data of $180°$ and $270°$, and for the reflection about the central horizontal axis of the input matrix. Show that the reflection about the central vertical axis of the input matrix does not impose additional constraints on the filter coefficients.

4.4 Given a two-dimensional $N \times N$-coefficient quadratic filter, derive the isotropy constraint (4.47) imposed by rotation of the input matrix by $180°$, the constraint (4.48) imposed by a $270°$ rotation of the input matrix, and the constraint (4.49) imposed by reflection about the central horizontal axis utilizing the lexicographic ordering of the samples of the two-dimensional input data.

4.5 Show that the constraints in (4.29)–(4.32) correspond to the octagonal symmetry of the linear filter in both the data and the frequency domains.

4.6 Consider the input image in Example 4.5. Derive expressions for the constants C_1, C_2, and C_3 and of the step Δ in the output image as functions of the coefficients of a homogeneous, two-dimensional 3×3-coefficient, symmetric, and isotropic quadratic filter.

4.7 *Computing Assignment:* Design a two-dimensional, symmetric, and isotropic quadratic filter containing a linear and a homogeneous quadratic component for the enhancement of low-contrast images using the optimization approach described in Section 4.4.1 and the following three steps:

(a) Use the mean-square error between the output of the filter and the desired output image as the measure of deviation included in the objective function. Include the conditions that guarantee the preservation of a constant input level at the output of the filter as part of the objective function.

(b) Derive the objective function using a training set that consists of a desired output image in the form of a vertical edge, and a corresponding input signal obtained by adding a randomly generated input noise to the desired output image. Assume a dummy value equal to one for the input image dynamics.

(c) Apply the conjugate gradient descent algorithm to design the filter.
To illustrate the performance of the filter designed, use a test image artificially created by compressing to 16 levels an image with gray-scale dynamics in the range [0,255], corrupted by a Gaussian-distributed random noise sequence having zero mean value and variance equal to 700.

(a) Compare the results obtained by applying the designed filter with those obtained using a linear, 3×3-sample averaging filter and a 3×3 cross-shaped median filter. In all these cases, expand the outputs of the filters to have full dynamics, that is, the resulting output image must take gray-level values in the range [0,255].

(b) Study the effects of employing different noise sequences in the design of the reference filter.

4.8 Derive the independent bi-impulse responses of the homogeneous two-dimensional quadratic filter described by the matrix

$$
\mathbf{H_2} = \begin{bmatrix}
1 & \frac{1}{2} & 0 & \frac{1}{2} & 0 & -\frac{1}{2} & 0 & -\frac{1}{2} & -1 \\
\frac{1}{2} & 0 & \frac{1}{2} & 0 & 0 & 0 & -\frac{1}{2} & 0 & -\frac{1}{2} \\
0 & \frac{1}{2} & 1 & -\frac{1}{2} & 0 & \frac{1}{2} & -1 & -\frac{1}{2} & 0 \\
\frac{1}{2} & 0 & -\frac{1}{2} & 0 & 0 & 0 & \frac{1}{2} & 0 & -\frac{1}{2} \\
0 & 0 & 0 & 0 & 0 & 0 & 0 & 0 & 0 \\
-\frac{1}{2} & 0 & \frac{1}{2} & 0 & 0 & 0 & -\frac{1}{2} & 0 & \frac{1}{2} \\
0 & -\frac{1}{2} & -1 & \frac{1}{2} & 0 & -\frac{1}{2} & 1 & \frac{1}{2} & 0 \\
-\frac{1}{2} & 0 & -\frac{1}{2} & 0 & 0 & 0 & \frac{1}{2} & 0 & \frac{1}{2} \\
-1 & -\frac{1}{2} & 0 & -\frac{1}{2} & 0 & \frac{1}{2} & 0 & \frac{1}{2} & 1
\end{bmatrix}
$$

and the corresponding input impulse configurations.

4.9 Suppose that we wish to design a two-dimensional quadratic operator having a 3×3-element plane of support. The filter should be such that it is capable of smoothing low-level noise and attenuating pairs of adjacent input impulses above a preselected amplitude. Such a result can be obtained by designing a quadratic filter having a lowpass linear component and a homogeneous quadratic component defined by the following three components of its bi-impulse response:

(a) A type 0 (α) response that cancels the linear effect for a certain input amplitude.

(b) Type 1 (β and ϵ) responses that attenuates pairs of adjacent impulses.

(c) Zero type 2 (δ, θ and μ) responses.

Assume as coefficients of the linear filter the following values:

$$h_{1L}(1) = h_{1L}(2) = 0.1 \quad h_{1L}(5) = 0.2.$$

Derive a set of possible values for the quadratic part on the basis of the constraints on the bi-impulse responses described above and the preservation of constant input levels on the filter support.

5

PARAMETER ESTIMATION

The main concerns in the previous chapters were the analysis of the properties of truncated Volterra filters and various methods for realizing such filters. In this chapter, we consider the important issue of estimating signals using truncated Volterra system models. For the sake of simplicity, we do not consider the problems associated with model selection and model order determination. In other words, we assume that we know the order of nonlinearity as well as the memory span of the system model prior to the estimation of the parameters. The reader may refer to Marple [162] and references cited therein for an overview of basic strategies employed for model order selection in estimation problems.

5.1 THE TRUNCATED VOLTERRA SERIES ESTIMATION PROBLEM

The problem of interest in this chapter is that of approximating a *desired response signal* $d(n)$ using a truncated Volterra series expansion of the *input signal* $x(n)$ as

$$\hat{d}(n) = h_0 + \sum_{p=1}^{P} \bar{h}_p[x(n)], \tag{5.1}$$

where

$$\bar{h}_p[x(n)] = \sum_{m_1=0}^{N-1} \sum_{m_2=m_1}^{N-1} \cdots \sum_{m_p=m_{p-1}}^{N-1} h_p(m_1, m_2, \ldots, m_p)x(n - m_1)$$
$$\times x(n - m_{2)} \cdots x(n - m_p). \tag{5.2}$$

This model incorporates the kernel symmetry, and consequently, the representation in (5.2) corresponds to the triangular form given in (2.38). Our objective is to identify the coefficients $h_p(m_1, m_2, \ldots, m_p)$ from either knowledge of the relevant statistics of the input and output signals or measurement of the input and output signals. The coefficients of the expression in (5.1) can be uniquely estimated in most cases.

5.1.1 Optimality of the Volterra Series Models

It is well-known [176] that the optimal minimum mean-square error (MMSE) estimator of a random variable x using N other random variables $y_1, y_2, \ldots, y_N$ is given by

$$\hat{x} = E\{x \,|\, y_1, y_2, \ldots, y_N\}. \tag{5.3}$$

Furthermore, when $x, y_1, y_2, \ldots, y_N$ are jointly Gaussian-distributed with zero mean values, this estimator takes the form of a linear system. In view of the preceding results, it is reasonable to ask the following question: *Is there a class of probability distributions of* $x, y_1, y_2, \ldots, y_N$ *for which the optimal MMSE estimator of* x *using* $y_1, y_2, \ldots, y_N$ *takes the form of a truncated Volterra filter?* Balakrishnan [11] has proved the following result that provides the answer to this question.

Theorem 5.1   Let $\phi_y(\omega_1, \omega_2, \ldots, \omega_N)$ denote the joint characteristic function of the random variables $y_1, y_2, \ldots, y_N$, i.e.,

$$\phi_y(\omega_1, \omega_2, \ldots, \omega_N) = E\left\{ e^{j \sum_{i=1}^{N} \omega_i y_i} \right\}. \tag{5.4}$$

Also, let $\phi_{x,y}(\omega_0, \omega_1, \ldots, \omega_N)$ denote the joint characteristic function of $x, y_1, y_2, \ldots, y_N$. We assume that the random variables have finite moments of all orders. Let D_k denote a differential operator such that

$$D_k \phi_y(\omega_1, \omega_2, \ldots, \omega_N) = \frac{\partial}{\partial(j\omega_k)} \phi_y(\omega_1, \omega_2, \ldots, \omega_N) \tag{5.5}$$

and let $g(\cdots)$ denote a polynomial in N variables. Then, a necessary and sufficient condition for

$$E\{x \,|\, y_1, y_2, \ldots, y_N\} = g(y_1, y_2, \ldots, y_N) \tag{5.6}$$

is

$$\frac{\partial}{\partial(j\omega_0)} \phi_{x,y}(\omega_0, \omega_1, \ldots, \omega_N)\bigg|_{\omega_0=0} = g(D_1, D_2, \ldots, D_N) \phi_y(\omega_1, \omega_2, \ldots, \omega_N). \tag{5.7}$$

This is a rather technical result and is included here for the sake of completeness. The proof may be found in Balakrishnan [11] and is beyond the scope of our discussion.

5.1.2 Sampling Requirements for Volterra System Identification

Suppose that the higher-order frequency response of a continuous-time, homogeneous pth-order Volterra system, $H_p(\omega_1, \omega_2, \ldots, \omega_p)$ is a band-limited function such that $H_p(\omega_1, \omega_2, \ldots, \omega_p) = 0$ outside the interval $0 < \omega_i \leq \omega_c$ for all i in the range $[1, p]$. Let the input signal be also band-limited to ω_s radians/s, where $\omega_s \geq \omega_c$. Even though the output signal for this system may have a bandwidth that is higher than ω_s, it is not necessary that the input and output signals be sampled at a rate larger than twice the bandwidth of the output signal to identify the system. In fact, we only need to sample the signals at a rate larger than twice the bandwidth of the input signal. This fact can be deduced from the multidimensional representation of a Volterra system. Recall that a linear, p-dimensional system can be identified by sampling the input and output signals at a rate larger than twice the bandwidth of the input signal in each dimension. Since in the multidimensional representation of Volterra systems, the input signal of the multidimensional linear system is created by evaluating the products of the one-dimensional input signal samples, it is clear that the bandwidth of the multidimensional input signal in each dimension is identical to the bandwidth of the actual input signal. The result follows immediately. Examples illustrating this idea may be found in the literature [71,316,341].

Remark 5.1 The preceding statements apply specifically to the identification of higher-order Volterra systems. If the output signal of a continuous-time Volterra system is sampled at a rate smaller than twice its bandwidth, aliasing will occur. The sampling requirements stated above implies that the system can be identified by digital techniques with aliased output signals if the sampling process does not introduce aliasing into the input signal.

5.2 DIRECT ESTIMATION OF THE PARAMETERS

It is convenient to represent the system of (5.1) using vector notation for our derivation. Let

$$\overline{h_p[x(n)]} = \mathbf{X}_{pr}^T(n)\vec{\mathbf{H}}_{pr} \tag{5.8}$$

as defined in Chapter 3. Let us also define the augmented input vector and coefficient vector as

$$\mathbf{X}(n) = [1, \mathbf{X}_1^T(n), \mathbf{X}_{2r}^T(n), \ldots, \mathbf{X}_{pr}^T(n)]^T \tag{5.9}$$

and

$$\vec{\mathbf{H}} = [h_0, \vec{\mathbf{H}}_1^T, \vec{\mathbf{H}}_{2r}^T, \ldots, \vec{\mathbf{H}}_{pr}^T]^T, \tag{5.10}$$

respectively. Now, we can rewrite (5.1) in a compact form as

$$\hat{d}(n) = \mathbf{X}^T(n)\vec{\mathbf{H}}. \tag{5.11}$$

We discuss the MMSE estimation of the coefficient vector $\vec{\mathbf{H}}$ first.

5.2.1   The MMSE Formulation

We consider the problem of estimating $\vec{\mathbf{H}}$ so that the mean-square estimation error is minimized, i.e., we select the coefficients $\vec{\mathbf{H}}_{\text{opt}}$ that minimizes

$$E\{e^2(n)\} = E\left\{(d(n) - \mathbf{X}^T(n)\vec{\mathbf{H}})^2\right\} \tag{5.12}$$

among all possible choices of $\vec{\mathbf{H}}$. In (5.12), $E\{(\cdot)\}$ denotes the statistical expectation of $(\cdot)$. The solution to this problem can be obtained from the following result.

Theorem 5.2: The Orthogonality Principle   The optimal estimation error $e_{\text{opt}}(n)$, defined as

$$e_{\text{opt}}(n) = d(n) - \mathbf{X}^T(n)\vec{\mathbf{H}}_{\text{opt}}, \tag{5.13}$$

is orthogonal to the input vector $\mathbf{X}(n)$, i.e.,

$$E\{\mathbf{X}(n)(d(n) - \mathbf{X}^T(n)\vec{\mathbf{H}}_{\text{opt}})\} = \mathbf{0}, \tag{5.14}$$

where $\mathbf{0}$ denotes a vector of zero elements and appropriate dimension.

PROOF   The objective function defined in (5.12) is a quadratic function in the elements of the coefficient vector. Consequently, it has a unique minimum in general, and the minimum point can be found by differentiating (5.12) with respect to $\vec{\mathbf{H}}$ and setting the result to zero for $\vec{\mathbf{H}} = \vec{\mathbf{H}}_{\text{opt}}$. This operation results in

$$\left.\frac{\partial E\{e^2(n)\}}{\partial \vec{\mathbf{H}}}\right|_{\vec{\mathbf{H}}=\vec{\mathbf{H}}_{\text{opt}}} = E\left\{2\left(d(n) - \vec{\mathbf{H}}_{\text{opt}}^T\mathbf{X}(n)\right)\mathbf{X}^T(n)\right\} = 0, \tag{5.15}$$

which proves the theorem.

5.2.1.1 Extension to Non-MMSE Estimation   The orthogonality principle stated above is valid for estimation problems other than those formulated in the MMSE sense. It is a straightforward task to show that the orthogonality principle holds for any properly formulated minimum-squared-norm estimation problem in well-defined inner-product spaces [93]. In particular, the orthogonality principle is valid for least-squares estimation problems in which the objective is to minimize a

weighted sum of squared estimation errors over a finite period of time.

5.2.1.2 The Optimal MMSE Coefficient Vector
The optimal coefficient vector can be evaluated from (5.15) as

$$\vec{\mathbf{H}}_{opt} = \mathbf{R}_{xx}^{-1}\mathbf{P}_{dx}, \tag{5.16}$$

where

$$\mathbf{R}_{xx} = E\{\mathbf{X}(n)\mathbf{X}^T(n)\} \tag{5.17}$$

is the statistical autocorrelation matrix of the input vector $\mathbf{X}(n)$ and

$$\mathbf{P}_{dx} = E\{d(n)\mathbf{X}(n)\} \tag{5.18}$$

is the statistical cross-correlation vector of the desired response signal $d(n)$ and the input vector $\mathbf{X}(n)$. This result assumes that the autocorrelation matrix is invertible. In using the time-independent notation of (5.16), we have also assumed that the signals $x(n)$ and $y(n)$ are jointly stationary at least up to order $2P$.

Remark 5.2 Equation (5.16) demonstrates some advantages and disadvantages of system modeling using truncated Volterra series expansions. Perhaps the single biggest advantage is that the underlying theory of parameter estimation as well as realization of such models is very similar to the corresponding theory for the linear case. However, Volterra series models require a large number of parameters to adequately model many nonlinear systems. Consequently, the matrix inversion necessary to compute the optimal coefficients can be a computationally complex and often numerically difficult task.

5.2.1.3 Minimum Mean-Square Error
We substitute (5.16) in (5.12) to obtain an expression for the minimum value of the mean-squared estimation error. This operation results in

$$\xi_{min} = E\left\{\left(d(n) - \vec{\mathbf{H}}_{opt}^T\mathbf{X}(n)\right)\right\}^2$$

$$= E\{d^2(n)\} - 2\vec{\mathbf{H}}_{opt}^T\mathbf{P}_{dx} + \vec{\mathbf{H}}_{opt}^T\mathbf{R}_{xx}\vec{\mathbf{H}}_{opt}$$

$$= E\{d^2(n)\} - \vec{\mathbf{H}}_{opt}^T\mathbf{P}_{dx}, \tag{5.19}$$

where we used the fact that $\mathbf{R}_{xx}\vec{\mathbf{H}}_{opt} = \mathbf{P}_{dx}$.

5.2.1.4 Estimation of Signal Statistics
Parameter estimation for a Pth-order Volterra system model requires knowledge of the input signal statistics of orders up to $2P$. It is only rarely that the statistics are known a priori in practice, and

one often has to estimate them from measurements of the input and output signals. The most common way of estimating the statistics required in estimation problems is to assume ergodicity of the input signals and estimate the statistics using time averages. For example, $E\{f(x(n))\}$ where $f(\cdot)$ is an arbitrary nonlinear function of $(\cdot)$, can be estimated as[1]

$$E\{\widehat{f(x(n))}\} = \frac{1}{M}\sum_{k=1}^{M} f(x(k)), \qquad (5.20)$$

where M denotes the number of samples employed in the estimate. One difficulty of estimating the parameters of the Volterra series representation of a nonlinear system is that the process requires estimation of the higher-order statistics of the input signals. Reliable estimation of the higher-order statistics requires long data records.

5.2.1.5 MMSE Estimation for Gaussian Input Signals

In general, all the required higher-order statistics of the input signal must be estimated before the coefficients of the nonlinear system model can be estimated. In the case of Gaussian signals, however, we can compute the higher-order statistics from knowledge of the second-order statistics. Let $x_1, x_2, \ldots, x_N$ denote N zero-mean, Gaussian random variables. Then, it is well-known that [319]

$$E\{x_1 x_2 \ldots x_N\} = \begin{cases} 0; & N \text{ odd} \\ \Sigma\Pi E\{x_i x_j\}; & N \text{ even,} \end{cases} \qquad (5.21)$$

where the summation and product signs together indicate a sum of all completely distinct ways of partitioning $x_1, x_2, \ldots x_N$ into pairs. There are $N!/((N/2)!2^{(N/2)})$ distinct ways in which N terms can be partitioned into $N/2$ pairs. Thus, the Nth-order expectations of Gaussian processes can be expressed as sums of products of second-order expectations. Consequently, the Volterra system parameters can be estimated using only knowledge of the second-order expectations of Gaussian input signals.

5.2.2 Least-Squares Formulation

As discussed in the previous section, the minimum mean-square estimation of the parameters is feasible only when the relevant statistics of the input signals are known. One approach to parameter estimation when the statistics are unknown is to estimate them from particular realizations of the input signals and substitute the estimated values for the true statistics in (5.16). An alternative is to formulate an estimation problem that can be solved exactly using measured values of input signal realizations. Least-squares algorithms belong to the latter class of methods.

---

[1] This book uses the same notation to denote random processes as well as particular realizations of random processes. The difference will, in general, be obvious from the context.

In the simplest formulation of the least-squares parameter estimation problem, we select the coefficient vector that minimizes.

$$J(M) = \frac{1}{M} \sum_{k=1}^{M} \left(d(k) - \mathbf{X}^T(k)\vec{\mathbf{H}} \right)^2. \tag{5.22}$$

It is straightforward to show that the optimal estimation error sequence satisfies the following orthogonality condition:

$$\frac{1}{M} \sum_{k=1}^{M} \left(d(k) - \vec{\mathbf{H}}_{\text{opt}}^T \mathbf{X}(k) \right) \mathbf{X}^T(k) = 0. \tag{5.23}$$

Consequently, the optimal least-squares solution is given by

$$\vec{\mathbf{H}}_{\text{opt}} = \hat{\mathbf{R}}_{\mathbf{xx}}^{-1} \hat{\mathbf{P}}_{d\mathbf{x}}, \tag{5.24}$$

where

$$\hat{\mathbf{R}}_{\mathbf{xx}} = \frac{1}{M} \sum_{k=1}^{M} \mathbf{X}(k)\mathbf{X}^T(k) \tag{5.25}$$

is the least-squares autocorrelation matrix of the input vector and $\hat{\mathbf{P}}_{d\mathbf{x}}$ is the least-squares cross-correlation vector of $d(n)$ and $\mathbf{X}(n)$ given by

$$\hat{\mathbf{P}}_{d\mathbf{x}} = \frac{1}{M} \sum_{k=1}^{M} d(k)\mathbf{X}(k). \tag{5.26}$$

Remark 5.3   The least-squares solution in (5.24) is identical to the approximate solution to the MMSE estimation problem suggested in Section 5.2.1 and obtained by substituting appropriate time-averaged estimates of the signal statistics in (5.16). The motivation for formulating the least-squares problem is the exact nature of the solution obtained.

5.2.2.1   *Data Windowing*   An implicit assumption employed in obtaining the solution given by (5.24) is that $x(n) = 0$ for $n < 1$. For noncausal system models, the solution also assumes that $x(n) = 0$ for $n > M$. This approach is known as *data windowing*. For large values of M, the bias introduced in the solution by data windowing is negligible. It is possible to avoid the errors caused by data windowing by formulating the least-squares estimation problem as the minimization of

$$J(M) = \frac{1}{M-N+1} \sum_{k=N}^{M} \left(d(k) - \mathbf{X}^T(k)\vec{\mathbf{H}} \right)^2. \tag{5.27}$$

This formulation employs only data samples available to the estimator. The derivation of the optimal estimator using this objective function is left as an exercise for the reader. Such approaches are used primarily in situations where the number of samples available to the estimator is not very large when compared with the memory span of the system model.

5.2.2.2 Variations of the Least-Squares Formulation
Several variations of the least-squares formulation are possible. One method that is particularly useful in adaptive filtering problems is the exponentially weighted least-squares method. In this approach, the optimal solution is found by minimizing the cost function

$$J(\lambda, M) = \sum_{k=1}^{M} \lambda^{M-k} \left(d(k) - \vec{\mathbf{H}}^T \mathbf{X}(k) \right)^2, \tag{5.28}$$

where λ is a positive parameter that is bounded by one. This problem is related to the adaptive estimation of the parameters, and is considered in detail in Chapter 7.

5.2.2.3 Convergence of Least-Squares Estimators
Suppose that the desired response signal and the input vector $\mathbf{X}(n)$ are related as

$$d(n) = \vec{\mathbf{H}}_{\text{true}}^T \mathbf{X}(n) + \eta(n). \tag{5.29}$$

Let the input signal $x(n)$ and the noise signal $\eta(n)$ be jointly ergodic processes so that

$$E\{ f(x(n), \eta(n)) \} = \lim_{M \to \infty} \frac{1}{M} \sum_{k=1}^{M} f(x(k), \eta(k)) \tag{5.30}$$

for an arbitrary function $f(\cdot)$. Then, it can be shown that the optimal least-squares coefficient vector given by (5.24) will converge to the true coefficient vector $\vec{\mathbf{H}}_{\text{true}}$ as the number of samples M approaches infinity if

$$E\{\mathbf{X}(n)\eta(n)\} = \mathbf{0} \tag{5.31}$$

Mendel has given proof of this result [176].

Example 5.1   We consider the identification of a memoryless Volterra system with input–output relationship

$$y(n) = 2 + x(n) - 0.5x^2(n) + x^3(n).$$

Assume that the input signal belongs to an independent, identically distributed (IID) Gaussian process with zero mean value and unit variance. Assume also that the output signal is observed in the presence of an additive, zero-mean IID Gaussian sequence with variance σ_η^2. That is, the desired response signal $d(n)$ is modeled as

$d(n) = y(n) + \eta(n)$, where $\eta(n)$ is the measurement noise that is independent of $x(n)$. The input vector for this problem is

$$\mathbf{X}(n) = [1\ x(n)\ x^2(n)\ x^3(n)]^T.$$

The autocorrelation matrix for this input vector can be calculated using (5.21) to be

$$\mathbf{R_{xx}} = \begin{bmatrix} 1 & 0 & 1 & 0 \\ 0 & 1 & 0 & 3 \\ 1 & 0 & 3 & 0 \\ 0 & 3 & 0 & 15 \end{bmatrix}.$$

The cross-correlation vector of $d(n)$ and $\mathbf{X}(n)$ is

$$\mathbf{P}_{dx} = E\{[1\ x(n)\ x^2(n)\ x^3(n)]^T (2 + x(n) - 0.5x^2(n) + x^3(n) + \eta(n))\}$$

$$= [1.5\ 4\ 0.5\ 18]^T.$$

The optimal MMSE solution to this problem is given by

$$\vec{\mathbf{H}}_{opt} = \mathbf{R_{xx}^{-1}} \mathbf{P}_{dx} = [2\ 1\ -0.5\ 1]^T,$$

which is identical to the coefficients of the unknown system.

We now consider the least-squares identification of the same system. Table 5.1 shows the ensemble averages of the coefficient values obtained over 100 independent estimates as well as the time-averaged, squared coefficient errors (mean-square deviation (MSD) of the coefficients) in the 100 estimates for two different values of the noise variance σ_η^2 and two different values of the number of samples M. The input signal was a pseudo-random sequence[2] with statistics identical to that of an IID Gaussian signal with zero mean value and unit variance. The noise sequence was independent of the input signal and also was an IID Gaussian signal with zero-mean value and appropriate variance. We can notice several things from the table. The mean values of the coefficients are very close to the corresponding values of the unknown system. The variance of the coefficient error increases with the noise variance and decreases with increasing sample size. In particular, the mean-square deviation of the estimates can be seen to be directly proportional to the measurement noise variance from the results.

Example 5.2 This example considers the identification of the same system as in the previous example. However, instead of directly estimating the autocorrelation matrix, we make use of (5.21) to estimate the higher-order statistics of the input signal from its second-order statistics estimated using measured values of the signal.

---

[2] We use pseudorandom sequences in all experiments, even though we will not make a distinction between random and pseudorandom processes in the rest of the book.

TABLE 5.1   Statistics of the Parameter Estimates in Example 5.1.

M	σ_η^2	True Value	Mean	MSD
100	0.01	2	1.998	0.165×10^{-3}
		1	0.999	0.221×10^{-3}
		−0.5	−0.499	0.617×10^{-4}
		1	1.000	0.248×10^{-4}
100	1.0	2	1.984	0.0165
		1	0.990	0.0211
		−0.5	−0.491	0.0062
		1	1.000	0.0025
1000	0.01	2	2.000	0.122×10^{-4}
		1	1.000	0.209×10^{-4}
		−0.5	−0.500	0.468×10^{-5}
		1	1.000	0.144×10^{-5}
1000	1.0	2	1.996	0.122×10^{-2}
		1	0.998	0.209×10^{-2}
		−0.5	−0.499	0.468×10^{-3}
		1	1.001	0.144×10^{-3}

Let $\hat{\sigma}_x^2$ denote an estimate of the input signal power. Then, we can estimate the autocorrelation matrix $\mathbf{R}_{xx}$ as

$$\hat{\mathbf{R}}_{xx} = \begin{bmatrix} 1 & 0 & \hat{\sigma}_x^2 & 0 \\ 0 & \hat{\sigma}_x^2 & 0 & 3\hat{\sigma}_x^4 \\ \hat{\sigma}_x^2 & 0 & 3\hat{\sigma}_x^4 & 0 \\ 0 & 3\hat{\sigma}_x^4 & 0 & 15\hat{\sigma}_x^6 \end{bmatrix}.$$

The cross-correlation vector $\mathbf{P}_{dx}$ is estimated in the usual manner. The mean values and error variances of the coefficients obtained from 100 independent estimates is tabulated in Table 5.2. We can see by comparing the results of Examples 5.1 and 5.2 that even though the mean values of the estimates were close to the true values in both instances, the variability of the estimates from one realization to the next was significantly higher for the method that assumes that the input signals are Gaussian-distributed. Furthermore, unlike the results of Example 5.1, the mean-square deviation of the estimates does not appear to depend significantly on the measurement noise variance in this example. The reason for the above-mentioned characteristics of the estimates can be attributed to the significant deviation of the higher-order statistics of the pseudo-random signals used in this example.

The reader should keep in mind that the technique used in this example can be applied *only when* the input signal is truly Gaussian-distributed. Otherwise, significant performance degradations will occur.

Example 5.3   A problem that occurs in several applications is the generation of harmonic signals. One example involves modeling of rotating machinery where the excitation is simple harmonic (sinusoidal) motion, but the measured vibrations

TABLE 5.2 Statistics of the Parameter Estimates in Example 5.2.

M	σ_η^2	True Value	Mean	MSD
100	0.01	2	1.978	1.2855
		1	0.957	5.676
		−0.5	−0.477	1.988
		1	1.013	0.816
100	1.0	2	1.964	1.312
		1	0.947	5.740
		−0.5	−0.470	1.994
		1	1.014	0.8271
1000	0.01	2	1.989	0.104
		1	0.962	0.7309
		−0.5	0.443	0.1721
		1	1.012	0.111
1000	1.0	2	1.985	0.105
		1	0.961	0.736
		−0.5	−0.482	0.174
		1	1.013	0.111

include higher-order harmonics [263]. We consider a problem in which the input signal is

$$x(n) = \sin(2\pi(0.01)n)$$

and the measured vibration is modeled as

$$d(n) = \sum_{k=1}^{3} \frac{1}{k}\sin(2\pi(0.01)kn) + \eta(n),$$

where $\eta(n)$ is a weak IID Gaussian measurement noise sequence with zero mean value and variance 10^{-5}. The measured spectrum of $d(n)$ is shown in Figure 5.1. The signal $d(n)$ was estimated using the least-squares approach with a truncated third-order Volterra system model with a memory span of two samples. It can be shown that the autocorrelation matrix of $x(n)$ is singular when the memory of the system model exceeds two samples. The measured spectrum of the estimate of $d(n)$ obtained using 2000 samples is plotted in Figure 5.2. A segment of $d(k)$ and its estimate are shown in Figure 5.3. We can see from these figures that a third-order nonlinearity is adequate to model the three harmonics in this example.

Example 5.4 In this example, we consider the estimation of the parameters of a homogeneous quadratic system with kernel given by

$$\mathbf{H}_2 = \begin{bmatrix} 1 & 0.5 & -0.25 & 0.10 \\ 0 & 0.5 & 0.2 & -0.15 \\ 0 & 0 & -0.25 & -0.20 \\ 0 & 0 & 0 & 0.5 \end{bmatrix}$$

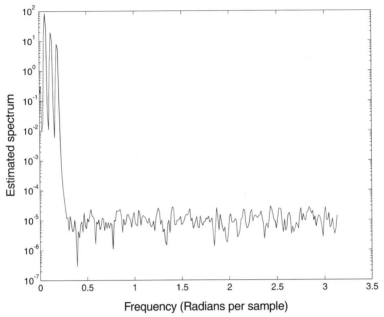

Figure 5.1 Estimated spectrum of $d(n)$ in Example 5.3.

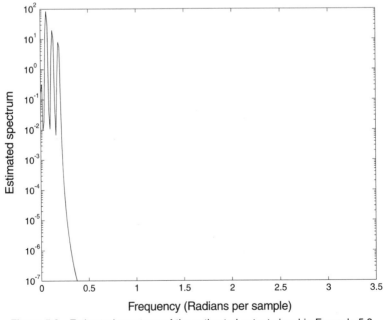

Figure 5.2 Estimated spectrum of the estimated output signal in Example 5.3.

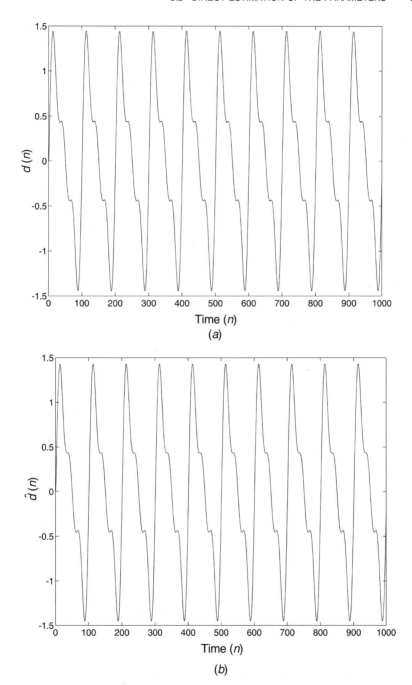

Figure 5.3 Desired response signal and its estimate in Example 5.3: (*a*) desired response signal *d*(*n*); (*b*) estimated *d*(*n*).

using least-squares techniques. The input signal is obtained by passing an IID, uniformly distributed pseudo-random signal in the range $[-1, 1]$ through a time-invariant IIR filter with input–output relationship

$$x(n) = 0.8x(n-1) + \xi(n).$$

Note that $x(n)$ has lowpass characteristics. The output of the system was observed through an additive zero-mean, IID, Gaussian white-noise sequence with variance 10^{-2} and independent of $x(n)$. Table 5.3 shows the mean values and the coefficient errors associated with each coefficient obtained over 100 independent estimates using 1000 samples. Once again, we can see that the coefficient estimates are reasonably close to the true values, indicating the usefulness of the parameter estimation techniques described in this chapter for colored and non-Gaussian input signals also.

5.2.3 Condition for Invertibility of the Autocorrelation Matrix

The derivations in the previous two subsections indicate that the optimal coefficient values can be uniquely estimated if the autocorrelation matrix is nonsingular. In general, it is difficult to determine conditions that guarantee the nonsingularity of the autocorrelation matrix. However, for the case of independent, identically distributed input signals, there is a simple but useful result. Extensions to correlated input signals and other variations of the following theorem are discussed in [23,201].

Theorem 5.3 $\mathbf{R_{xx}}$ defined in (5.17) is singular for an IID input if and only if $x(n)$ takes its values from a finite set of P or fewer values.

PROOF Consider a vector $\mathbf{X}_{\text{poly}}(n)$ defined as

$$\mathbf{X}_{\text{poly}}(n) = [1 \; x(n) \; x^2(n) \cdots x^P(n)]^T. \tag{5.32}$$

TABLE 5.3 Statistics of the Parameter Estimates in Example 5.4

Coefficient	True Value	Mean	MSD
$h_{0,0}$	1.00	1.000	0.409×10^{-3}
$h_{0,1}$	0.50	0.502	0.119×10^{-2}
$h_{0,2}$	-0.25	-0.248	0.107×10^{-2}
$h_{0,3}$	0.10	0.096	0.934×10^{-3}
$h_{1,1}$	0.50	0.498	0.583×10^{-3}
$h_{1,2}$	0.20	0.201	0.115×10^{-2}
$h_{1,3}$	-0.15	-0.147	0.107×10^{-2}
$h_{2,2}$	-0.25	-0.250	0.739×10^{-3}
$h_{2,3}$	-0.20	-0.204	0.143×10^{-2}
$h_{3,3}$	0.50	0.502	0.476×10^{-3}

We first show that

$$\mathbf{R}_{\text{poly}} = E\{\mathbf{X}_{\text{poly}}(n)\mathbf{X}_{\text{poly}}^T(n)\} \tag{5.33}$$

is singular if and only if $x(n)$ belongs to a finite set of P or less elements. To see this, recall that $\mathbf{R}_{\text{poly}}$ is singular if and only if there exists a vector $\mathbf{b} \neq \mathbf{0}$ such that

$$\mathbf{b}^T \mathbf{R}_{\text{poly}} \mathbf{b} = 0. \tag{5.34}$$

Now

$$\begin{aligned}
\mathbf{b}^T \mathbf{R}_{\text{poly}} \mathbf{b} &= E\{\mathbf{b}^T \mathbf{X}_{\text{poly}}(n)\mathbf{X}_{\text{poly}}^T(n)\mathbf{b}\} \\
&= E\{(\mathbf{b}^T \mathbf{X}_{\text{poly}}(n))^2\}.
\end{aligned} \tag{5.35}$$

The last equality is possible if and only if $\mathbf{b}^T \mathbf{X}_{\text{poly}}(n) = 0$ for all choices of $\mathbf{X}_{\text{poly}}(n)$. Since $\mathbf{b}^T \mathbf{X}_{\text{poly}}(n)$ defines a Pth-order polynomial in $x(n)$, it can be zero for all choices of $x(n)$ if and only if $x(n)$ takes at most P values, which are indeed the roots of the polynomial $\mathbf{b}^T \mathbf{X}_{\text{poly}}(n)$.

Let us now consider the autocorrelation matrix of $\mathbf{X}(n)$. Since all the elements of $\mathbf{X}_{\text{poly}}(n)$ belong to $\mathbf{X}(n)$, it can be shown that there exists a nonzero vector $\mathbf{b}'$ such that

$$\mathbf{X}^T(n)\mathbf{b}' = 0 \tag{5.36}$$

whenever $\mathbf{R}_{\text{poly}}$ is a singular matrix. To see this, let $\mathbf{C}$ be a permutation matrix such that the first $P + 1$ elements of

$$\mathbf{X}'(n) = \mathbf{C}\mathbf{X}(n) \tag{5.37}$$

form the vector $\mathbf{X}_{\text{poly}}(n)$. Since $\mathbf{X}_{\text{poly}}^T(n)\mathbf{b} = 0$, it is easy to show that

$$(\mathbf{X}'(n))^T \begin{pmatrix} \mathbf{b} \\ \mathbf{0} \end{pmatrix} = \mathbf{X}^T(n)\mathbf{C}^T \begin{pmatrix} \mathbf{b} \\ \mathbf{0} \end{pmatrix} = \mathbf{0}. \tag{5.38}$$

Comparing (5.36) and (5.38), we see that

$$\mathbf{b}' = \mathbf{C}^T \begin{pmatrix} \mathbf{b} \\ \mathbf{0} \end{pmatrix}. \tag{5.39}$$

Equation (5.38) shows that $\mathbf{R}_{\mathbf{xx}}$ is singular whenever $\mathbf{R}_{\text{poly}}$ is singular. Consequently, the condition in Theorem 5.3 is a sufficient condition for $\mathbf{R}_{\mathbf{xx}}$ to be singular for any input signal statistics. When the input signal is IID, we can show that the condition is also necessary. For this, consider the vector

$$\mathbf{X}_{\text{aug}}(n) = \mathbf{X}_{\text{poly}}(n) \otimes \mathbf{X}_{\text{poly}}(n-1) \otimes \cdots \otimes \mathbf{X}_{\text{poly}}(n-N+1). \tag{5.40}$$

By using the properties of the Kronecker product of matrices and the independence of $\mathbf{X}_{\text{poly}}(n)$ from $\mathbf{X}_{\text{poly}}(n-1)$, $\mathbf{X}_{\text{poly}}(n-2)$, ..., $\mathbf{X}_{\text{poly}}(n-N+1)$, we can show that

$$\mathbf{R}_{\text{aug}} = E\{\mathbf{X}_{\text{aug}}(n)\mathbf{X}_{\text{aug}}^T(n)\} = \underbrace{\mathbf{R}_{\text{poly}} \otimes \mathbf{R}_{\text{poly}} \otimes \cdots \otimes \mathbf{R}_{\text{poly}}}_{N_p \text{ times}} \qquad (5.41)$$

and that

$$\mathbf{R}_{\text{aug}}^{-1} = \mathbf{R}_{\text{poly}}^{-1} \otimes \mathbf{R}_{\text{poly}}^{-1} \otimes \cdots \otimes \mathbf{R}_{\text{poly}}^{-1}, \qquad (5.42)$$

which implies that $\mathbf{R}_{\text{aug}}$ is singular if and only if $\mathbf{R}_{\text{poly}}$ is singular. Since the condition in the theorem is both necessary and sufficient for $\mathbf{R}_{\text{poly}}$ to be singular, it follows that it is also a necessary and sufficient condition for $\mathbf{R}_{\text{aug}}$ to be singular when $x(n)$ is an IID signal.

Now, all the elements of $\mathbf{X}(n)$ are contained in $\mathbf{X}_{\text{aug}}(n)$. Suppose that $\mathbf{R}_{xx}$ is singular and the condition of the theorem does not hold. It is easy to show that $\mathbf{R}_{\text{aug}}$ is also singular. However, this contradicts our earlier result that $\mathbf{R}_{\text{aug}}$ is singular if and only if $x(n)$ belongs to a set of P or fewer elements. The results of the theorem follows immediately.

Remark 5.4   Theorem 5.3 Is Not a Necessary Condition for Correlated Input Signals   As described earlier, the condition in Theorem 5.3 is not a necessary one when $x(n)$ is not an IID signal. For example, one can show that a signal $x(n)$, defined as

$$x(n) = \sum_{i=1}^{K} a_i e^{j(\omega_i n + \theta_i)} \qquad (5.43)$$

has a singular autocorrelation matrix for both the linear and Volterra system identification problems whenever the system memory exceeds K samples. This is a consequence of the fact that a sum of K pure sinusoids can be predicted exactly using K consecutive samples.

5.2.3.1 *Singularity Condition for Least-Squares Criteria*   The result of Theorem 5.3 is valid for most least-squares formulations of the estimation problem also. It is left as an exercise for the reader to show that this statement is valid for the formulations in (5.22) and (5.28). The use of multilevel pseudorandom sequences as input signals for identification of Volterra filters is discussed in [200].

5.3 ORTHOGONAL METHODS FOR SYSTEM IDENTIFICATION

We have seen that a significant problem associated with representing nonlinear systems using Volterra series expansions is the large number of parameters required by such models. Consequently, parameter estimation is an even more complex problem for such models. Recall from (5.16) that the estimation of the parameters of a Pth-order Volterra system with N-sample memory requires the inversion of an $O(N^P) \times O(N^P)$-element autocorrelation matrix. For arbitrary input signals, the inversion of such matrices is a tedious task. We describe two methods for estimating the parameters of a Volterra system using orthogonal signals in this section. Both methods assume that the input signal is Gaussian-distributed. The first technique is applicable for white input signals, while the second method is useful for colored signals also.

5.3.1 Basic Definitions and Motivation

Suppose that we are interested in estimating a signal $d(n)$ as a linear combination of N other signals, $x_1(n), x_2(n), \ldots, x_N(n)$. The space spanned by $x_1(n), x_2(n), \ldots, x_N(n)$, specifically, the space consisting of all possible signals of the form $\sum_{i=1}^{N} a_i x_i(n)$, where the a_i terms are arbitrary scaling factors, is known as the *linear span* of $x_1(n), \ldots, x_N(n)$. A set of signals, $v_1(n), v_2(n), \ldots, v_N(n)$ is said to be an *orthogonal basis set* for this space if [3]

$$E\{v_i(n)v_k(n)\} = 0; \qquad i \neq k \tag{5.44}$$

and $v_1(n), v_2(n), \ldots, v_N(n)$ span the same space as $x_1(n), x_2(n), \ldots, x_N(n)$.

5.3.1.1 Efficiency of Using Orthogonal Signals in Estimation Problems
It can be seen from (5.44) that the autocorrelation matrix of a set of mutually orthogonal signals is diagonal. One complication that arises with Volterra system modeling is that even when the input signal is IID, the autocorrelation matrix is not diagonal in general. This is a consequence of the fact that the input vector contains various powers of the same signal samples. An illustration of such a situation can be found in Example 5.1.

There are several reasons why orthogonal signals are useful in estimation problems. We discuss two of the most important reasons here. Suppose that we are interested in finding the MMSE estimate of $d(n)$ as

$$\hat{d}(n) = \sum_{i=1}^{N} h_i v_i(n). \tag{5.45}$$

Since $v_1(n), v_2(n), \ldots, v_N(n)$ span the same space spanned by $x_1(n), x_2(n), \ldots, x_N(n)$, it should be clear that the MMSE estimate of the form (5.45) is

---

[3] This discussion assumes that the minimum mean-square error criterion is employed in our estimation problems. Definitions appropriate for alternate criteria such as least-squares can be made in a similar manner.

identical to the MMSE estimate of the form

$$\hat{d}(n) = \sum_{i=1}^{N} w_i x_i(n) \tag{5.46}$$

in the sense that the MMSE values are the same for both estimates. However, the coefficients $h_1, h_2, \ldots, h_N$ are different from $w_1, w_2, \ldots, w_N$ in general.

One key advantage of using orthogonal signals in estimation problems is that the coefficient of each signal can be evaluated independently of every other signal. Consequently, the estimation task can be broken up into several "smaller" sub-problems. For example, h_i can be evaluated as

$$h_i = \frac{E\{d(n)v_i(n)\}}{E\{v_i^2(n)\}}. \tag{5.47}$$

The simplicity of the estimation procedure is evident from this expression. A second advantage is that if we decide to add more signals to the set $v_1(n), v_2(n), \ldots, v_N(n)$ to obtain a better estimate, the coefficients $h_1, h_2, \ldots, h_N$ of the original set do not change. It is left as an exercise to show that the preceding statement is correct and that it does not hold when the signal set is not orthogonal.

5.3.2 Wiener G-Functionals

Wiener [332] developed the theory of using orthogonal signals for representing nonlinear systems. For certain technical reasons [264,275], it turns out that a larger class of nonlinear systems can be represented using this approach than with Volterra series expansions. We concentrate on the usefulness of Wiener's approach for characterizing and identifying truncated Volterra systems.

Let $y(n)$ represent the output of a possibly inhomogeneous Pth-order Volterra system. Wiener's method decomposes $y(n)$ into $(P + 1)$ G-functionals as

$$y(n) = g_0 + \sum_{i=1}^{P} \bar{g}_i[x(n)], \tag{5.48}$$

where $x(n)$ is the input signal to the system, g_0 is a constant corresponding to the zeroth-order G-functional, and $\bar{g}_i[x(n)]$ is the ith-order G-functional defined as the output of a ith-order, inhomogeneous Volterra system when its input is $x(n)$.

The G-functionals depend on the input and output signals of the nonlinear system. The property that distinguishes G-functionals from the outputs of arbitrary Volterra systems is their orthogonality when the input signal is an IID Gaussian sequence with zero mean value and finite variance σ_x^2. The G-functionals are constructed such that

$$E\{\bar{h}_l[x(n)]\bar{g}_p[x(n)]\} = 0; \qquad l < p. \tag{5.49}$$

whenever $\bar{h}_l[x(n)]$ is the output of any arbitrary, homogeneous, lth-order Volterra system and $x(n)$ is an IID Gaussian process with zero mean value and variance σ_x^2.

5.3.2.1 *Orthogonality of G-Functionals*   Since $\bar{g}_p[x(n)]$ is orthogonal to all homogeneous Volterra system outputs of order smaller than p, it is also orthogonal to all inhomogeneous Volterra system outputs of order smaller than p. In particular

$$E\{\bar{g}_l[x(n)]\bar{g}_p[x(n)]\} = 0; \qquad l \neq p. \tag{5.50}$$

The orthogonality described above, together with the discussion in the previous section, implies that the pth-order G-functional associated with a nonlinear system can be evaluated independently of the G-functionals of other orders. We now discuss how to constrain the structure of the G-functionals so that the orthogonality of (5.49) holds.

5.3.3   Structure of the *G*-Functionals

To simplify the presentation, we assume that the system of interest has finite order of nonlinearity P and is causal with system memory that spans the most recent N samples. Furthermore, for the sake of pedagogical simplicity, we assume throughout this discussion that the Volterra kernels are symmetric. The lth-order G-functional has the form

$$\bar{g}_l[x(n)] = g_{0,l} + \bar{g}_{1,l}[x(n)] + \cdots + \bar{g}_{l,l}[x(n)], \tag{5.51}$$

where $\bar{g}_{r,l}[x(n)]$ is the output of a homogeneous rth-order Volterra system of the form

$$\bar{g}_{r,l}[x(n)] = \sum_{k_1=0}^{N-1}\sum_{k_2=0}^{N-1}\cdots\sum_{k_r=0}^{N-1} g_{r,l}(k_1, k_2, \ldots, k_r)x(n - k_1)x(n - k_2)\cdots x(n - k_r).$$

$$\tag{5.52}$$

There are three characteristics that are common to all G-functionals:

1. The odd-ordered Volterra kernels of even-ordered G-functionals are zero. Similarly, the even-ordered Volterra kernels of odd-ordered G-functionals are zero. Since the odd-ordered moments of zero-mean Gaussian signals are zero,

it immediately follows that the odd-ordered G-functionals are uncorrelated with the outputs of even-ordered, homogeneous Volterra systems. Similarly, even-ordered G-functionals are uncorrelated with the outputs of odd-ordered, homogeneous Volterra systems.

2. All Volterra kernels of an lth-order G-functional are completely determined by the lth-order Volterra kernel of the functional.

3. The G-functionals depend on the variance σ_x^2 of the input signal.

The derivation of the structure of the G-functionals of arbitrary order is conceptually straightforward, but lengthy. Consequently, we derive the G-functionals up to order three, and then state the general result without proof.

5.3.3.1 *Zeroth-Order G-Functionals* The zeroth-order G-functional given by

$$\bar{g}_0[x(n)] = g_{0,0} \tag{5.53}$$

is a constant. The specific value of $g_{0,0}$ can be determined only from knowledge of the input and output signals. The steps for calculating this value are demonstrated in Example 5.5.

5.3.3.2 *First-Order G-Functional* The first-order G-functional is given by

$$\bar{g}_1[x(n)] = g_{0,1} + \sum_{k=0}^{N-1} g_{1,1}(k)x(n-k). \tag{5.54}$$

Because of the orthogonality property of the G-functionals, $\bar{g}_1[x(n)]$ should satisfy

$$E\{\bar{h}_0[x(n)]\bar{g}_1[x(n)]\} = 0 \tag{5.55}$$

for zero-mean and IID Gaussian input signals and any constant functional $\bar{h}_0[x(n)] = h_0$. This expectation can be evaluated as

$$E\{\bar{h}_0[x(n)]\bar{g}_1[x(n)]\} = h_0g_{0,1} + \sum_{k=0}^{N-1} h_0g_{1,1}(k)E\{x(n-k)\} = 0. \tag{5.56}$$

Since the second term on the right-hand side is always zero because $E\{x(n-k)\} = 0$, we must have that

$$g_{0,1} = 0 \tag{5.57}$$

Consequently, the first-order G-functional takes the form

$$\bar{g}_1[x(n)] = \sum_{k=0}^{N-1} g_{1,1}(k)x(n-k). \tag{5.58}$$

The specific value of the kernel $g_{1,1}(k)$ depends on the input and output signals. The details of the calculations are illustrated in Example 5.5.

5.3.3.3 The Second-Order G-Functional   The second-order G-functional is given by

$$\bar{g}_2[x(n)] = g_{0,2} + \sum_{k=0}^{N-1} g_{1,2}(k)x(n-k)$$

$$+ \sum_{k_1=0}^{N-1}\sum_{k_2=0}^{N-1} g_{2,2}(k_1, k_2)x(n-k_1)x(n-k_2). \tag{5.59}$$

Using the orthogonality of $\bar{g}_2[x(n)]$ with arbitrary $\bar{h}_0[x(n)]$ and $\bar{h}_1[x(n)]$, we get

$$E\{\bar{g}_2[x(n)]\bar{h}_0[x(n)]\} = g_{0,2}h_0 + \sum_{k=0}^{N-1} g_{1,2}(k)h_0E\{x(n-k)\}$$

$$+ \sum_{k_1=0}^{N-1}\sum_{k_2=0}^{N-1} g_{2,2}(k_1, k_2)h_0E\{x(n-k_1)x(n-k_2)\}$$

$$= 0 \tag{5.60}$$

and

$$E\{\bar{g}_2[x(n)]\bar{h}_1[x(n)]\} = g_{0,2} \sum_{k=0}^{N-1} h_1(k)E\{x(n-k)\}$$

$$+ \sum_{k=0}^{N-1}\sum_{k_1=0}^{N-1} h_1(k)g_{1,2}(k_1)E\{x(n-k)x(n-k_1)\}$$

$$+ \sum_{k=0}^{N-1}\sum_{k_1=0}^{N-1}\sum_{k_2=0}^{N-1} h_1(k)g_{2,2}(k_1, k_2)E\{x(n-k)x(n-k_1)x(n-k_2)\}$$

$$= 0. \tag{5.61}$$

We can simplify these relationships by recognizing that odd-ordered expectations of zero-mean Gaussian signals are zero and that $x(n)$ also belongs to an IID process. We get the following constraints on the G-functional of order two by making these simplifications:

$$g_{0,2}h_0 + \sigma_x^2 h_0 \sum_{k_1=0}^{N-1} g_{2,2}(k_1, k_1) = 0 \tag{5.62}$$

and

$$\sum_{k_1=0}^{N-1} \sigma_x^2 h_1(k_1)g_{1,2}(k_1) = 0. \tag{5.63}$$

Equation (5.62) implies that

$$g_{0,2} = -\sigma_x^2 \sum_{k_1=0}^{N-1} g_{2,2}(k_1, k_1), \tag{5.64}$$

Since $h_1(k_1)$ is arbitrary, the only way (5.63) can always be valid is if

$$g_{1,2}(k_1) = 0 \quad \text{for all } k_1. \tag{5.65}$$

This result simply illustrates the first characteristic of the G-functional discussed after (5.52). Equations (5.62) and (5.63) together imply that the second-order G-functional takes the form

$$\bar{g}_2[x(n)] = \sum_{k_1=0}^{N-1}\sum_{k_2=0}^{N-1} g_{2,2}(k_1, k_2)x(n - k_1)x(n - k_2) - \sigma_x^2 \sum_{k_1=0}^{N-1} g_{2,2}(k_1, k_1). \tag{5.66}$$

As before, we delay the illustration of the steps for calculating $g_{2,2}(k_1, k_2)$ to Example 5.5.

5.3.3.4 Third-Order-G-Functional
To simplify the discussion, we utilize the fact that the third-order G-functional has the form

$$\bar{g}_3[x(n)] = \sum_{k_1=0}^{N-1}\sum_{k_2=0}^{N-1}\sum_{k_3=0}^{N-1} g_{3,3}x(k_1, k_2, k_3)x(n - k_1)x(n - k_2)x(n - k_3)$$

$$+ \sum_{k_1=0}^{N-1} g_{1,3}(k_1)x(n - k_1) \tag{5.67}$$

in this discussion. Since the outputs of all homogeneous second and zeroth-order Volterra systems are uncorrelated with $\bar{g}_3[x(n)]$, we only need to find $g_{1,3}(k_1)$ that makes $\bar{g}_3[x(n)]$ uncorrelated with the outputs of all linear systems when the input is a zero-mean, Gaussian, and IID process with variance σ_x^2. Now

$$E\{\bar{g}_3[x(n)]\bar{h}_1[x(n)]\} = \sum_{k=0}^{N-1} \sum_{k_1=0}^{N-1} \sum_{k_2=0}^{N-1} \sum_{k_3=0}^{N-1} g_{3,3}(k_1, k_2, k_3)h_1(k)$$

$$\times E\{x(n-k_1)x(n-k_2)x(n-k_3)x(n-k)\}$$

$$+ \sum_{k=0}^{N-1} \sum_{k_1=0}^{N-1} g_{1,3}(k_1)h_1(k)E\{x(n-k_1)x(n-k)\}. \qquad (5.68)$$

We can simplify this equation by utilizing the facts that the input signal is an IID and Gaussian process and $g_{3,3}(k_1, k_2, k_3)$ is symmetric in its variables. Since

$$E\{x(n-k_1)x(n-k_2)x(n-k_3)x(n-k_4)\} = \sigma_x^4\{\delta(k_1-k_2)\delta(k_3-k_4)$$

$$+ \delta(k_1-k_3)\delta(k_2-k_4)$$

$$+ \delta(k_1-k_4)\delta(k_2-k_3)\} \qquad (5.69)$$

for IID Gaussian signals, we can simplify (5.68) to get

$$E\{\bar{g}_3[x(n)]\bar{h}_1[x(n)]\} = \sum_{k=0}^{N-1} \sum_{k_1=0}^{N-1} 3\sigma_x^4 g_{3,3}(k_1, k_1, k)h_1(k)$$

$$+ \sum_{k=0}^{N-1} \sigma_x^2 g_{1,3}(k)h_1(k)$$

$$= \sum_{k=0}^{N-1} \sigma_x^2 h_1(k) \left\{ \left[\sum_{k_1=0}^{N-1} 3\sigma_x^2 g_{3,3}(k_1, k_1, k) \right] + g_{1,3}(k) \right\}$$

$$= 0. \qquad (5.70)$$

The last equality arose because of the orthogonality of $\bar{g}_3[x(n)]$ and $\bar{h}_1[x(n)]$. Since $h_1(k)$ is arbitrary, we need to have

$$\sum_{k_1=0}^{N-1} 3\sigma_x^2 g_{3,3}(k_1, k_1, k) + g_{1,3}(k) = 0 \qquad (5.71)$$

to guarantee orthogonality of the two signals. Solving for $g_{1,3}(k)$ from (5.67), we find the appropriate structure for $\bar{g}_3[x(n)]$ to be

$$\bar{g}_3[x(n)] = \sum_{k_1=0}^{N-1} \sum_{k_2=0}^{N-1} \sum_{k_3=0}^{N-1} g_{3,3}(k_1, k_2, k_3)x(n - k_1)x(n - k_2)x(n - k_3)$$

$$- 3\sigma_x^2 \sum_{k=0}^{N-1} \left\{ \sum_{k_1=0}^{N-1} g_{3,3}(k_1, k_1, k) \right\} x(n - k). \tag{5.72}$$

5.3.3.5 The lth-Order G-Functional
It can be shown that the lth-order G-functional satisfies the following form:

$$\bar{g}_l[x(n)] = \sum_{i=0}^{[l/2]} \frac{(-1)^i l! \sigma_x^{2i}}{2^i(l - 2i)!i!} \sum_{k_1=0}^{N-1} \sum_{k_2=0}^{N-1} \cdots \sum_{k_{l-2i}=0}^{N-1} \sum_{m_1=0}^{N-1} \sum_{m_2=0}^{N-1} \cdots \sum_{m_i=0}^{N-1}$$

$$g_{l,l}(k_1, k_2, \ldots, k_{l-2i}, m_1, m_1, m_2, m_2, \ldots, m_i, m_i)$$

$$\times x(n - k_1)x(n - k_2) \cdots x(n - k_{l-2i}), \tag{5.73}$$

where $[l/2]$ denotes the largest integer smaller than $l/2$. Rugh has provided sketch of the proof for the continuous-time case [264]. It is left as an exercise for the reader to show that the G-functionals of order 0,1,2,3 satisfy the general form in (5.73).

5.3.3.6 Evaluation of Volterra Kernels from System Representations Using G-Functionals
Recall from the definition of G-functionals that a Pth-order Volterra system can be represented as

$$h[x(n)] = \sum_{i=0}^{P} \bar{g}_i[x(n)], \tag{5.74}$$

where $\bar{g}_l[x(n)]$ contains nonlinear terms up to order l. Consequently, it is straightforward to show that the lth-order kernel of the nonlinear system is given by

$$h_l(k_1, k_2, \ldots, k_l) = \sum_{m=l}^{P} g_{l,m}(k_1, k_2, \ldots, k_l). \tag{5.75}$$

Example 5.5   Consider a third-order Volterra system with input–output relationship

$$y(n) = x^3(n) + x^2(n) + x(n)x(n - 1) + 0.5x(n) + 2.$$

In the symmetric representation, the input–output relationship for this system is

$$y(n) = x^3(n) + x^2(n) + 0.5x(n)x(n-1) + 0.5x(n-1)x(n) + 0.5x(n) + 2.$$

In order to represent $y(n)$ using G-functionals, we first note that G-functionals up to order 3 are adequate in this example. Since only $\bar{g}_3[x(n)]$ contains a third-order term, it follows from (5.75) that

$$\bar{g}_{3,3}[x(n)] = x^3(n).$$

Since $\bar{g}_3[x(n)]$ is completely specified by $\bar{g}_{3,3}[x(n)]$, we can evaluate $\bar{g}_3[x(n)]$ using the result in (5.72). This process yields the following expression:

$$\bar{g}_3[x(n)] = x^3(n) - 3\sigma_x^2 x(n).$$

We can assume without loss of generality that the input signal power $\sigma_x^2 = 1$. This gives

$$\bar{g}_3[x(n)] = x^3(n) - 3x(n).$$

To find $\bar{g}_2[x(n)]$, we note that $\bar{g}_3[x(n)]$ contains no second-order terms. Consequently, the only term in the G-functional representation that contains a second-order nonlinearity is $\bar{g}_{2,2}[x(n)]$. To match the second-order nonlinearity in the direct-form and G-functional representation, we must have

$$\bar{g}_{2,2}[x(n)] = x^2(n) + 0.5x(n)x(n-1) + 0.5x(n-1)x(n)$$

$$= x^2(n) + x(n)x(n-1).$$

Using (5.66), we can immediately see that

$$\bar{g}_2[x(n)] = x^2(n) + x(n)x(n-1) - 1.$$

Recall from (5.75) that

$$\bar{h}_1[x(n)] = 0.5x(n) = \bar{g}_{1,3}[x(n)] + \bar{g}_{1,2}[x(n)] + \bar{g}_{1,1}[x(n)]$$

Since $\bar{g}_{1,3}[x(n)] = -3x(n)$, and $\bar{g}_{1,2}[x(n)] = 0$, we see that

$$\bar{g}_{1,1}[x(n)] = 3.5x(n),$$

implying that

$$\bar{g}_1[x(n)] = 3.5x(n)$$

Similarly

$$\bar{h}_0[x(n)] = 2 = \bar{g}_{0,0}[x(n)] + \bar{g}_{0,2}[x(n)].$$

Since $\bar{g}_{0,2}[x(n)] = -1$, we get $\bar{g}_0[x(n)] = \bar{g}_{0,0}[x(n)] = 3$.

Figure 5.4 shows the decomposition of the original system into an equivalent system involving the G-functional representation. This decomposition is valid for any input signal. However, the output of each component corresponding to a G-functional of specific order is orthogonal to the outputs of other components only when the input signal is a zero-mean, IID, Gaussian signal with unit variance.

5.3.4 System Identification Using *G*-Functionals

Suppose that we are interested in representing a nonlinear system using the G-functional representation as

$$y(n) = \sum_{i=0}^{P} \bar{g}_i[x(n)], \qquad (5.76)$$

and wish to estimate the parameters of this model using the joint statistics of the input and output signals when the input signal $x(n)$ belongs to a zero mean and IID Gaussian process with variance σ_x^2. We assume, as usual, that the system is causal and has N-sample memory. We now show that the appropriate cross-correlation functions involving the input and output samples can be directly used to estimate the kernels of the G-functionals.

Recall from our earlier discussion that

$$E\{\bar{g}_l[x(n)]\bar{g}_k[x(n)]\} = 0 \quad \text{for} \quad l \neq k \qquad (5.77)$$

and that

$$E\{\bar{g}_P[x(n)]x(n - m_1)x(n - m_2) \cdots x(n - m_l)\} = 0; \qquad P > l. \qquad (5.78)$$

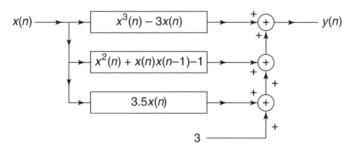

Figure 5.4 Decomposition of the system in Example 5.5 using *G*-functionals.

5.3.4.1 *Estimation of $\bar{g}_0[x(n)]$*   Since

$$E\{\bar{g}_k[x(n)]\} = 0; \qquad k \neq 0 \tag{5.79}$$

(see Problem 5.7), it follows that

$$E\{y(n)\} = E\{\bar{g}_0[x(n)]\} = g_{0,0}. \tag{5.80}$$

5.3.4.2 *Estimation of $\bar{g}_1[x(n)]$*   Even though it is not technically necessary, it is convenient to assume during the estimation of the ith-order G-functional that the estimates of G-functionals up to order $i-1$ have been removed from $y(n)$. Equivalently, we assume during the estimation of the ith-order G-functional that $y(n)$ can be represented as

$$y(n) = \sum_{k=i}^{P} \bar{g}_k[x(n)]. \tag{5.81}$$

For the first-order case, this simply means that we have removed the mean value from $y(n)$.

Now, consider a cross-correlation function of the form

$$E\{y(n)x(n-m_1)\} = E\left\{\sum_{k=1}^{P} \bar{g}_k[x(n)]x(n-m_1)\right\}, \tag{5.82}$$

Since $\bar{g}_2[x(n)], \bar{g}_3[x(n)], \ldots, \bar{g}_P[x(n)]$ are orthogonal to $x(n-m_1)$, this expectation simplifies to

$$E\{y(n)x(n-m_1)\} = E\{\bar{g}_1[x(n)]x(n-m_1)\}$$

$$= E\left\{\sum_{k_1=0}^{N-1} g_{1,1}(k_1)x(n-k_1)x(n-m_1)\right\}$$

$$= g_{1,1}(m_1)\sigma_x^2. \tag{5.83}$$

The last equality arose because $x(n)$ belongs to an IID process. We can solve for $g_{1,1}(m_1)$ from (5.83) to get

$$g_{1,1}(m_1) = \frac{1}{\sigma_x^2}E\{y(n)x(n-m_1)\}. \tag{5.84}$$

5.3.4.3 *Estimation of $\bar{g}_2[x(n)]$*   We proceed in a similar manner as before. Assuming that the zeroth- and first-order G-functionals have been removed from

$y(n)$ already, we get

$$E\{y(n)x(n - m_1)x(n - m_2)\} = E\{\bar{g}_2[x(n)]x(n - m_1)x(n - m_2)\}. \tag{5.85}$$

Expanding the right-hand side of (5.85) using the functional form of $\bar{g}_2[x(n)]$, we write the expectation as

$$
E\{\bar{g}_2[x(n)]x(n - m_1)x(n - m_2)\}
$$
$$
= E\left\{\sum_{k_1=0}^{N-1}\sum_{k_2=0}^{N-1} g_{2,2}(k_1, k_2)x(n - k_1)x(n - k_2)x(n - m_1)x(n - m_2)\right\}
$$
$$
- \sigma_x^2\left[\sum_{k_1=0}^{N-1} g_{2,2}(k_1, k_1)\right]E\{x(n - m_1)x(n - m_2)\}.
$$

$$\tag{5.86}$$

This expression can be simplified using the fact that $x(n)$ is a zero-mean, IID, Gaussian process. This operation results in

$$
E\{\bar{g}_2[x(n)x(n - m_1)x(n - m_2)\} = \sum_{k_1=0}^{N-1}\sum_{k_2=0}^{N-1} g_{2,2}(k_1, k_2)\sigma_x^4[\delta(k_1 - k_2)\delta(m_1 - m_2)
$$
$$
+ \delta(k_1 - m_1)\delta(k_2 - m_2) + \delta(k_1 - m_2)\delta(k_2 - m_1)]
$$
$$
- \sigma_x^4\sum_{k_1=0}^{N-1} g_{2,2}(k_1, k_1)\delta(m_1 - m_2). \tag{5.87}
$$

We can further simplify (5.87) by realizing that $g_{2,2}(k_1, k_2)$ is a symmetric kernel. The simplified form of (5.87) is

$$
E\{\bar{g}_2[x(n)]x(n - m_1)x(n - m_2)\} = \sum_{k_1=0}^{N-1} g_{2,2}(k_1, k_1)\sigma_x^4\delta(m_1 - m_2)
$$
$$
+ 2\sigma_x^4 g_{2,2}(m_1, m_2)
$$
$$
- \sigma_x^4\sum_{k_1=0}^{N-1} g_{2,2}(k_1, k_1)\delta(m_1 - m_2) \tag{5.88}
$$
$$
= 2\sigma_x^4 g_{2,2}(m_1, m_2). \tag{5.89}
$$

Combining (5.89) with (5.85), we see that $g_{2,2}(m_1, m_2)$ can be estimated using the relationship

$$g_{2,2}(m_1, m_2) = \frac{1}{2\sigma_x^4} E\{y(n)x(n - m_1)x(n - m_2)\}. \tag{5.90}$$

5.3.4.4 *Estimation of $\bar{g}_r[x(n)]$*
Extension of the preceding results to the general case is conceptually straightforward. However, the derivations are lengthy. We provide only a statement of the general result. For the general case, we assume that

$$y(n) = \sum_{k=r}^{P} \bar{g}_k[x(n)]. \tag{5.91}$$

We can then show that

$$E\{y(n)x(n - m_1)x(n - m_2) \cdots x(n - m_r)\} = r!\sigma_x^{2r} g_{r,r}(m_1, m_2, \ldots, m_r), \tag{5.92}$$

implying that the rth order G-functional can be estimated using the relationship

$$g_{r,r}(m_1, m_2, \ldots, m_r) = \frac{1}{r!\sigma_x^{2r}} E\{y(n)x(n - m_1)x(n - m_2) \cdots x(n - m_r)\}. \tag{5.93}$$

5.3.5 Orthogonalization of Correlated Gaussian Signals

The use of G-functionals to identify truncated Volterra systems is useful when the input signal is IID and Gaussian. Schetzen has extended the result to correlated Gaussian signals using the concept of L-functionals, and to the case of certain types of non-Gaussian signals using the concept of S-functionals [275]. In this section, we discuss a somewhat different approach for orthogonalizing correlated Gaussian signals. The approach presented here is based on the recent work in [170] and is useful for identifying truncated Volterra systems with finite memory.

5.3.5.1 *Orthogonalization for Second-Order Volterra Systems*
In this section we develop an efficient technique for creating an orthogonal basis set for the entries of the input vector $[1, \mathbf{X}_{1r}^T(n), \mathbf{X}_{2r}^T(n)]$, given by

$$\{1, x(n), x(n - 1), \ldots, x(n - N + 1), x^2(n), x(n)x(n - 1), \ldots,$$
$$x(n)x(n - N + 1), \ldots, x^2(n - N + 1)\} \tag{5.94}$$

when the input signal is zero-mean, Gaussian, and stationary. This process results in the derivation of a lattice filter structure that is applicable only to Gaussian input signals and second-order Volterra system models. These concepts are later extended to higher-order nonlinearities. The algorithm for orthogonalization of correlated Gaussian signals for identification of second-order Volterra systems was first developed by Koh and Powers [115].

We start with finding an orthogonal basis set for the entries of $\mathbf{X}_{1r}(n)$, the vector that contains only the linear components in the input signal set in (5.94). Let

$$\mathbf{R}_1 = E\{\mathbf{X}_{1r}(n)\mathbf{X}_{1r}^T(n)\} \qquad (5.95)$$

denote the $N \times N$ autocorrelation matrix of $\mathbf{X}_{1r}$. Let $\mathbf{Q}$ be a lower triangular matrix such that

$$\mathbf{Q}^T\mathbf{R}_1\mathbf{Q} = \mathbf{I}. \qquad (5.96)$$

Consider the transformation

$$\mathbf{U}(n) = \mathbf{Q}^T\mathbf{X}_{1r}(n). \qquad (5.97)$$

Direct calculation will show that

$$E\{\mathbf{U}(n)\mathbf{U}^T(n)\} = \mathbf{I}, \qquad (5.98)$$

and consequently, the elements of $\mathbf{U}(n)$ are mutually orthogonal. Let $u_i(n)$; $i = 0, 1, \ldots, N-1$ represent the orthogonal signals that make up $\mathbf{U}(n)$. Then

$$E\{u_i(n)u_j(n)\} = \delta(i-j). \qquad (5.99)$$

Remark 5.5   The Elements of U(n) Are Independent   The elements of the set $\{u_0(n), u_1(n), \ldots, u_{N-1}(n)\}$ are Gaussian, zero-mean, and uncorrelated with each other. Since all of them have unit variance, they also have identical distribution functions. Furthermore, since uncorrelated Gaussian processes are also independent processes, $u_0(n), u_1(n), \ldots, u_{N-1}(n)$ are mutually independent random processes. In particular

$$E\{f(u_i(n))g(u_l(n))\} = E\{f(u_i(n))\}E\{g(u_l(n))\} \qquad (5.100)$$

whenever $i \neq l$ for arbitrary functions f and g.

Remark 5.6:   Orthogonalization Using Backward Prediction   An efficient method to orthogonalize the elements of $\mathbf{X}_{1r}(n)$ is to employ a lattice predictor. It is shown in Appendix C that the backward prediction error signals defined as

$$b_0(n) = x(n)$$

$$b_i(n) = x(n - i) - \sum_{k=0}^{i-1} c_{k,i} x(n - k); \qquad i = 1, 2, \ldots, N - 1, \qquad (5.101)$$

where $b_i(n)$ is the optimal MMSE error for estimating $x(n - i)$ using $x(n), x(n - 1), \ldots, x(n - i + 1)$, form an orthogonal basis set for the span of $x(n), x(n - 1), \ldots, x(n - N + 1)$. The coefficients $c_{k,i}$ in (5.101) correspond to the optimal minimum mean-square error coefficients for the ith order backward predictor. A computationally efficient method, known as the Levinson–Durbin algorithm [90], for evaluating the backward prediction errors is described in Appendix C.

The elements of the vector $\mathbf{U}(n)$ may be obtained by normalizing the backward prediction errors by their respective root-mean-squared values, i.e.,

$$u_i(n) = \frac{b_i(n)}{\sigma_i}, \qquad (5.102)$$

where

$$\sigma_i^2 = E\{b_i^2(n)\}. \qquad (5.103)$$

Now, consider a product signal of the form

$$u_{i,j}^{(2)}(n) = u_i(n)u_j(n) - \delta(i - j). \qquad (5.104)$$

Proposition 5.1:   The elements of the vector

$$\mathbf{U}(n) = [1, u_0(n), u_1(n), \ldots, u_{N-1}(n), u_{0,0}^{(2)}(n), u_{0,1}^{(2)}(n), \ldots,$$

$$u_{0,N-1}^{(2)}(n), u_{1,1}^{(2)}(n), u_{1,2}^{(2)}(n), \ldots, u_{1,N-1}^{(2)}(n), \ldots, u_{N-1,N-1}^{(2)}(n)]^T \qquad (5.105)$$

are mutually orthogonal.

PROOF   Obviously

$$E\{u_i(n)u_j(n)\} = \delta(i - j). \qquad (5.106)$$

Since third-order correlations of all zero-mean Gaussian variables are zero, we see that

$$E\{u_i(n)u_{l,m}^{(2)}(n)\} = 0 \quad \text{for all } i, l, m. \tag{5.107}$$

Furthermore, since $E\{u_i(n)\} = 0$ and $E\{u_{i,j}^{(2)}(n)\} = 0$ for all i and j, the only issue that needs to be solved for proving the proposition is to show that

$$E\{u_{i,j}^{(2)}(n)u_{k,l}^{(2)}(n)\} = 0 \tag{5.108}$$

whenever $(i, j) \neq (k, l)$. First, it is easy to show that

$$E\{(u_{i,j}^{(2)}(n))^2\} = \begin{cases} 2; & i = j \\ 1; & i \neq j. \end{cases} \tag{5.109}$$

Let $(i, j) \neq (k, l)$. Several possibilities exist:

1. $i = j$ and $k = l$

$$\begin{aligned} E\{u_{i,i}^{(2)}(n)u_{k,k}^{(2)}(n)\} &= E\{(u_i^2(n) - 1)(u_k^2(n) - 1)\} \\ &= E\{u_i^2(n)\}E\{u_k^2(n)\} - E\{u_i^2(n)\} - E\{u_k^2(n)\} + 1 \\ &= 0. \end{aligned} \tag{5.110}$$

2. $i \neq j$ and $k = l$ or $i = j$ and $k \neq l$. It is a straightforward task to show that

$$E\{u_{i,j}^{(2)}(n)u_{k,l}^{(2)}(n)\} = 0 \tag{5.111}$$

in this case.

3. $i \neq j$ and $k \neq l$ with the possibility that either $i = k$ or $j = l$. Once again one can show by direct evaluation that $u_{i,j}^{(2)}(n)$ and $u_{k,l}^{(2)}(n)$ are uncorrelated in this case.

All of these results together prove our proposition. Consequently, we can design an orthogonalizer for the input signal as shown in Figure 5.5. Note that the orthogonalization of $x(n), x(n-1), \ldots, x(n-N+1)$ is performed using a lattice predictor structure.

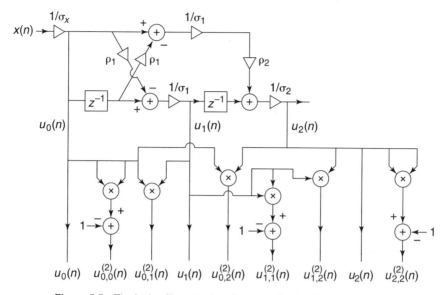

Figure 5.5   The lattice filter structure for second-order Volterra systems.

5.3.5.2   Estimation of a Second-Order Volterra System Using the Orthogonalized Input Signals

Let the optimal MMSE second-order Volterra estimate of $d(n)$ using $x(n)$ be given by

$$\hat{d}(n) = w_0 + \sum_{m_1=0}^{N-1} w_{m_1}^{(1)} u_{m_1}(n) + \sum_{m_1=0}^{N-1} \sum_{m_2=m_1}^{N-1} w_{m_1,m_2}^{(2)} u_{m_1,m_2}^{(2)}(n). \tag{5.112}$$

The coefficients of the orthogonal input signals are, in general, different from those of the direct-form estimator.

Estimation of w_0. Since $u_{m_1}(n)$ and $u_{m_1,m_2}^{(2)}(n)$ terms are all zero-mean quantities, the mean values of all the terms involving them are zero. Consequently, we can estimate w_0 as the mean value of $d(n)$:

$$w_0 = E\{d(n)\} \tag{5.113}$$

Estimation of $w_{m_1}^{(1)}$ and $w_{m_1,m_2}^{(2)}$. Even though it is technically not necessary, we remove its mean value from $d(n)$ before we compute the necessary expectations. Let

$$d'(n) = d(n) - E\{d(n)\}. \tag{5.114}$$

We can show that

$$E\{d(n)u_{m_1}(n)\} = E\{d'(n)u_{m_1}(n)\} \tag{5.115}$$

and

$$E\{d(n)u_{m_1,m_2}^{(2)}(n)\} = E\{d'(n)u_{m_1,m_2}^{(2)}(n)\} \tag{5.116}$$

for all m_1 and m_2. While using $d'(n)$ or $d(n)$ produces identical results in MMSE estimates, using $d'(n)$ may be advantageous in situations where one has to estimate the statistics of the signals. Now, since all the elements of $\mathbf{U}(n)$ defined in (5.105) are mutually orthogonal, it is easy to show that

$$w_{m_1}^{(1)} = \frac{E\{d'(n)u_{m_1}(n)\}}{E\{u_{m_1}^2(n)\}} \tag{5.117}$$

and

$$w_{m_1,m_2}^{(2)} = \frac{E\{d'(n)u_{m_1,m_2}^{(2)}(n)\}}{E\{(u_{m_1,m_2}^{(2)}(n))^2\}} \tag{5.118}$$

for $m_1 = 0, 1, \ldots, N-1$ and $m_2 = m_1, m_1 + 1, \ldots, N-1$. Recall from (5.99) that

$$E\{u_{m_1}^2(n)\} = 1 \tag{5.119}$$

and from (5.109) that

$$E\{(u_{m_1,m_2}^{(2)}(n))^2\} = \begin{cases} 2; \ m_1 = m_2 \\ 1; \ m_1 \neq m_2. \end{cases} \tag{5.120}$$

We can develop systematic methods to evaluate the cross-correlation functions in the numerators of (5.117) and (5.118) from knowledge of all the cross-correlation values of the type $E\{d(n)x(n-m_1)\}$ and $E\{d(n)x(n-m_1)x(n-m_2)\}$. However, in most practical situations, we will measure the cross-correlation functions involving the orthogonalized signals directly.

5.3.5.3 Orthogonalization for Higher-Order Volterra Systems The extension of the ideas presented for the second-order Volterra systems to higher-order cases is only slightly more complicated. Let us define vectors $\mathbf{U}_{P,i}(n)$ as

$$\mathbf{U}_{P,i}(n) = [1 \ u_i(n) \ u_i^2(n) \ \cdots \ u_i^P(n)]^T \tag{5.121}$$

for $i = 0, 1, \ldots, N-1$.

Let $\mathbf{Q}_P$ be a lower triangular, $(P+1) \times (P+1)$ element matrix that orthogonalizes $\mathbf{U}_{P,i}(n)$. Since $u_i(n)$ for all values of i have identical distributions, the same $\mathbf{Q}_P$ orthogonalizes $\mathbf{U}_{P,i}(n)$ for all values of i. Furthermore, since the statistics of $\mathbf{U}_{P,i}(n)$

are known a priori, we can pre-compute $\mathbf{Q}_P$. As an example, for $P = 2$, one choice of $\mathbf{Q}_2$ is given by

$$\mathbf{Q}_2 = \begin{bmatrix} 1 & 0 & 0 \\ 0 & 1 & 0 \\ -1 & 0 & 1 \end{bmatrix}. \tag{5.122}$$

Thus

$$\mathbf{Q}_2 \mathbf{U}_{2,i}(n) = \begin{bmatrix} 1 \\ u_i(n) \\ u_i^2(n) - 1 \end{bmatrix}. \tag{5.123}$$

Example 5.6 For $P = 5$, one choice of $\mathbf{Q}_5$ is given by

$$\mathbf{Q}_5 = \begin{bmatrix} 1 & 0 & 0 & 0 & 0 & 0 \\ 0 & 1 & 0 & 0 & 0 & 0 \\ -1 & 0 & 1 & 0 & 0 & 0 \\ 0 & -3 & 0 & 1 & 0 & 0 \\ 3 & 0 & -6 & 0 & 1 & 0 \\ 0 & 15 & 0 & -10 & 0 & 1 \end{bmatrix}.$$

This matrix can be obtained by direct substitution of the higher-order statistics of the signals from (5.21) in each of the five estimation problems for creating the orthogonal signal set. The reader should also verify by direct calculation that the variances of the orthogonal signals are not unity and also not necessarily identical in this case.

Let $\mathbf{V}_{P,i}(n)$ be an orthogonal vector obtained as

$$\mathbf{V}_{P,i}(n) = \mathbf{Q}_P \mathbf{U}_{P,i}(n). \tag{5.124}$$

Theorem 5.4 The elements of

$$\mathbf{V}(n) = \mathbf{V}_{P,0}(n) \otimes \mathbf{V}_{P,1}(n) \otimes \cdots \otimes \mathbf{V}_{P,N-1}(n) \tag{5.125}$$

are mutually orthogonal.

PROOF   Let $v_{P,i,k}(n)$ denote the kth element of $\mathbf{V}_{P,i}(n)$. Recall that $v_{P,i,k}(n)$ and $v_{P,l,k}(n)$ are independent random processes if $i \neq l$. Any element of $\mathbf{V}(n)$ has the form

$$v_{p,0,m_0}(n)v_{p,1,m_1}(n)\cdots v_{P,N-1,m_{N-1}}(n). \tag{5.126}$$

Now, let us evaluate the cross-correlation of any two elements of $\mathbf{V}(n)$. The expectation has the form

$$E\{(v_{P,0,m_0}(n)v_{P,1,m_1}(n)\cdots v_{P,N-1,m_{N-1}}(n))(v_{P,0,l_0}(n)v_{P,1,l_1}(n)\cdots v_{P,N-1,l_{N-1}}(n))\}$$

$$= E\{v_{P,0,m_0}(n)v_{P,0,l_0}(n)\}E\{v_{P,1,m_1}(n)v_{P,1,l_1}(n)\}\cdots E\{v_{P,N-1,m_{N-1}}(n)v_{P,N-1,l_{N-1}}(n)\}. \tag{5.127}$$

The expectations can be separated as on the right-hand side of these equations because of the independence of the various elements involved. The expectations of the form $E\{v_{P,i,m_i}(n)v_{P,i,l_i}(n)\}$ are zero whenever $m_i \neq l_i$ since $v_{p,i,m_i}(n)$ and $v_{p,i,l_i}(n)$ are zero-mean and uncorrelated processes. Consequently, the only situation under which (5.127) is nonzero is when $m_0 = l_0, m_1 = l_1, \dots$ and $m_{N-1} = l_{N-1}$. This implies that the elements of $\mathbf{V}(n)$ are mutually orthogonal, proving the theorem.

It is straightforward to show that a linear transformation exists between the elements of $\mathbf{V}(n)$ and those of $\mathbf{X}_{\text{aug}}(n)$ defined as

$$\mathbf{X}_{\text{aug}}(n) = \mathbf{X}_P(n) \otimes \mathbf{X}_P(n-1) \otimes \cdots \otimes \mathbf{X}_P(n-N+1), \tag{5.128}$$

where

$$\mathbf{X}_P(n) = [1 \ x(n) \ x^2(n) \ \cdots \ x^P(n)]^T. \tag{5.129}$$

It follows immediately that the elements of $\mathbf{V}(n)$ is an orthogonal basis set for the elements of $\mathbf{X}_{\text{aug}}(n)$.

While the above result is satisfying in many ways, one should not overlook the fact that the number of elements contained in $\mathbf{X}_{\text{aug}}(n)$ and $\mathbf{V}(n)$ far exceeds the number of terms in a Pth-order Volterra series expansion with N-sample memory. It would be useful to derive an orthogonal basis set for the signals involved in the general Pth-order Volterra series expansion. The next theorem provides a solution to this problem.

Theorem 5.5

$$\{v_{P,0,m_0}(n)v_{P,1,m_1}(n)\cdots v_{P,N-1,m_{N-1}}(n)|m_0 + m_1 + \cdots + m_{N-1} \le P\}$$

is an orthogonal basis set for

$$\{x^{m_0}(n)x^{m_1}(n-1)x^{m_2}(n-2)\cdots x^{m_{N-1}}(n-N+1)|m_0 + m_1 + \cdots + m_{N-1} \le P\}.$$

Note that $v_{P,i,0}(n) = 1$ for all i and that each m_i takes values from $0 \le m_i \le P$.

PROOF The proof is a consequence of the fact that $v_{P,i,m_i}(n)$ can be written as a linear combination of $1, x(n), x(n-2), \ldots, x(n-i+1)$ and their products of order up to m_i. Consequently, there exists a one-to-one linear transformation between the elements of the two sets defined in the theorem. This implies that the elements of both sets span the same space. The result follows immediately.

5.3.5.4 Orthogonalization of IID Non-Gaussian Signals

The procedure described above for orthogonalizing the input signal set is also applicable to IID non-Gaussian signals. The procedure is even simpler for such signals, since the elements of $\mathbf{X}_1(n)$ are mutually orthogonal. Consequently, there is no need for a linear lattice predictor in the orthogonalization procedure.

Example 5.7 Consider a Gaussian input signal $x(n)$ generated according to the model

$$x(n) = 0.8x(n-1) + 0.6\xi(n),$$

where $\xi(n)$ is a zero-mean and white Gaussian signal with unit variance. We are interested in identifying a homogeneous third-order Volterra system with three-sample memory using this input signal. Since the 3×3-element autocorrelation matrix of $x(n)$ is

$$\mathbf{R}_{xx} = \begin{bmatrix} 1.0 & 0.8 & 0.64 \\ 0.8 & 1.0 & 0.8 \\ 0.64 & 0.8 & 1.0 \end{bmatrix},$$

we can show that

$$b_0(n) = x(n),$$

$$b_1(n) = \frac{1}{0.6}(x(n-1) - 0.8x(n))$$

and

$$b_2(n) = \frac{1}{0.6}(x(n-2) - 0.8x(n-1))$$

constitutes an orthogonal basis signal set for $x(n)$, $x(n-1)$ and $x(n-2)$. Furthermore, each basis signal has unit variance and zero mean value. By direct evaluation of the coefficient matrix, we can show that $\mathbf{Q}_3$ is given by

$$\mathbf{Q}_3 = \begin{bmatrix} 1 & 0 & 0 & 0 \\ 0 & 1 & 0 & 0 \\ -1 & 0 & 1 & 0 \\ 0 & -3 & 0 & 1 \end{bmatrix}.$$

Let us define the following signals for $i = 0, 1$ and 2:

$$v_{3,i,0}(n) = 1, \; v_{3,i,1}(n) = b_i(n), \; v_{3,i,2}(n) = b_i^2(n) - 1$$

and

$$v_{3,i,3}(n) = b_i^3(n) - 3b_i(n).$$

Even though Theorem 5.5 discusses arbitrary truncated Volterra systems, we can use similar ideas to get the orthogonal basis set for homogeneous Volterra systems also. The orthogonal basis set for $\{x^3(n), x^2(n)x(n-1), \; x^2(n)x(n-2), \; x(n)x^2(n-1), \; x(n)x(n-1)x(n-2), \; x(n)x^2(n-2), x^3(n-1), \; x^2(n-1)x(n-2), \; x(n-1)x^2(n-2), x^3(n-2)\}$ is given by the set

$$\begin{aligned} \{ & v_{3,0,0}(n)v_{3,1,0}(n)v_{3,2,3}(n), \;\; v_{3,0,0}(n)v_{3,1,1}(n)v_{3,2,2}(n), \;\; v_{3,0,0}(n)v_{3,1,2}(n)v_{3,2,1}(n), \\ & v_{3,0,0}(n)v_{3,1,3}(n)v_{3,2,0}(n), \;\; v_{3,0,1}(n)v_{3,1,0}(n)v_{3,2,2}(n), \;\; v_{3,0,1}(n)v_{3,1,1}(n)v_{3,2,1}(n), \\ & v_{3,0,1}(n)v_{3,1,2}(n)v_{3,2,0}(n), \;\; v_{3,0,2}(n)v_{3,1,0}(n)v_{3,2,1}(n), \;\; v_{3,0,2}(n)v_{3,1,1}(n)v_{3,2,0}(n), \\ & v_{3,0,3}(n)v_{3,1,0}(n)v_{3,2,0}(n) \}. \end{aligned}$$

5.4 LATTICE ORTHOGONALIZATION FOR ARBITRARY INPUT SIGNALS

In this section[4] we discuss the situation in which the input signals to a nonlinear system cannot be characterized as Gaussian processes. The orthogonalization procedure that we described in the previous section will not work when the input signal is non-Gaussian. We now describe a general lattice orthogonalization procedure that is useful for all input signals. The price we pay for the generality is the increased complexity of the structure. For ease of presentation, we restrict our discussion to truncated second-order Volterra systems. However, the approach is valid for arbitrary-order Volterra systems with finite memory.

---

[4] This section may be omitted without loss of continuity.

5.4.1 Multichannel Representation of Nonlinear Systems

In order to develop the lattice parameterization of Volterra filters, it is convenient to visualize the nonlinear filtering problem as a linear, multichannel filtering problem. This characterization is depicted in Figure 5.6 for the second-order Volterra filter. The multichannel characterization is somewhat different from many traditional multichannel adaptive filtering problems in the sense that each channel uses a different number of delay elements and coefficients when compared with the rest of the channels. To overcome this difficulty, many lattice realizations of Volterra filters [141] use additional coefficients and delay elements in each "channel" to make the number of coefficients the same for every "channel." Such realizations correspond to special shapes for the region of support of the Volterra kernels. However, there are lattice structures available for truncated Volterra systems as given in (5.2). We now discuss one such structure that is based on a multichannel lattice filter developed by Ling and Proakis [145] and a nonlinear lattice predictor developed by Zarzycki [339]. This structure has been described for a second-order Volterra system model in [165] and for the bilinear system model in [8].

5.4.2 The Nonlinear Lattice Prediction

For simplicity, we consider the case when $N = 3$ and $P = 2$. A block diagram of the nonlinear lattice predictor is shown in Figure 5.7. The basic idea employed in the derivation of the lattice Volterra filter is very similar to the lattice orthogonalization using linear prediction techniques described in Appendix C.

5.4.2.1 Orthogonalization Using Backward Prediction   Consider the arrangement shown in Figure 5.8 for the signal set used in the estimation process. Let us define three vectors, $\mathbf{x}_0^b(n)$, $\mathbf{x}_1^b(n)$, and $\mathbf{x}_2^b(n)$, as

$$\mathbf{x}_0^b(n) = [x(n)\ x^2(n)]^T, \tag{5.130}$$

$$\mathbf{x}_1^b(n) = [x(n-1)\ x^2(n-1)\ x(n)x(n-1)]^T, \tag{5.131}$$

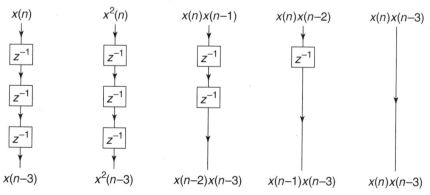

Figure 5.6 Characterization of the Volterra filtering problem as a multichannel estimation problem.

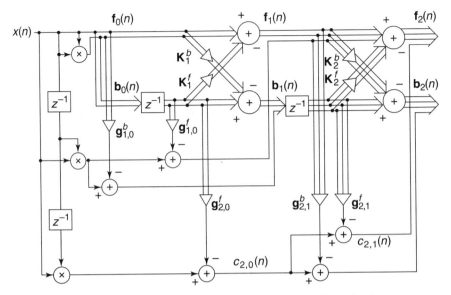

Figure 5.7 A nonlinear lattice predictor for arbitrary input signals.

and

$$\mathbf{x}_2^b(n) = [x(n-2) \ x^2(n-2) \ x(n-1)x(n-2) \ x(n)x(n-2)]^T. \tag{5.132}$$

Note that

$$\mathbf{x}_1^b(n) = [(\mathbf{x}_0^b(n-1))^T \ x(n)x(n-1)]^T \tag{5.133}$$

and

$$\mathbf{x}_2^b(n) = [(\mathbf{x}_1^b(n-1))^T \ x(n)x(n-2)]^T. \tag{5.134}$$

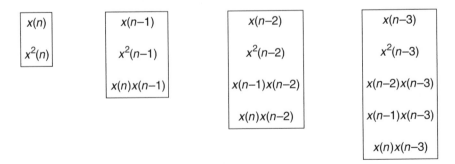

Figure 5.8 Signal set used in the backward prediction process.

Let us estimate the vector $\mathbf{x}_1^b(n)$ using $\mathbf{x}_0^b(n)$ as

$$\hat{\mathbf{x}}_1^b(n) = \mathbf{B}_{01}\mathbf{x}_0^b(n), \tag{5.135}$$

where

$$\mathbf{B}_{01} = (E\{\mathbf{x}_0^b(n)\mathbf{x}_0^b(n)^T\})^{-1}E\{\mathbf{x}_1^b(n)(\mathbf{x}_0^b(n))^T\} \tag{5.136}$$

is the optimal MMSE coefficient matrix of size 2×3 elements. It follows from the orthogonality principle that the estimation error vector

$$\mathbf{b}_1(n) = \mathbf{x}_1^b(n) - \hat{\mathbf{x}}_1^b(n) \tag{5.137}$$

is orthogonal to $\mathbf{x}_0^b(n)$. Similarly, the error in estimating $\mathbf{x}_2^b(n)$ using $\mathbf{x}_0^b(n)$ and $\mathbf{x}_1^b(n)$

$$\mathbf{b}_2(n) = \mathbf{x}_2^b(n) - \mathbf{B}_{02}\mathbf{x}_0^b(n) - \mathbf{B}_{12}\mathbf{x}_1^b(n) \tag{5.138}$$

is orthogonal to $\mathbf{x}_0^b(n)$ and $\mathbf{x}_1^b(n)$. The matrices $\mathbf{B}_{02}$ and $\mathbf{B}_{12}$ in (5.138) represent the optimal MMSE coefficient matrices for estimating $\mathbf{x}_2^b(n)$ using $\mathbf{x}_0^b(n)$ and $\mathbf{x}_1^b(n)$, respectively. Since $\mathbf{b}_1(n)$ is a linear combination of $\mathbf{x}_0^b(n)$ and $\mathbf{x}_1^b(n)$, the estimation error vector $\mathbf{b}_2(n)$ is also orthogonal to $\mathbf{b}_1(n)$. It is now straightforward to show that $\mathbf{b}_0(n) = \mathbf{x}_0^b(n)$, $\mathbf{b}_1(n)$ and $\mathbf{b}_2(n)$ are mutually orthogonal vectors. Furthermore, since the three vectors can be expressed as linear combinations of the elements of $\mathbf{x}_0^b(n)$, $\mathbf{x}_1^b(n)$, and $\mathbf{x}_2^b(n)$, the elements of $\mathbf{b}_0(n)$, $\mathbf{b}_1(n)$, and $\mathbf{b}_2(n)$ span the space spanned by the signal set in the original estimation problem.

Remark 5.7:   The Elements of $\mathbf{b}_i(n)$ Are Not Mutually Orthogonal   Although it is true that

$$E\{\mathbf{b}_i(n)\mathbf{b}_j^T(n)\} = \mathbf{0}; \qquad i \neq j, \tag{5.139}$$

it is not the case that

$$E\{\mathbf{b}_i(n)\mathbf{b}_i^T(n)\} = \mathbf{I}. \tag{5.140}$$

This is because we did not orthogonalize the elements within each vector. Complete orthogonalization can be achieved by performing a Gram–Schmidt orthogonalization of the elements of each vector separately. However, our discussion will not delve into this issue.

Remark 5.8:   Backward Nonlinear Prediction   Appendix C discusses *linear lattice filters* and *backward linear prediction*. Since $\mathbf{x}_2^b(n)$ and $\mathbf{x}_1^b(n)$ can be partially obtained by delaying $\mathbf{x}_0^b(n)$ by 2 or 1 samples, respectively, the process of creating $\mathbf{b}_0(n)$, $\mathbf{b}_1(n)$, and $\mathbf{b}_2(n)$ is similar to that of backward prediction. Consequently, we refer to this process as *backward nonlinear prediction*.

5.4.2.2 *Updating Backward Prediction Error Vectors* Suppose that we have already computed $\mathbf{b}_0(n)$ and $\mathbf{b}_1(n)$ and would like to compute $\mathbf{b}_2(n)$. Is there an efficient method for updating the prediction error vectors using the information we already have about $\mathbf{b}_0(n)$ and $\mathbf{b}_1(n)$? The answer to this question is in the affirmative. To see this, we first define $\overline{\mathbf{b}}_2(n)$ as the top three elements of $\mathbf{b}_2(n)$ and consider the estimation of $\overline{\mathbf{b}}_2(n)$ using $\mathbf{x}_0^b(n)$ and $\mathbf{x}_1^b(n)$. Recall that the top three elements of $\mathbf{x}_2^b(n)$ represent a delayed version of $\mathbf{x}_1^b(n)$. Because of the stationarity of the signals involved, it should be immediately clear that the optimal error in estimating the top three elements of $\mathbf{x}_2^b(n)$ using $\mathbf{x}_0^b(n-1) = \mathbf{b}_0(n-1)$ is $\mathbf{b}_1(n-1)$. This follows from

$$
\begin{aligned}
\mathbf{b}_1(n-1) &= \mathbf{x}_1^b(n-1) - \hat{\mathbf{x}}_1^b(n-1) \\
&= \mathbf{x}_1^b(n-1) - \mathbf{B}_{0,1}\mathbf{x}_0^b(n-1).
\end{aligned}
\tag{5.141}
$$

Note that the coefficient matrix is time-invariant because of the stationarity of the signals. For consistency of notation in the remainder of the discussion of the lattice predictor, we define $\mathbf{K}_1^b = \mathbf{B}_{0,1}$.

What (5.141) provides for us is a partial solution to solving the problem of estimating $\mathbf{x}_2^b(n)$ with $\mathbf{x}_1^b(n)$ and $\mathbf{x}_0^b(n)$. It tells us that $\mathbf{b}_1(n-1)$ is the error in estimating the top three elements of $\mathbf{x}_2^b(n)$ using $x(n-1)$ and $x^2(n-1)$. The additional elements we need to use in the estimation process are contained in the vector $[x(n) \quad x^2(n) \quad x(n)x(n-1)]^T$.

The "new" information contained in this vector is present in that component of $[x(n) \quad x^2(n) \quad x(n)x(n-1)]^T$ that is orthogonal to $[x(n-1) \quad x^2(n-1)]^T$. This component is precisely the estimation error vector obtained when the three-element vector defined above is estimated using the two-element vector that is also defined above. The estimation error vector that is produced in the process is nothing but the first-order forward prediction error vector. In general, the ith-order forward prediction error vector $\mathbf{f}_i(n)$ is defined as the error vector produced when the data vector

$$
[x(n) \quad x^2(n) \quad x(n)x(n-1) \quad x(n)x(n-2) \quad \cdots \quad x(n)x(n-i)]^T
\tag{5.142}
$$

is estimated using all possible linear and quadratic terms formed using the elements of the set $\{x(n-1), x(n-2), \ldots, x(n-i)\}$. From the discussion above, we can express $\overline{\mathbf{b}}_2(n)$ as

$$
\overline{\mathbf{b}}_2(n) = \mathbf{b}_1(n-1) - \mathbf{K}_2^b \mathbf{f}_1(n),
\tag{5.143}
$$

where $\mathbf{K}_2^b$ is the appropriate coefficient matrix. Similarly, one can show that the top three elements of $\mathbf{f}_2(n)$ can be evaluated as

$$
\overline{\mathbf{f}}_2(n) = \mathbf{f}_1(n) - \mathbf{K}_2^f \mathbf{b}_1(n-1),
\tag{5.144}
$$

where the notation is similar to that used in (5.143).

The last element of $\mathbf{b}_2(n)$, which is the error in estimating $x(n)x(n-2)$ using the same five input elements in the first two columns of Figure 5.8, has to be computed separately. This element can be computed by subtracting a linear combination of all the elements of $\mathbf{b}_0(n)$ and $\mathbf{f}_1(n)$ from $x(n)x(n-2)$ since the components of $\mathbf{b}_0(n)$ and $\mathbf{f}_1(n)$ do span the same space spanned by the element of the first two columns of Figure 5.8. In general, the last element of $\mathbf{b}_i(n)$ can be obtained by subtracting an appropriate linear combination of the elements of the vectors $\mathbf{b}_0(n-1)$, $\mathbf{b}_1(n-1), \ldots, \mathbf{b}_{i-2}(n-1)$, and $\mathbf{f}_{i-1}(n)$ from $x(n)x(n-i)$. Similarly, the last element of $\mathbf{f}_i(n)$ can be obtained by subtracting an estimate of $x(n)x(n-i)$ obtained as a linear combination of the elements of $\mathbf{b}_0(n-1), \mathbf{b}_1(n-1), \ldots, \mathbf{b}_{i-1}(n-1)$ from $x(n)x(n-i)$. The basic lattice predictor algorithm for a second-order Volterra system with $N-1$ delays is given in Table 5.4.

5.4.2.3 Complexity of the Nonlinear Lattice Predictor

The ith stage of the lattice involves estimating $i+2$ signals using $i+1$ signals. Consequently, we require $O(i^2)$ coefficients in the ith stage. A second-order Volterra predictor with N-sample memory therefore requires $\sum_{i=1}^{N} O(i^2) = O(N^3)$ coefficients. Consequently, the lattice predictor described above is an order of magnitude more complex than the direct form realization of a second-order Volterra system with N-sample memory.

Example 5.8: **Design of a Nonlinear Predictor** We consider the design of a nonlinear lattice predictor for $N=3$ and nonlinearity order $P=2$. The autocorrelation matrix of the input vector $[x(n), x^2(n), x(n-1), x^2(n-1), x(n)x(n-1), x(n-2), x^2(n-2), x(n-1)x(n-2), x(n)x(n-2)]^T$ is

$$\mathbf{R}_{xx} = \begin{bmatrix} 1.0 & 0.0 & 0.8 & 0.0 & 0.0 & 0.6 & 0.0 & 0.0 & 0.0 \\ 0.0 & 2.0 & 0.0 & 1.6 & 0.0 & 0.0 & 1.2 & 0.2 & 0.0 \\ 0.8 & 0.0 & 1.0 & 0.0 & 0.0 & 0.8 & 0.0 & 0.0 & 0.0 \\ 0.0 & 1.6 & 0.0 & 2.0 & 0.0 & 0.0 & 1.6 & 0.0 & 0.6 \\ 0.0 & 0.0 & 0.0 & 0.0 & 1.6 & 0.0 & 0.2 & 0.6 & 0.2 \\ 0.6 & 0.0 & 0.8 & 0.0 & 0.0 & 1.0 & 0.0 & 0.0 & 0.0 \\ 0.0 & 1.2 & 0.0 & 1.6 & 0.2 & 0.0 & 2.0 & 0.0 & 0.0 \\ 0.0 & 0.2 & 0.0 & 0.0 & 0.6 & 0.0 & 0.0 & 1.6 & 0.2 \\ 0.0 & 0.0 & 0.0 & 0.6 & 0.2 & 0.0 & 0.0 & 0.2 & 1.2 \end{bmatrix}.$$

Using (5.136) we find that the coefficient vector for estimating $[x(n-1), x^2(n-1)]^T$ using $[x(n), x^2(n)]^T$ is given by

$$\mathbf{K}_1^b = \mathbf{B}_{01}$$

$$= \begin{bmatrix} 1.0 & 0.0 \\ 0.0 & 2.0 \end{bmatrix}^{-1} \begin{bmatrix} 0.8 & 0.0 \\ 0.0 & 1.6 \end{bmatrix} = \begin{bmatrix} 0.8 & 0.0 \\ 0.0 & 0.8 \end{bmatrix}.$$

TABLE 5.4   The Lattice Volterra Filter Structure for Arbitrary Input Signals[a]

Initialization

$$f_0(n) = b_0(n) = \begin{pmatrix} x(n) \\ x^2(n) \end{pmatrix}$$

$$e_0(n) = d(n)$$

$$b_1(n) = \begin{pmatrix} b_0(n-1) - (K_1^b)^T f_0(n) \\ x(n)x(n-1) - (g_{1,0}^b)^T f_0(n) \end{pmatrix}$$

$$f_1(n) = \begin{pmatrix} f_0(n) - (K_1^f)^T b_0(n-1) \\ x(n)x(n-1) - (g_{1,0}^f)^T b_0(n-1) \end{pmatrix}$$

$$c_{i,0}(n) = x(n)x(n-i) - (g_{i,0}^f)^T b_0(n-1); \; i = 2, 3, \ldots, N-1$$
$$e_1(n) = e_0(n) - (k_0^d)^T b_0(n)$$

Lattice sections 2 through $N-1$
  Do for $i = 2, 3, \ldots, N-1$

Backward prediction error update

$$b_i(n) = \begin{pmatrix} b_{i-1}(n-1) - (K_i^b)^T f_{i-1}(n) \\ c_{i,i-2}(n) - (g_{i,i-1}^b)^T f_{i-1}(n) \end{pmatrix}$$

Forward prediction error update

$$f_i(n) = \begin{pmatrix} f_{i-1}(n) - (K_i^f)^T b_{i-1}(n-1) \\ c_{i,i-2}(n) - (g_{i,i-1}^f)^T b_{i-1}(n-1) \end{pmatrix}$$

Auxiliary variable update

$$c_{j,i-1}(n) = c_{j,i-2}(n) - (g_{j,i-1}^f)^T b_{i-1}(n-1); \; j = i+1, i+2, \ldots, N-1$$

Joint process estimation error update

$$e_i(n) = e_{i-1}(n) - (k_{i-1}^d)^T b_{i-1}(n)$$

Final joint process estimation error

$$e(n) = e_N(n) = e_{N-1}(n) - (k_{N-1}^d)^T b_{N-1}(n)$$

[a]Notes: K_i^f and K_i^b are $(i+1) \times (i+1)$-element matrices; $g_{i,j}^f$ and $g_{i,j}^b$ are vectors with $(j+1)$ elements; k_i^d has $i+2$ elements; $c_{i,j}(n)$ are scalar signals and are used to compute the last elements of $f_i(n)$ and $b_i(n)$. The coefficient matrices and vectors are assumed to be known.

Similarly, the coefficients of the forward predictor of order one is given by

$$K_1^f = \begin{bmatrix} 1.0 & 0.0 \\ 0.0 & 2.0 \end{bmatrix}^{-1} \begin{bmatrix} 0.8 & 0.0 \\ 0.0 & 1.6 \end{bmatrix} = \begin{bmatrix} 0.8 & 0.0 \\ 0.0 & 0.8 \end{bmatrix}.$$

For the backward prediction problem, we still need to estimate $x(n)x(n-1)$ using $[x(n), x^2(n)]^T$.

The coefficient vector for this problem is

$$
\mathbf{g}_{1,0}^b = \left(E\left\{ \begin{bmatrix} x(n) \\ x^2(n) \end{bmatrix} [x(n) \quad x^2(n)] \right\} \right)^{-1} E\left\{ x(n)x(n-1) \begin{bmatrix} x(n) \\ x^2(n) \end{bmatrix} \right\}
$$

$$
= \begin{bmatrix} 1.0 & 0.0 \\ 0.0 & 2.0 \end{bmatrix}^{-1} \begin{bmatrix} 0.0 \\ 0.0 \end{bmatrix} = \begin{bmatrix} 0.0 \\ 0.0 \end{bmatrix}.
$$

Similarly, the coefficients for estimating $x(n)x(n-1)$ using $[x(n-1), x^2(n-1)]^T$ can be shown to be

$$
\mathbf{g}_{1,0}^f = \begin{bmatrix} 0.0 \\ 0.0 \end{bmatrix}.
$$

The preceding steps conclude the evaluation of the coefficients of the first-order forward and backward nonlinear predictors. To complete our analysis, we require the evaluation of $\mathbf{K}_2^b$ and $\mathbf{K}_2^f$, as well as the entries of $\mathbf{g}_{2,1}^b$ and those of $\mathbf{g}_{2,0}^f$ and $\mathbf{g}_{2,1}^f$. Recall that $\mathbf{K}_2^b$ estimates $[x(n-2), x^2(n-2), x(n)x(n-1)]^T$ using $\mathbf{f}_1(n)$, and that $\mathbf{K}_2^f$ estimates $[x(n), x^2(n), x(n)x(n-1)]^T$ using $\mathbf{b}_1(n-1)$.

In order to evaluate the coefficient matrix $\mathbf{K}_2^f$, we must first calculate the correlation matrix of $\mathbf{b}_1(n)$. Since the signals involved are stationary, their statistics do not change when the time index is changed to $n-1$. Recall that $\mathbf{b}_1(n)$ is the optimal estimation error when the vector $\mathbf{x}_1^b(n)$ defined in (5.131) is estimated as a linear combination of the elements of $\mathbf{x}_0^b(n)$ as defined in (5.130). Consequently, we can evaluate the correlation matrix of $\mathbf{b}_1(n)$ to be

$$
E\{\mathbf{b}_1(n)\mathbf{b}_1^T(n)\} = E\{\mathbf{x}_1^b(n)\,(\mathbf{x}_1^b(n))^T\} - \begin{bmatrix} (\mathbf{K}_1^b)^T \\ (\mathbf{g}_{1,0}^b)^T \end{bmatrix} E\{\mathbf{x}_0^b(n)(\mathbf{x}_1^b(n))^T\}
$$

$$
= \begin{bmatrix} 1.0 & 0.0 & 0.0 \\ 0.0 & 2.0 & 0.0 \\ 0.0 & 0.0 & 1.6 \end{bmatrix} - \begin{bmatrix} 0.8 & 0.0 \\ 0.0 & 0.8 \\ 0.0 & 0.0 \end{bmatrix} \begin{bmatrix} 0.8 & 0.0 & 0.0 \\ 0.0 & 1.6 & 0.0 \end{bmatrix}
$$

$$
= \begin{bmatrix} 0.36 & 0.00 & 0.00 \\ 0.00 & 0.72 & 0.00 \\ 0.00 & 0.00 & 1.60 \end{bmatrix}.
$$

The cross-correlation matrix of $\mathbf{b}_1(n-1)$ and $[x(n), x^2(n), x(n)x(n-1)]^T$ can be evaluated as

$$
E\left\{ \mathbf{b}_1(n-1) \begin{bmatrix} x(n) \\ x^2(n) \\ x(n)x(n-1) \end{bmatrix}^T \right\}
$$

$$
= E\left\{ \left(\begin{bmatrix} x(n-2) \\ x^2(n-2) \\ x(n-1)x(n-2) \end{bmatrix} - \begin{bmatrix} (\mathbf{K}_1^b)^T \\ (\mathbf{g}_{1,0}^b)^T \end{bmatrix} \begin{bmatrix} x(n-1) \\ x^2(n-1) \end{bmatrix} \right) \begin{bmatrix} x(n) \\ x^2(n) \\ x(n)x(n-1) \end{bmatrix}^T \right\}
$$

$$
= \begin{bmatrix} 0.6 & 0.0 & 0.0 \\ 0.0 & 1.2 & 1.0 \\ 0.0 & 0.6 & 1.0 \end{bmatrix} - \begin{bmatrix} 0.8 & 0.0 \\ 0.0 & 0.8 \\ 0.0 & 0.0 \end{bmatrix} \begin{bmatrix} 0.8 & 0.0 & 0.0 \\ 0.0 & 1.6 & 0.0 \end{bmatrix}
$$

$$
= \begin{bmatrix} -0.04 & 0.00 & 0.00 \\ 0.00 & -0.08 & 1.00 \\ 0.00 & 0.60 & 1.00 \end{bmatrix}.
$$

We are now able to evaluate $\mathbf{K}_2^f$ as

$$
\mathbf{K}_2^f = (E\{\mathbf{b}_1(n)\mathbf{b}_1^T(n)\})^{-1} E\left\{ \mathbf{b}_1(n-1) \begin{bmatrix} x(n) \\ x^2(n) \\ x(n)x(n-1) \end{bmatrix}^T \right\}
$$

$$
= \begin{bmatrix} 0.36 & 0.00 & 0.00 \\ 0.00 & 0.72 & 0.00 \\ 0.00 & 0.00 & 1.60 \end{bmatrix}^{-1} \begin{bmatrix} -0.04 & 0.00 & -0.64 \\ 0.00 & -0.08 & 0.00 \\ 0.00 & 0.60 & 1.00 \end{bmatrix}
$$

$$
= \begin{bmatrix} -0.11 & 0.00 & -1.78 \\ 0.00 & -0.11 & 0.00 \\ 0.00 & 0.38 & 0.63 \end{bmatrix}.
$$

In a similar manner, we can show that

$$
E\{\mathbf{f}_1(n)\mathbf{f}_1^T(n)\} = E\{\mathbf{b}_1(n)\mathbf{b}_1^T(n)\} = \begin{bmatrix} 0.36 & 0.00 & 0.00 \\ 0.00 & 0.72 & 0.00 \\ 0.00 & 0.00 & 1.60 \end{bmatrix}
$$

and that

$$E\left\{\mathbf{f}_1(n)\begin{bmatrix} x(n-2) \\ x^2(n-2) \\ x(n-1)x(n-2) \end{bmatrix}^T\right\} = \begin{bmatrix} -0.04 & 0.00 & 0.00 \\ 0.00 & -0.08 & 0.60 \\ 0.00 & 1.00 & 1.00 \end{bmatrix}.$$

We can evaluate the coefficient matrix $\mathbf{K}_2^b$ using the preceding two results to be

$$\mathbf{K}_2^b = \begin{bmatrix} 0.36 & 0.00 & 0.00 \\ 0.00 & 0.72 & 0.00 \\ 0.00 & 0.00 & 1.60 \end{bmatrix}^{-1} \begin{bmatrix} -0.04 & 0.00 & 0.00 \\ 0.00 & -0.08 & 0.60 \\ 0.00 & 1.00 & 1.00 \end{bmatrix}$$

$$= \begin{bmatrix} -0.11 & 0.00 & 0.00 \\ 0.00 & -0.11 & 0.38 \\ -1.78 & 0.00 & 0.63 \end{bmatrix}.$$

Finally, we need to evaluate the auxiliary coefficients $\mathbf{g}_{2,0}^f, \mathbf{g}_{2,1}^f$, and $\mathbf{g}_{2,1}^b$. These coefficient vectors estimate $x(n)x(n-2)$ using $\mathbf{b}_0(n-1), \mathbf{b}_1(n-1)$, and $\mathbf{f}_1(n)$, respectively. Since

$$E\{\mathbf{b}_0(n-1)x(n)x(n-2)\} = [0.0\ 0.0]^T,$$

it is clear that $\mathbf{g}_{2,0}^f = [0.0\ 0.0]^T$. Now,

$$E\{\mathbf{b}_1(n-1)x(n)x(n-2)\}$$

$$= E\left\{\left(\begin{bmatrix} x(n-2) \\ x^2(n-2) \\ x(n-1)x(n-2) \end{bmatrix} - \begin{bmatrix} (\mathbf{K}_1^b)^T \\ (\mathbf{g}_{1,0}^b)^T \end{bmatrix} \begin{bmatrix} x(n-1) \\ x^2(n-1) \end{bmatrix}\right)x(n)x(n-2)\right\}$$

$$= [0.0\ 0.0\ 0.2]^T.$$

We can now evaluate $\mathbf{g}_{2,1}^f$ as

$$\mathbf{g}_{2,1}^f = (E\{\mathbf{b}_1(n-1)\mathbf{b}_1^T(n-1)\})^{-1}E\{\mathbf{b}_1(n-1)x(n)x(n-2)\}$$

$$= \begin{bmatrix} 0.36 & 0.00 & 0.00 \\ 0.00 & 0.72 & 0.00 \\ 0.00 & 0.00 & 1.60 \end{bmatrix}^{-1} \begin{bmatrix} 0.0 \\ 0.0 \\ 0.2 \end{bmatrix} = \begin{bmatrix} 0.00 \\ 0.00 \\ 0.32 \end{bmatrix}.$$

Similarly

$$E\{\mathbf{f}_1(n)x(n)x(n-2)\}$$

$$= E\left\{ \left(\begin{bmatrix} x(n) \\ x^2(n) \\ x(n)x(n-1) \end{bmatrix} - \begin{bmatrix} (\mathbf{K}_1^f)^T \\ (\mathbf{g}_{1,0}^f)^T \end{bmatrix} \begin{bmatrix} x(n-1) \\ x^2(n-1) \end{bmatrix} \right) x(n)x(n-2) \right\}$$

$$= [0.0 \ -0.48 \ 0.2]^T.$$

Therefore

$$\mathbf{g}_{2,1}^b = (E\{\mathbf{f}_1(n)\mathbf{f}_1^T(n)\})^{-1} E\{\mathbf{f}_1(n)x(n)x(n-2)\}$$

$$= \begin{bmatrix} 0.36 & 0.00 & 0.00 \\ 0.00 & 0.72 & 0.00 \\ 0.00 & 0.00 & 1.60 \end{bmatrix}^{-1} \begin{bmatrix} 0.0 \\ -0.48 \\ 0.2 \end{bmatrix} = \begin{bmatrix} 0.00 \\ -0.35 \\ 0.32 \end{bmatrix}.$$

This completes the derivation of the nonlinear lattice predictor.

5.4.3   Joint Process Estimation Using the Orthogonal Basis Signal Set

Let $\hat{d}(n)$ given by

$$\hat{d}(n) = \sum_{i=0}^{N-1} (\mathbf{k}_i^d)^T \mathbf{b}_i(n) \tag{5.145}$$

represent the optimal estimate of $d(n)$ using the orthogonal vectors. As described in previous sections, we can find the coefficient vectors $\mathbf{k}_i^d$ independently for each $\mathbf{b}_i(n)$ since the $\mathbf{b}_i(n)$ terms are mutually orthogonal. It is easy to show that the $(i+2)$-element, optimal coefficient vector $\mathbf{k}_i^d$ is given by

$$\mathbf{k}_i^d = (E\{\mathbf{b}_i(n)\mathbf{b}_i^T(n)\})^{-1} E\{d(n)\mathbf{b}_i(n)\}. \tag{5.146}$$

5.5   IDENTIFICATION OF CASCADE NONLINEAR SYSTEMS

Identification of cascade systems poses several difficulties that do not exist for identification of direct-form structures. For example, in general, the cascade representation of nonlinear systems are not unique. Even if the structure is constrained in such a way that the resulting representation is unique, the performance surface (e.g., the mean-square error surface) may not be quadratic, implying that the performance surface may have local minima. Because of such difficulties, most methods for estimating cascade structures [123] are iterative techniques. We do not describe them here. Adaptive filtering using cascade structures is discussed in Chapter 7. Iterative batch estimation schemes for such structures use ideas similar to those employed by the adaptive filters.

5.6 EXERCISES

5.1 State and prove the appropriate orthogonality principles for the least-squares estimation problems formulated as in (**a**) (5.22) and (**b**) (5.28). In each case, obtain an expression for the least-squares estimation error.

5.2 *Computing Assignment:* The purpose of this assignment is to investigate the effect of variations in the input signal spectrum on the performance of the least-squares estimator of (5.24). Consider the identification of a second-order truncated Volterra system of the form

$$y(n) = x(n) + 0.5x(n-1) + 0.3x^2(n) - 0.5x^2(n-1) + 0.25x(n)x(n-1)$$

from measurements of the input and output signals. For this purpose, generate 1000 samples of an input sequence using the difference equation

$$x(n) = bx(n-1) + \sqrt{(1-b^2)}\zeta(n),$$

where $\zeta(n)$ is a zero-mean and white Gaussian process with unit variance, for different values of $b = 0$, 0.5, 0.9, 0.99, and 0.999. Verify that the input signals so generated have unit variances. Generate the measured output sequences by processing the input sequences presented above with the second-order Volterra system described above and adding to each output a zero-mean and white Gaussian noise sequence that is uncorrelated with the input sequence and has variance $\sigma_\eta^2 = 0.01$.

(**a**) Assuming that we know the model order of the system to be identified, find the eigenvalues of the autocorrelation matrices of the input vector for the system identification problem. Characterize each matrix using its *condition number,* which is defined as the ratio of the maximum and minimum eigenvalues of the matrix.

(**b**) Find the least-squares estimates of the coefficients of the unknown system using (5.24) for each signal set. Repeat the experiments using 50 independent data sets, and find the mean values and the variances of the estimated coefficients for each choice of b over the ensemble of the 50 experiments. What conclusions can you reach regarding any relationship between the condition number of the autocorrelation matrix and the performance of the least-squares estimator? Since the condition number of the autocorrelation matrix is large for narrowband signals and small for wideband signals, we can also make inferences about the dependence of the estimator on the input signal spectrum.

(c) Explain what might happen if the input signal consists of a finite number of pure sinusoids.

5.3 Show that Theorem 5.3 is valid for the least-squares formulations of the estimation problem as in (5.22) and (5.28).

5.4 Extend Theorem 5.3 to find a condition for the invertibility of $E\{\mathbf{X}(n)\mathbf{X}^T(n)\}$ where

$$\mathbf{X}(n) = [x^L(n), x^{L+1}(n), \ldots, x^P(n)]^T.$$

5.5 Consider an IID, two-valued sequence $x(n)$ that takes its values from the set $[1,0]$. We are interested in identifying a memoryless nonlinear system with input–output relationship

$$y(n) = 1 + 0.5x(n) + x^2(n) - 0.5x^3(n)$$

from the measurement of its output signals when the input is $x(n)$.

(a) Calculate the autocorrelation matrix and the cross-correlation vector for this estimation problem when the order of nonlinearity is 3.

(b) Find the optimal coefficient vectors and the MMSE error values when the system identification is performed using nonlinearity orders 1, 2, and 3. Assume that the desired response signal is obtained by corrupting the output $y(n)$ of the unknown system with an additive IID noise sequence that is independent of the input signal and has zero mean value and variance $\sigma_\eta^2 = 0.01$. Use the pseudoinverse of the autocorrelation matrix whenever it is singular.[5]

(c) Comment on the results in part **b**.

5.6 Suppose that we wish to estimate a random variable d as a linear combination of N random variables $x_1, x_2, \ldots, x_N$ using the minimum mean-square error criterion. Let the optimal MMSE coefficients be denoted by $w_1, w_2, \ldots, w_N$. Let x_{N+1} be another random variable that is uncorrelated with each of $x_1, x_2, \ldots, x_N$. Show that the first N coefficients of the MMSE estimator of d obtained using $x_1, x_2, \ldots, x_N$ and x_{N+1} are $w_1, w_2, \ldots, w_N$ obtained for the first estimator.

5.7 Show that $E\{\bar{g}_k[x(n)]\} = 0$ for all $k \neq 0$.

---

[5] If the singular value decomposition of a matrix $\mathbf{A}$ is given by $\mathbf{A} = \mathbf{USV}$, where $\mathbf{U}$ and $\mathbf{V}$ are orthonormal matrices of appropriate dimensions and $\mathbf{S}$ is a diagonal matrix that contains the singular values of $\mathbf{A}$, its pseudoinverse is $\mathbf{A}^\# = \mathbf{V}^T\mathbf{S}^\#\mathbf{U}^T$, where $\mathbf{S}^\#$ is obtained by replacing all nonzero singular values in $\mathbf{S}$ with their respective reciprocals and leaving the zero singular values without change.

5.8 Show that the G-functionals of orders 1, 2, and 3 satisfy the form of (5.73).

5.9 Find a representation of the following nonlinear system using the Wiener G-functionals when the input signal is zero-mean and white Gaussian-distributed:

$$y(n) = 0.5x^3(n) + 0.2x^2(n)x(n-1) + 0.1x^2(n)x(n-2)$$
$$- 0.5x(n)x(n-1)x(n-2) + 0.4x^2(n-1)x(n-2)$$
$$+ x^3(n-1) - 0.3x^2(n-2)x(n) - 0.7x^3(n-3).$$

5.10 Consider the input signal employed in Exercise 5.2. Show that the output of the system

$$y(n) = \frac{1}{\sqrt{(1-b^2)}}x(n) - \frac{b}{\sqrt{(1-b^2)}}x(n-1)$$

is a white, Gaussian signal with zero mean value and unit variance. Extend the ideas related to the G-functionals to develop a system identification technique that whitens the input signal as above and then develops the G-functional representation of an appropriately modified nonlinear system. Note that the input to your system is still $x(n)$ and not the whitened signal.

5.11 *Computing Assignment:* In this problem, we consider the identification of a truncated Volterra system using zero-mean and white Gaussian input signals. Create a 2000-sample-long pseudorandom sequence that matches the statistics of a zero-mean and unit variance Gaussian signal. Process this signal with the system in Exercise 5.9, and create a desired response signal by adding an IID Gaussian sequence to the output of the system. The noise sequence should have zero mean value and variance $\sigma_\eta^2 = 0.01$, and should be independent of the input signal. Using time averages to evaluate the signal statistics, estimate the coefficients of the unknown system using the method of G-functionals, as well as directly using higher-order statistics. Repeat the experiments 50 times using independent data sets and find the mean-square deviation of the coefficients from their optimal values as ensemble averages over the 50 experiments. Comment on the results of your experiments. The performance of the estimator will depend quite heavily on the quality of the pseudo-Gaussian signals you generate.

5.12 Derive a transformation that relates the coefficients of the orthogonal basis set of (5.105) to that of the direct form coefficients of a second-order Volterra system using the autocorrelation values $r_{xx}(m)$ for $m = 0, 1, \ldots, N - 1$.

5.13 Consider the orthogonalization of a correlated, zero-mean, Gaussian input signal for identifying a homogeneous Pth-order truncated Volterra system with N-sample memory. Show that Theorem 5.5 can be modified as follows for the problem:

The signal set defined by

$$\{v_{P,0,m_0}(n)v_{P,1,m_1}(n)\cdots v_{P,N-1,m_{N-1}}(n)|m_0 + m_1 + \cdots + m_{N-1} = P\}$$

is an orthogonal basis set for

$$\{x^{m_0}(n)x^{m_1}(n-1)x^{m_2}(n-2)\cdots x^{m_{N-1}}(n-N+1)|m_0 + m_1 + \cdots + m_{N-1} = P\}.$$

5.14 *Computing Assignment:* Repeat Exercise 5.11 using a colored Gaussian signal obtained as

$$x(n) = 0.8x(n-1) + 0.6\xi(n)$$

where $\xi(n)$ is a zero-mean, white Gaussian sequence with unit variance. Instead of using the method of G-functionals, use the method described in Section 5.3.5 for orthogonalizing correlated Gaussian input signals. Note that, in order to evaluate the mean-square deviation of the coefficients, you will have to transform the coefficients of the orthogonal signal set to that of the direct-form structure. The performance of the estimator will depend quite heavily on the quality of the pseudo-Gaussian signals you generate.

5.15 Show that the orthogonalization technique for correlated Gaussian signals also applies to IID non-Gaussian signals.

5.16 Show that $\mathbf{Q}_5$ in Example 5.6 is an appropriate choice for orthogonalizing $\mathbf{U}_{5,i}(n)$. Suppose now that it is also required that the elements of the vector

$$\mathbf{V}_{5,i}(n) = \mathbf{Q}_5\mathbf{U}_{5,i}(n)$$

should have unit variance. How would you modify the solution in Example 5.7 to achieve this objective?

5.17 Develop a lattice predictor algorithm for orthogonalizing arbitrary input signals in quadratic system identification problems such that $E\{\mathbf{b}_i(n)\mathbf{b}_i^T(n)\} = \mathbf{I}$, that is, that the elements of each backward prediction error vector are mutually orthogonal.

6

FREQUENCY-DOMAIN METHODS
FOR VOLTERRA SYSTEM
IDENTIFICATION

It was shown in Chapter 5 that parameter estimation using pth-order truncated Volterra system models requires knowledge of the higher-order statistics of the input signals up to order $2p$. Higher-order spectra relate the statistical expectations of higher-order products of time-domain processes to frequency-domain concepts. This chapter describes the concepts of higher-order spectra, also known as *polyspectra*, and explains how polyspectra and related frequency-domain ideas can be used to estimate the parameters of a truncated Volterra system.

6.1 HIGHER-ORDER STATISTICS

We start our discussion by defining the important concepts of *moments* and *cumulants*. The joint moments of order $r = k_1 + k_2 + \cdots + k_p$ of a set of p real random variables $x_1, x_2, \ldots, x_p$ are defined as

$$\text{mom}(x_1^{k_1}, x_2^{k_2}, \ldots, x_p^{k_p}) \equiv E\{x_1^{k_1} x_2^{k_2} \cdots x_p^{k_p}\}$$

$$= \frac{\partial^r \Phi(\omega_1, \omega_2, \ldots, \omega_p)}{\partial^{k_1}(j\omega_1) \partial^{k_2}(j\omega_2) \cdots \partial^{k_p}(j\omega_p)}\bigg|_{\omega_1 = \omega_2 = \cdots = \omega_p = 0}, \quad (6.1)$$

where

$$\Phi(\omega_1, \omega_2, \ldots, \omega_p) = E\{e^{j(\omega_1 x_1 + \omega_2 x_2 + \cdots + \omega_p x_p)}\} \quad (6.2)$$

is the joint *characteristic function* of $x_1, x_2, \ldots, x_p$ [218]. The joint moments are the coefficients of the Taylor series expansion of the joint characteristic function $\Phi(\omega_1, \omega_2, \ldots, \omega_p)$ of $x_1, x_2, \ldots, x_p$ about zero.

The joint *cumulants* of order r of the same set of random variables are defined as the coefficients of the Taylor series expansion of the natural logarithm of the joint characteristic function [29,260]:

$$\text{cum}(x_1^{k_1}, x_2^{k_2}, \ldots, x_p^{k_p}) = \frac{\partial^r \ln[\Phi(\omega_1, \omega_2, \ldots, \omega_p)]}{\partial^{k_1}(j\omega_1)\partial^{k_2}(j\omega_2)\cdots\partial^{k_p}(j\omega_p)}\Bigg|_{\omega_1=\omega_2=\cdots=\omega_p=0}. \tag{6.3}$$

Example 6.1 Consider the exponential probability density function given by

$$f_x(x) = \begin{cases} \lambda e^{-\lambda x}; & x \geq 0, \lambda > 0 \\ 0; & \text{otherwise.} \end{cases}$$

The characteristic function $\phi(\omega_1)$ for a random variable x with this probability density function is given by

$$\phi(\omega_1) = \frac{\lambda}{\lambda - j\omega_1}.$$

The kth-order moment of x is given by

$$m_k = \frac{\partial^k}{\partial^k(j\omega_1)}\Phi(\omega_1)\Bigg|_{\omega_1=0}$$

$$= \frac{\lambda k!}{(\lambda - j\omega_1)^{k+1}}\Bigg|_{\omega_1=0} = \frac{k!}{\lambda^k}.$$

The kth-order cumulant of x can be evaluated by finding the kth derivative of $\ln[\Phi(\omega_1)]$ with respect to $j\omega_1$ and evaluating the results when $\omega_1 = 0$. This operation results in

$$c_k = \frac{\partial^k}{\partial(j\omega_1)^k}\left\{\ln\left(\frac{\lambda}{\lambda - j\omega_1}\right)\right\}\Bigg|_{\omega_1=0}$$

$$= \frac{(k-1)!}{(\lambda - j\omega_1)^k}\Bigg|_{\omega_1=0} = \frac{(k-1)!}{\lambda^k}.$$

6.1.1 Cumulants of Gaussian Random Variables

Consider a Gaussian random variable x with mean value μ and variance σ^2. It can be shown by direct evaluation that the characteristic function for this random variable is given by

$$\Phi(\omega_1) = e^{j\omega_1\mu - (\omega_1^2\sigma^2/2)}. \tag{6.4}$$

Since

$$\ln \Phi(\omega_1) = j\omega_1\mu - \frac{\omega_1^2\sigma^2}{2}, \tag{6.5}$$

it is straightforward to see that

$$\frac{\partial^r}{\partial^r(j\omega_1)} \ln \Phi(\omega_1) = 0; \qquad r > 2, \tag{6.6}$$

implying that all cumulants of order larger than two are zero for Gaussian random variables. Note that the higher-order moments of Gaussian random variables are not necessarily zero. The first and second-order cumulants of Gaussian variables can be evaluated from (6.5) as

$$c_1 = \text{cum}(x) = \frac{\partial}{\partial(j\omega_1)} \ln \Phi(\omega_1)\bigg|_{\omega_1=0} = \mu \tag{6.7}$$

and

$$c_2 = \text{cum}(x^2) = \frac{\partial^2}{\partial^2(j\omega_1)} \ln \Phi(\omega_1)\bigg|_{\omega_1=0} = \sigma^2. \tag{6.8}$$

6.1.1.1 *Test for Gaussianity and Nonlinearity*   The fact that cumulants of order larger than two of a Gaussian random variable are zero can be used to test whether a random variable is Gaussian. Since the output of a linear system is Gaussian whenever its input is Gaussian, it is also possible to devise a test if a system is nonlinear by finding the higher-order cumulants of the output of the system. If its input is Gaussian-distributed and if at least one cumulant of order larger than two is nonzero, it must be inferred that the system is nonlinear.

6.1.1.2 *Robust Processing in the Presence of Gaussian Noise*   The fact that the higher-order cumulants of a Gaussian process are zero can also be utilized in processing non-Gaussian signals corrupted by additive Gaussian noise using higher-order cumulant-based approaches. If the necessary statistics are exactly

known, they will have no contribution from the Guassian noise components. Consequently, we can design systems that are robust to additive Gaussian noise in their inputs using higher-order cumulants.

6.1.2 Relationship between Cumulants and Moments

The general relationship between joint moments and cumulants of order $r = p$ of a set of p real random variables $x_1, x_2, \ldots, x_p$ is given by [29,142,260]

$$\text{cum}(x_1, x_2, \ldots, x_p) = \sum_{l=1}^{p} (-1)^{l-1}(l-1)! \sum E\left\{ \prod_{i \in s_1} x_i \right\} E\left\{ \prod_{i \in s_2} x_i \right\} \cdots E\left\{ \prod_{i \in s_l} x_i \right\},$$

(6.9)

where the inner summation is over all partitions $s_1, s_2, \ldots, s_l$ of the indices such that they are mutually exclusive and their union is the set $\{1, 2, \ldots, p\}$. Therefore, the computation of the joint cumulants of order r requires the knowledge of all the joint moments up to order r.

A similar formula exists for the conversion from cumulants to moments also. To develop this formula, let $S_{l,j}$ denote a partition of $\{x_1, x_2, \ldots, x_p\}$ containing l mutually exclusive subsets such that

$$\bigcup_{j=1}^{l} S_{l,j} = \{x_1, x_2, \ldots, x_p\}.$$

(6.10)

The parameter l can take values in the range $\{1, 2, \ldots, p\}$. The relationship between the moments and cumulants are then given by

$$\text{mom}(x_1, x_2, \ldots, x_p) = \sum_{l=1}^{p} \sum_{j} \text{cum}(x_i \in S_{l,j}),$$

(6.11)

where the notation $\text{cum}(x_i \in S_{l,j})$ denotes the joint cumulant of the variables that belong to $S_{l,j}$.

Example 6.2 We wish to find the relationship between the fourth-order moment $\text{mom}(x_1, x_2, x_3, x_4)$ and the fourth-order cumulant $\text{cum}(x_1, x_2, x_3, x_4)$ of the four random variables x_1, x_2, x_3, and x_4 using (6.9). In order to accomplish this, we must first determine the number of ways in which we can partition x_1, x_2, x_3, and x_4 into 1,

2, 3, and 4 mutually exclusive sets. These partitions are tabulated below:

$$l = 1: \quad \{x_1, x_2, x_3, x_4\}; \qquad (-1)^{l-1}(l-1)! = 1$$

$$l = 2: \quad \{x_1\}, \{x_2, x_3, x_4\}; \qquad (-1)^{l-1}(l-1)! = -1$$
$$\{x_2\}, \{x_1, x_3, x_4\}$$
$$\{x_3\}, \{x_1, x_2, x_4\}$$
$$\{x_4\}, \{x_1, x_2, x_3\}$$
$$\{x_1, x_2\}, \{x_3, x_4\}$$
$$\{x_1, x_3\}, \{x_2, x_4\}$$
$$\{x_1, x_4\}, \{x_2, x_3\}$$

$$l = 3: \quad \{x_1\}, \{x_2\}, \{x_3, x_4\}; \qquad (-1)^{l-1}(l-1)! = 2$$
$$\{x_1\}, \{x_3\}, \{x_2, x_4\}$$
$$\{x_1\}, \{x_4\}, \{x_2, x_3\}$$
$$\{x_2\}, \{x_3\}, \{x_1, x_4\}$$
$$\{x_2\}, \{x_4\}, \{x_1, x_3\}$$
$$\{x_3\}, \{x_4\}, \{x_1, x_2\}$$

$$l = 4: \quad \{x_1\}, \{x_2\}, \{x_3\}, \{x_4\}; \qquad (-1)^{l-1}(l-1)! = -6$$

Now, we can express the fourth-order cumulant as follows:

$$\begin{aligned}
\mathrm{cum}(x_1, x_2, x_3, x_4) = {}& E\{x_1 x_2 x_3 x_4\} - E\{x_1\}E\{x_2 x_3 x_4\} \\
& - E\{x_2\}E\{x_1 x_3 x_4\} - E\{x_3\}E\{x_1 x_2 x_4\} \\
& - E\{x_4\}E\{x_1 x_2 x_3\} - E\{x_1 x_2\}E\{x_3 x_4\} \\
& - E\{x_1 x_3\}E\{x_2 x_4\} - E\{x_1 x_4\}E\{x_2 x_3\} \\
& + 2E\{x_1\}E\{x_2\}E\{x_3 x_4\} + 2E\{x_1\}E\{x_3\}E\{x_2 x_4\} \\
& + 2E\{x_1\}E\{x_4\}E\{x_2 x_3\} + 2E\{x_2\}E\{x_3\}E\{x_1 x_4\} \\
& + 2E\{x_2\}E\{x_4\}E\{x_1 x_3\} + 2E\{x_3\}E\{x_4\}E\{x_1 x_2\} \\
& - 6E\{x_1\}E\{x_2\}E\{x_3\}E\{x_4\}.
\end{aligned}$$

If the random variables have zero mean values, this relationship simplifies to

$$\begin{aligned}
\mathrm{cum}(x_1, x_2, x_3, x_4) = {}& E\{x_1 x_2 x_3 x_4\} - E\{x_1 x_2\}E\{x_3 x_4\} \\
& - E\{x_1 x_3\}E\{x_2 x_4\} - E\{x_1 x_4\}E\{x_2 x_3\}.
\end{aligned}$$

Derivation of an expression relating the fourth-order moment to the cumulants of order up to four is left as an exercise for the reader.

Example 6.3   In this example, we derive the relationships between the first four cumulants and the first four moments of a single random variable x. For this, we first express the cumulant generating function $\tilde{\Phi}(\omega) = \ln \Phi(\omega)$ using the Taylor series expansion. Using the definition of cumulants in (6.3), we can write

$$\tilde{\Phi}(\omega) = c_1(j\omega) + \frac{c_2}{2}(j\omega)^2 + \cdots + \frac{c_k}{k!}(j\omega)^k + \cdots.$$

The reader should verify that the constant term in this expansion is indeed zero. We find the first four derivatives of $\Phi(\omega) = e^{\tilde{\Phi}(\omega)}$ with respect to $j\omega$ and then evaluate the results at $\omega = 0$ to find the first four moments of x. The results are as follows:

$$m_1 = \left. \frac{\partial}{\partial(j\omega)} \Phi(\omega) \right|_{\omega=0}$$

$$= \left. \left\{ \frac{\partial}{\partial(j\omega)} \tilde{\Phi}(\omega) \right\} e^{\tilde{\Phi}(\omega)} \right|_{\omega=0} = c_1,$$

$$m_2 = \left. \frac{\partial^2}{\partial(j\omega)^2} \Phi(\omega) \right|_{\omega=0}$$

$$= \left. \left\{ \frac{\partial^2}{\partial(j\omega)^2} \tilde{\Phi}(\omega) + \left(\frac{\partial}{\partial(j\omega)} \tilde{\Phi}(\omega) \right)^2 \right\} e^{\tilde{\Phi}(\omega)} \right|_{\omega=0},$$

$$= c_2 + c_1^2,$$

$$m_3 = \left. \frac{\partial^3}{\partial(j\omega)^3} \Phi(\omega) \right|_{\omega=0}$$

$$= \left. \left\{ \frac{\partial^3}{\partial(j\omega)^3} \tilde{\Phi}(\omega) + 3 \left(\frac{\partial^2}{\partial(j\omega)^2} \tilde{\Phi}(\omega) \right) \left(\frac{\partial}{\partial(j\omega)} \tilde{\Phi}(\omega) \right) + \left(\frac{\partial}{\partial(j\omega)} \tilde{\Phi}(\omega) \right)^3 \right\} e^{\tilde{\Phi}(\omega)} \right|_{\omega=0},$$

$$= c_3 + 3c_1 c_2 + c_1^3$$

and

$$m_4 = \left. \frac{\partial^4}{\partial(j\omega)^4} \Phi(\omega) \right|_{\omega=0}$$

$$= \left\{ \frac{\partial^4}{\partial(j\omega)^4} \tilde{\Phi}(\omega) + 4 \left(\frac{\partial^3}{\partial(j\omega)^3} \tilde{\Phi}(\omega) \right) \left(\frac{\partial}{\partial(j\omega)} \tilde{\Phi}(\omega) \right) + 3 \left(\frac{\partial^2}{\partial(j\omega)^2} \tilde{\Phi}(\omega) \right)^2 \right.$$

$$\left. + 6 \left(\frac{\partial}{\partial(j\omega)} \tilde{\Phi}(\omega) \right)^2 \left(\frac{\partial^2}{\partial(j\omega)^2} \tilde{\Phi}(\omega) \right) + \left(\frac{\partial}{\partial(j\omega)} \tilde{\Phi}(\omega) \right)^4 \right\} e^{\tilde{\Phi}(\omega)} \Bigg|_{\omega=0},$$

$$= c_4 + 4c_3 c_1 + 3c_2^2 + 6c_2 c_1^2 + c_1^4.$$

We can solve for c_1, c_2, c_3 and c_4 from the above equations to obtain the following results:

$$c_1 = m_1$$

$$c_2 = m_2 - m_1^2$$

$$c_3 = m_3 - 3m_2 m_1 + 2m_1^3$$

$$c_4 = m_4 - 4m_3 m_1 - 3m_2^2 + 12m_2 m_1^2 - 6m_1^4$$

6.1.3 Additional Properties of Cumulants

We briefly summarize some additional properties of moments and cumulants below [29,175,197,260]. These properties can be easily derived from the definitions of moments and cumulants.

1. Given a set of random variables $x_1, x_2, \ldots, x_p$ and a set of constants $a_1, a_2, \ldots, a_p$,

$$\text{mom}(a_1 x_1, a_2 x_2, \ldots, a_p x_p) = a_1 a_2 \cdots a_p \text{mom}(x_1, x_2, \ldots, x_p) \tag{6.12}$$

and

$$\text{cum}(a_1 x_1, a_2 x_2, \ldots, a_p x_p) = a_1 a_2 \cdots a_p \text{cum}(x_1, x_2, \ldots, x_p). \tag{6.13}$$

2. Moments and cumulants are symmetric functions of their arguments; that is they remain unchanged for any permutation of the variables $x_1, x_2, \ldots, x_p$.
3. If a subset of the p random variables $x_1, x_2, \ldots, x_p$ is independent of the remaining ones, the cumulant of order p is identically equal to zero, i.e.,

$$\text{cum}(x_1, x_2, \ldots, x_p) = 0. \tag{6.14}$$

In general

$$\text{mom}(x_1, x_2, \ldots, x_p) \neq 0. \tag{6.15}$$

4. Let $x_1, x_2, \ldots, x_p$ and $y_1, y_2, \ldots, y_p$ be independent sets of random variables. Then

$$\begin{aligned} \text{cum}(x_1 + y_1, x_2 + y_2, \ldots, x_p + y_p) \\ = \text{cum}(x_1, x_2, \ldots, x_p) + \text{cum}(y_1, y_2, \ldots, y_p) \end{aligned} \tag{6.16}$$

Such a property does not hold in general for the moments.

5. For a set of random variables $x_1, x_2, \ldots, x_p$, and another arbitrary random variable y_1,

$$\text{cum}(x_1 + y_1, x_2, \ldots, x_p) = \text{cum}(x_1, x_2, \ldots, x_p) + \text{cum}(y_1, x_2, \ldots, x_p). \quad (6.17)$$

As a consequence, $\text{cum}(x_1 + y_1, x_2 + y_2, \ldots, x_p + y_p)$ can be expressed as the sum of 2^p terms for two dependent sets of random variables $x_1, x_2, \ldots, x_p$ and $y_1, y_2, \ldots, y_p$.

6. The cumulants are blind to an additive constant, i.e.,

$$\text{cum}(\alpha + x_1, x_2, \ldots, x_p) = \text{cum}(x_1, x_2, \ldots, x_p), \quad (6.18)$$

where α is an arbitrary constant.

6.1.4 Higher-Order Statistics of Discrete-Time Stationary Processes

Let $\{x(n)\}$ be a real and stationary, discrete-time random signal. Then, the pth-order moments and cumulants are defined as $(p - 1)$-dimensional functions that depend only on the time differences $l_i, i = 1, 2, \ldots, p - 1$ as

$$m_p^x(l_1, l_2, \ldots, l_{p-1}) = \text{mom}(x(n), x(n - l_1), \ldots, x(n - l_{p-1}))$$
$$\equiv E\{x(n), x(n - l_1), \ldots, x(n - l_{p-1}))\} \quad (6.19)$$

and

$$c_p^x(l_1, l_2, \ldots, l_{p-1}) \equiv \text{cum}(x(n), x(n - l_1), \ldots, x(n - l_{p-1})). \quad (6.20)$$

Example 6.4   Some familiar examples of cumulants of stationary random processes and their relationship to the corresponding moments include the mean value defined as

$$c_1^x = m_1^x = E\{x(n)\} \quad (6.21)$$

and the covariance function given by

$$c_2^x(l_1) = m_2^x(l_1) - (m_1^x)^2,$$

where $m_2^x(l_1)$ is the autocorrelation function. If the discrete-time random signal $x(n)$ is a zero-mean process,

$$c_2^x(l_1) = m_2^x(l_1).$$

For zero-mean processes, we can derive the following simplified relationships between the third- and fourth-order cumulants and the moments of order up to four using (6.9):

$$c_3^x(l_1, l_2) = m_3^x(l_1, l_2)$$

and

$$c_4^x(l_1, l_2, l_3) = m_4^x(l_1, l_2, l_3) - m_2^x(l_1)m_2^x(l_3 - l_2)$$
$$- m_2^x(l_2)m_2^x(l_3 - l_1) - m_2^x(l_3)m_2^x(l_2 - l_1).$$

While the third-order moments and cumulants are identical for a zero-mean random signal, the knowledge of the second- and fourth-order moments are required to compute the fourth-order cumulant.

6.1.4.1 *Variance, Skewness, and Kurtosis* The *variance* of a stationary random process is the second-order cumulant evaluated at lag $l_1 = 0$:

$$\text{Variance} = \sigma_x^2 = c_2^x(0) = E\{x^2(n)\} - [E\{x(n)\}]^2. \tag{6.22}$$

The *skewness* and *kurtosis* [60] parameters are used to further characterize the statistical distribution of the signal samples using the third and fourth-order cumulants at zero lags as

$$\text{Skewness} = c_3^x(0, 0)$$
$$= E\{x^3(n)\} - 3E\{x^2(n)\}E\{x(n)\} + 2[E\{x(n)\}]^3 \tag{6.23}$$

and

$$\text{Kurtosis} = c_4^x(0, 0, 0)$$
$$= E\{x^4(n)\} - 4E\{x^3(n)\}E\{x(n)\} - 3[E\{x^2(n)\}]^2$$
$$+ 12E\{x^2(n)\}[E\{x(n)\}]^2 - 6[E\{x(n)\}]^4. \tag{6.24}$$

The normalized forms of skewness and kurtosis defined as $c_3^x(0, 0)/\sigma_x^3$ and $c_4^x(0, 0, 0)/\sigma_x^4$ are often referred to as the *coefficient of skewness* and the *coefficient of kurtosis*, respectively.

6.1.4.2 Symmetry of Cumulants   The estimation of cumulants and analyses involving cumulants can be simplified for real and stationary signals by making use of the symmetry properties of the cumulants of such signals. Since $\text{cum}(x(n), x(n - l_1), \ldots, x(n - l_{p-1}))$ does not change for any permutation of the samples, it is straightforward to see that

$$c_p^x(l_1, l_2, \ldots, l_{p-1}) = c_p^x(\pi(l_1, l_2, \ldots, l_{p-1})), \tag{6.25}$$

where $\pi(\cdots)$ represents a permutation of the variables within the parentheses. Since the signal $x(n)$ is stationary, we can show that

$$\begin{aligned} c_p^x(l_1, l_2, \ldots, l_{p-1}) &= \text{cum}(x(n), x(n - l_1), \ldots, x(n - l_{p-1})) \\ &= \text{cum}(x(n + l_1), x(n), \ldots, x(n - l_{p-1} + l_1)) \\ &= \text{cum}(x(n), x(n + l_1), \ldots, x(n - l_{p-1} + l_1)) \\ &= c_p^x(-l_1, l_2 - l_1, \ldots, l_{p-1} - l_1). \end{aligned} \tag{6.26}$$

This analysis can be repeated for each of the lag variables, implying p different symmetries for the pth-order cumulant. The $(p - 1)!$ symmetries provided by (6.25) can be applied to each of the symmetries presented above, resulting in $p!$ different symmetries for the pth-order cumulant of a real, stationary random signal.

Example 6.5   The following symmetries are valid for the third-order cumulants:

$$\begin{aligned} c_3^x(l_1, l_2) &= c_3^x(l_2, l_1) \\ &= c_3^x(-l_1, l_2 - l_1) = c_3^x(l_2 - l_1, -l_1) \\ &= c_3^x(-l_2, l_1 - l_2) = c_3^x(l_1 - l_2, -l_2) \end{aligned}$$

Figure 6.1 shows the regions of symmetry on the $l_1 - l_2$ plane.

6.1.4.3 Estimation of Cumulants   The definition of cumulants in (6.3) requires the knowledge of the cumulant generating function in order to evaluate the cumulants. This is not a practical means for computing the cumulants since the probability density functions or the cumulant generating functions are seldom known a priori. One practical way of estimating the pth-order cumulants of an ergodic random process is to compute the moments up to order p as time averages of a single realization, and then to apply the formula in (6.9) that relates the moments and cumulants. The pth-order moment of a stationary random process $x(n)$ can be estimated from M samples of a realization of the process as

$$\hat{m}_p^x(l_1, l_2, \ldots, l_{p-1}) = \frac{1}{M} \sum_{n=1}^{M} x(n)x(n - l_1) \cdots x(n - l_{p-1}). \tag{6.27}$$

6.1.4.4 Cross-Cumulants   Let $x_1(n), x_2(n), \ldots, x_p(n)$ be p real and jointly stationary discrete-time random signals. Their pth-order cross-cumulant is defined as

$$c_{x_1 x_2 \cdots x_p}(l_1, l_2, \ldots, l_{p-1}) = \text{cum}(x_1(n), x_2(n - l_1), \ldots, x_p(n - l_{p-1})). \tag{6.28}$$

Cross-cumulants play an important role in nonlinear system identification from input and output measurements.

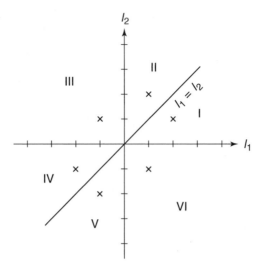

Figure 6.1 Regions of symmetry of the third-order cumulant. The values of the cumulant at the points marked are identical.

6.1.5 Cumulant Spectra

Let $x(n)$ be a real and stationary, discrete-time random signal. The pth-order cumulant spectrum of $x(n)$, denoted by $C_p^x(\omega_1, \omega_2, \ldots, \omega_{p-1})$, is defined as the $(p-1)$-dimensional Fourier transform of its pth-order cumulant $c_p^x(l_1, l_2, \ldots, l_{p-1})$ [29,260], i.e.,

$$
C_p^x(\omega_1, \omega_2, \ldots, \omega_{p-1}) = \sum_{l_1=-\infty}^{\infty} \sum_{l_2=-\infty}^{\infty} \cdots \sum_{l_{p-1}=-\infty}^{\infty} c_p^x(l_1, l_2, \ldots, l_{p-1})
$$
$$
\times e^{-j(\omega_1 l_1 + \omega_2 l_2 + \cdots + \omega_{p-1} l_{p-1})}. \tag{6.29}
$$

A sufficient condition for the existence of the pth-order cumulant spectrum is that $c_p^x(l_1, l_2, \ldots, l_{p-1})$ is absolutely summable, i.e.,

$$
\sum_{l_1=-\infty}^{\infty} \sum_{l_2=-\infty}^{\infty} \cdots \sum_{l_{p-1}=-\infty}^{\infty} |c_p^x(l_1, l_2, \ldots, l_{p-1})| < \infty. \tag{6.30}
$$

The implication of this condition is that the cumulant spectrum exists if the samples of the signals that are far apart are sufficiently independent of each other [29].

In general, $C_p^x(\omega_1, \omega_2, \ldots, \omega_{p-1})$ is a complex function. It is also periodic with period 2π in each dimension. Consequently,

$$
C_p^x(\omega_1, \omega_2, \ldots, \omega_{p-1}) = C_p^x(\omega_1 + 2\pi r_1, \omega_2 + 2\pi r_2, \ldots, \omega_{p-1} + 2\pi r_p) \tag{6.31}
$$

for arbitrary integer values $r_1, r_2, \ldots, r_p$.

Cross-cumulant spectra may also be defined in a similar manner. The cumulant spectra are also called *higher-order* spectra or *polyspectra*. Three special cases of the higher-order spectra are the power spectrum, the bispectrum and the trispectrum.

The Power Spectrum. The power spectrum of a stationary, discrete-time random signal is defined as

$$C_2^x(\omega_1) = \sum_{l_1=-\infty}^{\infty} c_2^x(l_1)e^{-j\omega_1 l_1}. \tag{6.32}$$

Since the covariance $c_2^x(l_1)$ of a real process is a real symmetric function, $C_2^x(\omega_1)$ is a real nonnegative symmetric function [218].

The Bispectrum. The bispectrum of a stationary, discrete-time random signal is defined as

$$C_3^x(\omega_1, \omega_2) = \sum_{l_1=-\infty}^{\infty} \sum_{l_2=-\infty}^{\infty} c_3^x(l_1, l_2)e^{-j(\omega_1 l_1 + \omega_2 l_2)}. \tag{6.33}$$

Using this definition and the symmetry properties of the third-order moments and cumulants, it can be shown that the bispectrum of a real process is in general a complex function with 12 symmetry regions [197]. The symmetries arise from the six sets of symmetries of the third-order cumulant of real processes, and the fact that the Fourier transform of real functions is a conjugate even function. As a consequence of the symmetries, the triangular region $\omega_2 \geq 0$, $\omega_1 \geq \omega_2$, $\omega_1 + \omega_2 \leq \pi$ completely specifies the bispectrum. The regions of symmetry of the bispectrum are shown in Figure 6.2.

The Trispectrum. The trispectrum of a discrete-time stationary random signal is defined as

$$C_4^x(\omega_1, \omega_2, \omega_3) = \sum_{l_1=-\infty}^{\infty} \sum_{l_2=-\infty}^{\infty} \sum_{l_3=-\infty}^{\infty} c_4^x(l_1, l_2, l_3)e^{-j(\omega_1 l_1 + \omega_2 l_2 + \omega_3 l_3)}. \tag{6.34}$$

Using a similar analysis as was done for the bispectrum, we can show that there are 48 regions of symmetry for the trispectrum [224][1]

6.1.5.1 *Estimation of Polyspectra*   Nikias and colleagues have described methods for the estimation of the higher-order spectra [196–198]. One straightforward method for estimating the cumulant spectra is to first estimate the cumulants using appropriate time averages, and to then compute the Fourier transform of the cumulant estimates.

---

[1] The fundamental period of the trispectrum is divided into 96 pyramidal regions in this work. However, knowledge of the trispectrum in two of the 96 regions are required for completely specifying the trispectrum.

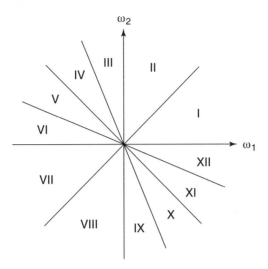

Figure 6.2 Regions of symmetry of the bispectrum.

6.2 IDENTIFICATION OF TRUNCATED VOLTERRA SYSTEMS USING CUMULANT SPECTRA

In this section, we derive analytical results that will enable us to develop formal methods of identifying truncated Volterra systems using cumulant spectra. These methods are primarily derived for Gaussian input signals since the cumulant spectra of order larger than two are zero for such signals, and therefore, the methods simplify considerably in such situations. All the methods described in this section make use of a useful result in the theory of cumulant spectra derived by Leonov and Shiryaev [142]. We state the result without proof next.

6.2.1 A Useful Result

Consider a two-dimensional array

$$
\begin{array}{cccc}
(1,1) & (1,2) & \cdots & (1,l_1) \\
(2,1) & (2,2) & \cdots & (2,l_2) \\
\vdots & \vdots & \vdots & \vdots \\
(p,1) & (p,2) & \cdots & (p,l_p)
\end{array}
\tag{6.35}
$$

A partition $S_1, S_2, \ldots, S_M$ of this array is said to be *indecomposable* [125,142] if there are no sets $S_{m_1}, S_{m_2}, \ldots, S_{m_N}$; $N < M$ and rows of the array $R_{i_1}, R_{i_2}, \ldots, R_{i_s}$; $s < p$ such that

$$
S_{m_1} \cup S_{m_2} \cup \cdots \cup S_{m_N} = R_{i_1} \cup R_{i_2} \cup \cdots \cup R_{i_s}.
$$

In this definition, $1 \leq m_1, m_2, \ldots, m_N \leq M$ and $1 \leq i_1, i_2, \ldots, i_s \leq p$.

Example 6.6   Consider the array

$$
\begin{array}{ccc}
(1, 1) & (1, 2) & (1, 3) \\
(2, 1) & (2, 2) & (2, 3) \\
(3, 1) & (3, 2) & (3, 3)
\end{array}
$$

Then

$$
\begin{aligned}
S_1 &= \{(1, 1), (1, 2), (1, 3)\} \\
S_2 &= \{(2, 1), (2, 2), (2, 3)\} \\
S_3 &= \{(3, 1), (3, 2), (3, 3)\}
\end{aligned}
$$

is not an indecomposable partition since each set corresponds to a separate row. On the other hand, the partition given by

$$
\begin{aligned}
S_1 &= \{(1, 1), (2, 1), (3, 1)\} \\
S_2 &= \{(1, 2), (2, 2), (3, 2)\} \\
S_3 &= \{(1, 3), (2, 3), (3, 3)\}
\end{aligned}
$$

is an indecomposable partition since no two sets alone contain the elements of one or two rows of the original array.

The following result was proved by Leonov and Shiryaev [142].

Theorem 6.1   Let $S = \{x_{i,j}; 1 \leq j \leq l_i, 1 \leq i \leq p\}$ denote a two-dimensional array of random variables. Let

$$
y_i = \prod_{j=1}^{l_i} x_{i,j}. \tag{6.36}
$$

Then

$$
\mathrm{cum}(y_1, y_2, \ldots, y_p) = \sum_r \mathrm{cum}(x_{i,j}; x_{i,j} \in S_{r,1}) \cdots \mathrm{cum}(x_{i,j}; x_{i,j} \in S_{r,M_r}), \tag{6.37}
$$

where the summation is over all possible indecomposable partitions of S such that $S_{r,1} \cup S_{r,2} \cup \cdots \cup S_{r,M_r} = S$.

Example 6.7   Consider a homogeneous quadratic system with input–output relationship

$$
y(n) = \sum_{m_1=0}^{N-1} \sum_{m_2=0}^{N-1} h_2(m_1, m_2) x(n - m_1) x(n - m_2).
$$

We wish to evaluate the joint cumulant $\text{cum}(y(n), x(n - l_1), x(n - l_2))$ and the corresponding joint cross-cumulant spectrum $C_{y,x,x}(\omega_1, \omega_2)$ when the input signal $x(n)$ is a Gaussian process.

The analysis can be considerably simplified by using some of the properties of cumulants described in Section 6.1.3. Using (6.17), we see that

$$\text{cum}(y(n), x(n - l_1), x(n - l_2))$$

$$= \text{cum}\left(\sum_{m_1=0}^{N-1} \sum_{m_2=0}^{N-1} h_2(m_1, m_2) x(n - m_1) x(n - m_2), x(n - l_1), x(n - l_2)\right)$$

$$= \sum_{m_1=0}^{N-1} \sum_{m_2=0}^{N-1} h_2(m_1, m_2) \text{cum}(x(n - m_1) x(n - m_2), x(n - l_1), x(n - l_2)).$$

Now we can use Theorem 6.1 to evaluate the cross-cumulant functions in this expression. Let us define the array S as

$$S = \begin{pmatrix} x(n - m_1) & x(n - m_2) \\ x(n - l_1) & \\ x(n - l_2) & \end{pmatrix}.$$

The indecomposable partitions of this array are

(1) $\{x(n - m_1), \ x(n - m_2), \ x(n - l_1), \ x(n - l_2)\}$

(2) $\{x(n - m_1), \ x(n - l_1), \ x(n - l_2)\} \ \{x(n - m_2)\}$

(3) $\{x(n - m_2), \ x(n - l_1), \ x(n - l_2)\} \ \{x(n - m_1)\}$

(4) $\{x(n - m_1), \ x(n - l_1)\} \ \{x(n - m_2), \ x(n - l_2)\}$

(5) $\{x(n - m_1), \ x(n - l_2)\} \ \{x(n - m_2), \ x(n - l_1)\}$

Since the third- and fourth-order cumulants of Gaussian processes are zero, we can express $\text{cum}(x(n - m_1) x(n - m_2), x(n - l_1), x(n - l_2))$ using Theorem 6.1 as

$$\text{cum}(x(n - m_1) x(n - m_2), x(n - l_1), x(n - l_2))$$
$$= c_2^x(l_1 - m_1) c_2^x(l_2 - m_2) + c_2^x(l_1 - m_2) c_2^x(l_2 - m_1).$$

Substituting this result in the expression for $\text{cum}(y(n), x(n - l_1), x(n - l_2))$, we get

$$\text{cum}(y(n), x(n - l_1), x(n - l_2)) =$$

$$\sum_{m_1=0}^{N-1} \sum_{m_2=0}^{N-1} h_2(m_1, m_2) \{c_2^x(l_1 - m_1) c_2^x(l_2 - m_2) + c_2^x(l_1 - m_2) c_2^x(l_2 - m_1)\}.$$

We can now evaluate the cross-cumulant spectrum $C_{yxx}(\omega_1, \omega_2)$ as

$$C_{yxx}(\omega_1, \omega_2) = \sum_{l_1=-\infty}^{\infty} \sum_{l_2=-\infty}^{\infty} \sum_{m_1=0}^{N-1} \sum_{m_2=0}^{N-1} h_2(m_1, m_2)$$

$$\times \{c_2^x(l_1 - m_1)c_2^x(l_2 - m_2) + c_2^x(l_1 - m_2)c_2^x(l_2 - m_1)\}e^{-j(\omega_1 l_1 + \omega_2 l_2)}$$

$$= 2H_2(\omega_1, \omega_2)C_2^x(\omega_1)C_2^x(\omega_2).$$

The two-dimensional transfer function $H_2(\omega_1, \omega_2)$ can be evaluated from knowledge of the power spectrum and the cross-cumulant spectrum as

$$H_2(\omega_1, \omega_2) = \frac{C_{yxx}(\omega_1, \omega_2)}{2C_2^x(\omega_1)C_2^x(\omega_2)}.$$

It is important to realize that this expression is valid only when the input signal is Gaussian-distributed.

6.2.2 Identification of Homogeneous Volterra Systems

The results of Example 6.7 can be extended to include all homogeneous Volterra systems [29]. We can show that the joint $(p + 1)$th-order cumulant spectrum $C_{yx\cdots x}(\omega_1, \omega_2, \ldots, \omega_p)$ of the output and input signals of a homogeneous pth order Volterra system is related to the power spectrum of its input signal as

$$H_p(\omega_1, \omega_2, \ldots, \omega_p) = \frac{C_{yx\cdots x}(\omega_1, \omega_2, \ldots, \omega_p)}{p!C_2^x(\omega_1)C_2^x(\omega_2)\ldots C_2^x(\omega_p)} \tag{6.38}$$

when the input signal is Gaussian-distributed.

Remark 6.1 Equation (6.38) often appears in the literature as

$$H_p(\omega_1, \omega_2, \ldots, \omega_p) = \frac{C_{yx\cdots x}(-\omega_1, -\omega_2, \ldots, -\omega_p)}{p!C_2^x(\omega_1)C_2^x(\omega_2)\cdots C_2^x(\omega_p)}. \tag{6.39}$$

The difference arises from the different definitions used for the cross-cumulants. In this book, we have used

$$c_{yx\cdots x}(l_1, l_2, \ldots, l_p) = \text{cum}(y(n), x(n - l_1), \ldots, x(n - l_p)), \tag{6.40}$$

whereas

$$c_{yx\cdots x}(l_1, l_2, \ldots, l_p) = \text{cum}(y(n), x(n + l_1), \ldots, x(n + l_p)), \tag{6.41}$$

is also used in the literature. The second definition results in (6.39), while our definition results in (6.38). Since the power spectrum of a real signal is an even function of the frequency, both definitions result in identical values of $C_2^x(\omega_1)$.

The derivation of (6.38) is similar to that worked out in Example 6.7. We first define an array S as

$$
S = \begin{pmatrix} x(n-m_1) & x(n-m_2) & \cdots & x(n-m_p) \\ x(n-l_1) & & & \\ & \vdots & & \\ x(n-l_p) & & & \end{pmatrix}. \tag{6.42}
$$

Since the cumulants of all orders other than two are zero for zero-mean Gaussian signals, we only need to consider indecomposable partitions of S that takes two samples per set. Furthermore, since pairing of two samples from the second through $(p+1)$th rows will cause two rows of the arrays to be completely specified by a subset of the partition, and thus cause the partition to be not indecomposable, we require the pairs to be formed by one sample from the first row, and another sample from one of the other rows. There are $p!$ ways of forming such pairs. According to the results of Theorem 6.1, $\mathrm{cum}(y(n)x(n-l_1)x(n-l_2)\ldots x(n-l_p))$ contains $p!$ terms of the form

$$
\sum_{m_1=0}^{N-1} \sum_{m_2=0}^{N-1} \cdots \sum_{m_p=0}^{N-1} h_p(m_1, m_2, \ldots, m_p) c_2^x(l_1 - m_1) c_2^x(l_2 - m_2) \cdots c_2^x(l_p - m_p). \tag{6.43}
$$

The $p!$ terms are formed by different combinations of the lag indices l_k and m_r in the expressions for the second-order cumulants. Such permutations do not affect the overall sum, and therefore

$$
\mathrm{cum}(y(n)x(n-l_1)x(n-l_2)\cdots x(n-l_p))
$$
$$
= p! \sum_{m_1=0}^{N-1} \sum_{m_2=0}^{N-1} \cdots \sum_{m_p=0}^{N-1} h_p(m_1, m_2, \ldots, m_p) c_2^x(l_1 - m_1) c_2^x(l_2 - m_2) \cdots c_2^x(l_p - m_p).
$$
$$
\tag{6.44}
$$

Taking the p-dimensional Fourier transform of both sides of this expression gives

$$
C_{yxx\cdots x}(\omega_1, \omega_2, \ldots, \omega_p) = p! H_p(\omega_1, \omega_2, \ldots, \omega_p) C_{xx}(\omega_1) C_{xx}(\omega_2) \cdots C_{xx}(\omega_p),
$$
$$
\tag{6.45}
$$

from which we can easily obtain (6.38).

6.2.3 Identification of Quadratic Systems

We now consider a more general quadratic system with input–output relationship

$$y(n) = \sum_{m_1=0}^{N-1} h_1(m_1)x(n - m_1)$$

$$+ \sum_{m_1=0}^{N-1} \sum_{m_2=0}^{N-1} h_2(m_1, m_2)x(n - m_1)x(n - m_2). \tag{6.46}$$

We assume that the input signal is Gaussian-distributed. Using an analysis similar to that used in Example 6.7, we can show that

$$\text{cum}(y(n), x(n - l_1)) = \sum_{m_1=0}^{N-1} h_1(m_1)c_2^x(l_1 - m_1) \tag{6.47}$$

and

$$\text{cum}(y(n), x(n - l_1), x(n - l_2)) = \sum_{m_1=0}^{N-1} \sum_{m_2=0}^{N-1} h_2(m_1, m_2)\{c_2^x(l_1 - m_1)c_2^x(l_2 - m_2)$$

$$+ c_2^x(l_1 - m_2)c_2^x(l_2 - m_1)\}. \tag{6.48}$$

Taking the Fourier transform of both sides of the above expressions, we get

$$C_{yx}(\omega_1) = H_1(\omega_1)C_2^x(\omega_1) \tag{6.49}$$

and

$$C_{yxx}(\omega_1, \omega_2) = 2H_2(\omega_1, \omega_2)C_2^x(\omega_1)C_2^x(\omega_2). \tag{6.50}$$

This analysis shows that the contributions of the homogeneous quadratic and linear systems can be decoupled when the input signal is Gaussian-distributed. Consequently, the identification of the linear and homogeneous quadratic kernels can be performed independent of each other. The solution to the identification problem is

$$H_1(\omega_1) = \frac{C_{yx}(\omega_1)}{C_2^x(\omega_1)} \tag{6.51}$$

and

$$H_2(\omega_1, \omega_2) = \frac{C_{yxx}(\omega_1, \omega_2)}{2C_2^x(\omega_1)C_2^x(\omega_2)}. \tag{6.52}$$

Remark 6.2 Decoupling of linear and homogeneous quadratic components seldom occurs when the input signal is not Gaussian. Such a decoupling also does not occur for higher-order Volterra systems. Equations (6.51) and (6.52) were first derived by Tick [313].

Example 6.8 The system model used in this example is the same as the one presented by Kim and Powers [112]. We consider the identification of a system with input–output relationship given by

$$y(n) = -0.64x(n) + x(n-2) + 0.9x^2(n) + x^2(n-1),$$

where the first two terms represent the linear component and the last two terms represent the homogeneous quadratic response. Consequently, the linear and quadratic transfer functions are given by

$$H_1(e^{j\omega_1}) = -0.64 + e^{-j2\omega_1}$$

and

$$H_2(e^{j\omega_1}, e^{j\omega_2}) = 0.9 + e^{-j(\omega_1+\omega_2)}.$$

The actual transfer functions of the system are plotted in Figures 6.3 and 6.4.

The input signal was obtained by processing a white Gaussian signal $\xi(n)$ with zero mean value and unit variance through a linear system with input–output relationship given by

$$x(n) = 0.6x(n-1) + 0.8\xi(n).$$

The output signal was obtained by processing the input signal $x(n)$ with the nonlinear system and then corrupting its output with a zero-mean and white Gaussian noise sequence with variance 0.001. The noise sequence was independent of the input signals. In total, 50 independent experiments using 2048 samples each were conducted and the results presented were obtained by averaging over the 50 experiments. The cumulant spectra were estimated by first estimating the cumulants for lag values ranging from -32 to 31 and then taking the Fourier transform of the estimates. The power spectrum was computed using 64-point DFTs, and the cross-cumulant spectrum was computed using 64×64-point two-dimensional DFTs. The linear and quadratic components of the unknown system were then estimated using (6.51) and (6.52), respectively.

Figure 6.5 shows the average estimates of the amplitude and phase function of the linear component obtained using the method presented in this section. The corresponding results for the quadratic components are given in Figure 6.6. Comparing the average estimates with the true transfer functions, we can see that the estimation technique works well in this example.

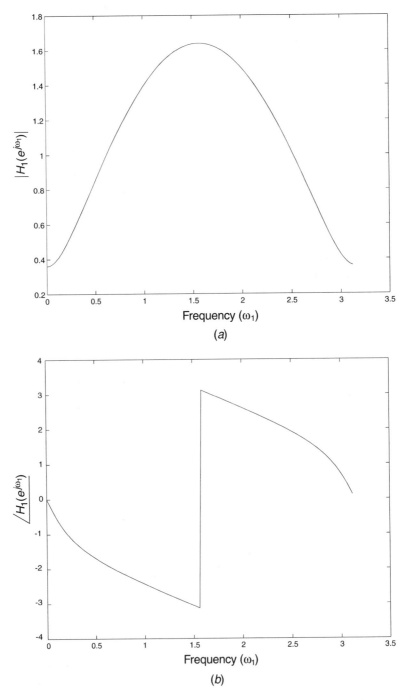

Figure 6.3 Amplitude and phase response of the linear component of the system in Example 6.8: (a) amplitude response; (b) phase response.

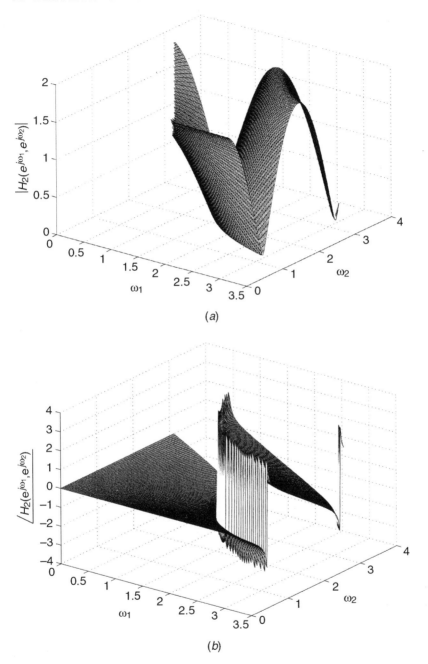

Figure 6.4 Amplitude and phase response of the quadratic component of the system in Example 6.8: (a) amplitude response; (b) phase response.

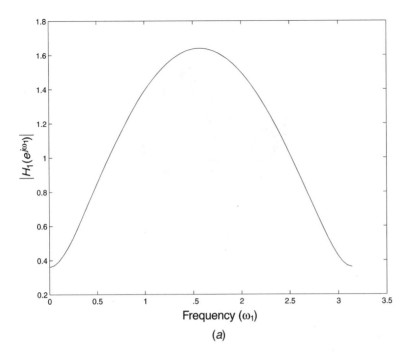

(a)

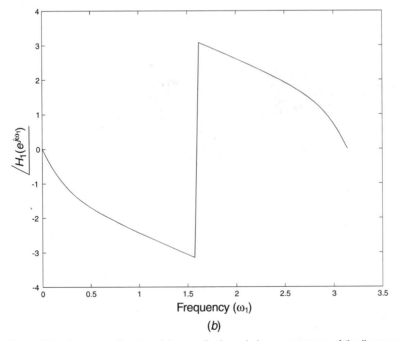

(b)

Figure 6.5 Average estimates of the amplitude and phase responses of the linear component of the system in Example 6.8: (a) amplitude response; (b) phase response.

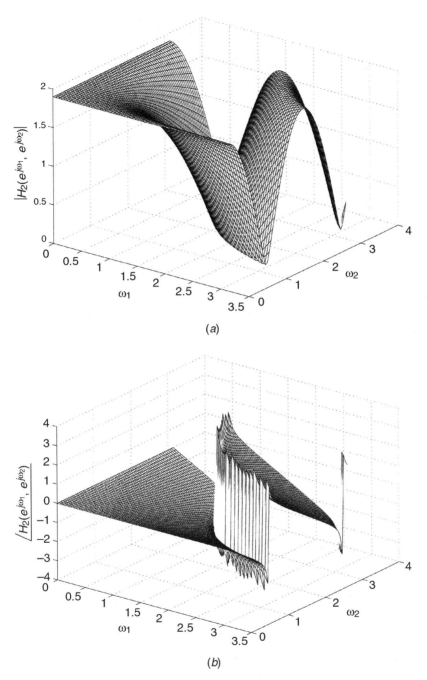

Figure 6.6 Average estimates of the amplitude and phase responses of the quadratic component of the system in Example 6.8: (a) amplitude response; (b) phase response.

6.2.4   Identification of the Kernels of a *p*th-Order Volterra System

Koukoulas and Kalouptsidis [125] extended the results of the previous sections to inhomogeneous truncated Volterra systems of order larger than two. Their analysis is based on the derivation of an expression for the cross-cumulant of the output sequence of $y(n)$ with several samples of the input signal. The result is described in the following theorem.

Theorem 6.2   Let

$$y(n) = h_0 + \sum_{m_1=0}^{N-1} h_1(m_1)x(n-m_1) + \sum_{m_1=0}^{N-1}\sum_{m_2=0}^{N-1} h_2(m_1,m_2)x(n-m_1)x(n-m_2)$$

$$+ \cdots + \sum_{m_1=0}^{N-1}\sum_{m_2=0}^{N-1} \cdots \sum_{m_p=0}^{N-1} h_p(m_1,m_2,\ldots,m_p)x(n-m_1)x(n-m_2)\cdots x(n-m_p)$$

$$+ \eta(n), \tag{6.53}$$

where $x(n)$ is a zero-mean, Gaussian random process and $\eta(n)$ is a noise process that is independent of $x(n)$. Then, the $(K+1)$th-order cross-cumulant of $y(n)$ with K samples of the input signal

$$\text{cum}(y(n), x(n-l_1), \ldots, x(n-l_K)) = 0 \tag{6.54}$$

whenever $K > p$. For $K \leq p$,

$$\text{cum}(y(n), x(n-l_1), \ldots, x(n-l_K))$$

$$= K! \sum_{m_1=0}^{N-1}\sum_{m_2=0}^{N-1} \cdots \sum_{m_K=0}^{N-1} h_K(m_1, m_2, \ldots, m_K)c_2^x(l_1-m_1)$$

$$\times c_2^x(l_2-m_2)\cdots c_2^x(l_K-m_K)$$

$$+ \sum_{v=1}^{\lfloor\frac{p-K}{2}\rfloor} \frac{(K+2v)!}{v!2^v} \sum_{m_1=0}^{N-1}\sum_{m_2=0}^{N-1} \cdots \sum_{m_{K+2v}=0}^{N-1} h_{K+2v}(m_1, m_2, \ldots, m_{K+2v})$$

$$\times c_2^x(l_1-m_1)c_2^x(l_2-m_2)\cdots c_2^x(l_K-m_K)$$

$$\times c_2^x(m_{K+2}-m_{K+1})c_2^x(m_{K+4}-m_{K+3})\cdots c_2^x(m_{K+2v}-m_{K+2v-1}), \tag{6.55}$$

where $\lfloor(\cdot)\rfloor$ denotes the integer part of $(\cdot)$.

A proof for this theorem is provided later.

6.2.4.1   *Frequency-Domain Result*   By taking the K-dimensional Fourier transform of both sides of (6.55) and solving for $H_K(\omega_1, \omega_2, \ldots, \omega_K)$, we get the

following result:

$$H_K(\omega_1, \omega_2, \ldots, \omega_K) = \frac{C_{yx\cdots x}(\omega_1, \omega_2, \ldots, \omega_K)}{K! C_2^x(\omega_1) C_2^x(\omega_2) \cdots C_2^x(\omega_K)}$$

$$- \sum_{v=1}^{\lfloor \frac{p-K}{2} \rfloor} \frac{(K + 2v)!}{v! 2^v K!} \left(\frac{1}{2\pi}\right)^v \int_{-\pi}^{\pi} \int_{-\pi}^{\pi} \cdots \int_{-\pi}^{\pi}$$

$$H_{K+2v}(\omega_1, \ldots, \omega_K, \omega_{K+1}, -\omega_{K+1}, \ldots, \omega_{K+v}, -\omega_{K+v})$$

$$\times C_2^x(\omega_{K+1}) C_2^x(\omega_{K+2}) \cdots C_2^x(\omega_{K+v})$$

$$d\omega_{K+1} d\omega_{K+2} \cdots d\omega_{K+v}. \tag{6.56}$$

From this result, it is clear that an effective way of identifying a pth-order Volterra system using Gaussian input signals is to first identify the pth-order kernel, then the $(p - 1)$th-order kernel and so on until we identify the zeroth-order kernel. This approach is described for a fourth-order truncated Volterra system identification problem in the following example.

Example 6.9   In this example, we show how we can apply Theorem 6.2 to identify the kernels of a fourth-order Volterra system. According to (6.56), we have

$$C_{yxxxx}(\omega_1, \omega_2, \omega_3, \omega_4) = 4! H_4(\omega_1, \omega_2, \omega_3, \omega_4) C_2^x(\omega_1) C_2^x(\omega_2) C_2^x(\omega_3) C_2^x(\omega_4).$$

Consequently, we can estimate the fourth order kernel as

$$H_4(\omega_1, \omega_2, \omega_3, \omega_4) = \frac{C_{yxxxx}(\omega_1, \omega_2, \omega_3, \omega_4)}{4! C_2^x(\omega_1) C_2^x(\omega_2) C_2^x(\omega_3) C_2^x(\omega_4)}.$$

Again, using (6.56) for the third-order kernel, we see that

$$C_{yxxx}(\omega_1, \omega_2, \omega_3) = 3! H_3(\omega_1, \omega_2, \omega_3) C_2^x(\omega_1) C_2^x(\omega_2) C_2^x(\omega_3),$$

implying that the third-order kernel can be estimated using

$$H_3(\omega_1, \omega_2, \omega_3) = \frac{C_{yxxx}(\omega_1, \omega_2, \omega_3)}{3! C_2^x(\omega_1) C_2^x(\omega_2) C_2^x(\omega_3)}.$$

A similar calculation for the estimation of the second-order kernel will show that the third-order cross-cumulant spectrum depends on the second- and fourth-order kernels. However, we have already identified the fourth-order kernel, and therefore, the contributions of the fourth-order kernel can be removed. Thus, we can estimate the second-order kernel as

$$H_2(\omega_1, \omega_2) = \frac{C_{yxx}(\omega_1, \omega_2)}{2!C_2^x(\omega_1)C_2^x(\omega_2)} - \frac{6}{2\pi}\int_{-\pi}^{\pi} H_4(\omega_1, \omega_2, \omega_3, -\omega_3)C_2^x(\omega_3)d\omega_3$$

The linear kernel can be estimated in a similar way as

$$H_1(\omega_1) = \frac{C_{yx}(\omega_1)}{C_2^x(\omega_1)} - \frac{3}{2\pi}\int_{-\pi}^{\pi} H_3(\omega_1, \omega_2, -\omega_2)C_2^x(\omega_2)d\omega_2.$$

Finally, the constant bias term h_0 may be evaluated using c_1^y, the mean value of $y(n)$, as

$$h_0 = c_1^y - \frac{1}{2\pi}\int_{-\pi}^{\pi} H_2(\omega_1, -\omega_1)C_2^x(\omega_1)d\omega_1$$

$$- 3\left(\frac{1}{2\pi}\right)^2 \int_{-\pi}^{\pi}\int_{-\pi}^{\pi} H_4(\omega_1, -\omega_1, \omega_2, -\omega_2)C_2^x(\omega_1)C_2^x(\omega_2)d\omega_1 d\omega_2.$$

Remark 6.3   One drawback of the method described above is that we must know the order of the nonlinear system before we can begin the process of identifying the kernel. Koukoulas and Kalouptsidis [125] have developed an iterative algorithm that begins with the estimation of h_0, then estimates the linear kernel and then the second-order kernel and so on, thus avoiding the drawback just mentioned. They have also shown that the pth-order system identification technique designed according to the discussion surrounding Theorem 6.2 is in fact the minimum error estimate of an arbitrary nonlinear system using the pth-order Volterra system model.

6.2.4.2 Proof of Theorem 6.2

To prove the theorem, we first note using the linearity property of cumulants that

$$\text{cum}(y(n), x(n - l_1), x(n - l_2), \ldots, x(n - l_K))$$

$$= \sum_{m_1=0}^{N-1} \sum_{m_2=0}^{N-1} \cdots \sum_{m_p=0}^{N-1} h_p(m_1, m_2, \ldots, m_p) \tag{6.57}$$

$$\times \text{cum}(x(n - m_1)x(n - m_2)\cdots x(n - m_p), x(n - l_1), x(n - l_2)\cdots x(n - l_K)).$$

Consequently, we consider the computation of the cumulant

$\text{cum}(x(n - m_1) \, x(n - m_2) \cdots x(n - m_p), x(n - l_1), x(n - l_2) \cdots x(n - l_K))$. We proceed as we did in Example 6.7 and Section 6.2.2 by defining an array S as

$$S = \begin{pmatrix} x(n - m_1) & x(n - m_2) & \cdots & x(n - m_i) \\ x(n - l_1) & & & \\ \vdots & & & \\ x(n - l_K) & & & \end{pmatrix}. \tag{6.58}$$

As stated in Section 6.2.2, the only indecomposable partitions that contribute nonzero values of the cumulants are those involving sets containing only pairs of the samples. Furthermore, such a set should not contain more than one element from rows 2 through $K + 1$. We consider several cases:

$K + i$ *is odd:* The partition of S must contain at least one set with only one element. Consequently, the cumulant in this case is zero.

$K > i$: After pairing each element in rows 2 through $i + 1$, we still have $K - i$ rows left. Therefore, every partition of S involving two samples must contain at least one set involving two of rows 2 through $K + 1$. Thus, we have no valid indecomposable partitions involving only pairs of samples from S, implying that the cumulant in this case also is zero.

$K = i$: This case was considered in Section 6.2.2. The cumulant is given by (6.44) with $p = K = i$.

$K < i$: If $K + i$ is odd, the cumulant is zero. If $K + i$ is even, let us denote $i - K$ by $2v$ since $i - K$ is an even number. After pairing the elements of the last K rows with an element each from the first row, we have $2v$ samples left in the first row. These elements must be partitioned into v sets of pairs of samples in every indecomposable partition of S that contributes nonzero value in the cumulant calculation. There are $\begin{pmatrix} i \\ K \end{pmatrix} = i!/(K!2v!)$ ways in which we can choose K elements from the top row of S. Having chosen the K elements, we have $K!$ ways in which to combine each element with the one element from the bottom K rows of S. Then we have $2v!/(v!2^v)$ different ways to pair the remaining $2v$ samples in the first row of S. Thus, we have a total of

$$\frac{i!}{K!2v!} K! \frac{2v!}{v!2^v} = \frac{i!}{v!2^v} \tag{6.59}$$

distinctly different indecomposable sets in S. Consequently,

$$\begin{aligned} \text{cum}(x(n - m_1)x(n - m_2) & \cdots x(n - m_i), x(n - l_1), x(n - l_2), \ldots, x(n - l_K)) \\ &= \sum c_2^x(l_1 - m_1)c_2^x(l_2 - m_2) \cdots c_2^x(l_K - m_K)c_2^x(m_{K+2} - m_{K+1}) \\ &\quad \cdots c_2^x(m_{K+2v} - m_{K+2v-1}), \end{aligned} \tag{6.60}$$

where the summation is over all possible $i!/(v!2^v)$ partitions as derived above.

Now, let us consider evaluating the cumulant of the output $y(n)$ of the pth-order truncated Volterra system with K samples of the input signals. Let $p - K = 2v$. Using the analysis above, we can show in a straightforward manner that

$$\text{cum}(y(n), x(n - l_1), x(n - l_2) \cdots x(n - l_K))$$

$$= \sum_{m_1=0}^{N-1} \sum_{m_2=0}^{N-1} \cdots \sum_{m_K=0}^{N-1} h_K(m_1, m_2, \ldots, m_K)$$

$$\times \text{cum}(x(n - m_1)x(n - m_2) \cdots x(n - m_K), x(n - l_1), x(n - l_2) \cdots x(n - l_K))$$

$$+ \sum_{m_1=0}^{N-1} \sum_{m_2=0}^{N-1} \cdots \sum_{m_{K+2}=0}^{N-1} h_{K+2}(m_1, m_2, \ldots, m_{K+2})$$

$$\times \text{cum}(x(n - m_1)x(n - m_2) \cdots x(n - m_{K+2}), x(n - l_1), x(n - l_2) \cdots x(n - l_K))$$

$$+ \cdots + \sum_{m_1=0}^{N-1} \sum_{m_2=0}^{N-1} \cdots \sum_{m_{K+2v}=0}^{N-1} h_{K+2v}(m_1, m_2, \ldots, m_{K+2v})$$

$$\times \text{cum}(x(n - m_1)x(n - m_2) \cdots x(n - m_{K+2v}), x(n - l_1), x(n - l_2) \cdots x(n - l_K)).$$

$$(6.61)$$

Substituting (6.60) in this equation results in (6.55). Computing the p-dimensional Fourier transform of both sides of (6.55) gives (6.56), completing the proof.

6.3 ESTIMATION OF QUADRATIC SYSTEMS USING ARBITRARY INPUT SIGNALS

We saw in the previous sections that the computational complexity of estimating the parameters of a Volterra system can be significantly reduced when the input signal is Gaussian-distributed. However, there are many applications in which the assumption that the input signal is Gaussian-distributed is not accurate. Examples of real-life applications in modeling drift oscillations of moored vessels in sea waves and electromagnetic scattering in which the input signals are non-Gaussian have been described in [112,192]. In this section, we describe a frequency-domain method for estimating the parameters of a quadratic system when the input signal is not Gaussian-distributed. This method was developed by Kim and Powers [112], and later extended to third-order nonlinearities in [192]. We will see that frequency-domain processing results in considerable reduction in the computational complexity of the estimation algorithm when compared with the time-domain techniques described in Chapter 5.

6.3.1 The System Model

We consider the identification of a quadratic nonlinear system with input–output relationship

$$y(n) = \sum_{m_1=0}^{N-1} h_1(m_1)x(n - m_1) + \sum_{m_1=0}^{N-1}\sum_{m_2=0}^{N-1} h_2(m_1, m_2)x(n - m_1)x(n - m_2). \quad (6.62)$$

For simplicity of presentation, we have assumed the symmetric realization of the system model. Note that there is no constant bias term corresponding to the zeroth-order kernel in the system model. This case is left as an exercise for the reader. Our objective is to develop a parameter estimation algorithm for this model from measurements of the input and output signals. The output signal is measured in the presence of additive zero-mean noise that is independent of the input signal, and is modeled as

$$d(n) = y(n) + \eta(n). \quad (6.63)$$

6.3.2 The Parameter Estimation Algorithm

Recall from (2.85) that we can express the M-point discrete Fourier transform of the measured output of a quadratic system as

$$D(k) = \vec{\eta}(k) + H_1(k)X(k)$$

$$+ \frac{1}{M}\sum_{k_1=0}^{M-1}\sum_{k_2=0}^{M-1} H_2(k_1, k_2)X(k_1)X(k_2)(u_0(k_1 + k_2 - k) + u_0(k_1 + k_2 - M - k)),$$

$$(6.64)$$

where $u_0(k)$ represents the Dirac delta function and $\vec{\eta}(k)$ is the DFT of the measurement noise sequence, and we have assumed that M is larger than or equal to N. Furthermore, since the DFTs are computed over a finite number of samples M, they exist as long as the time-domain signals are finite. We can also perform the expectation operations on the DFT coefficients whenever the higher-order expectations of the time-domain signals can be performed.

The key to parameter estimation in the DFT domain is the fact that the kth coefficient of the DFT of the output sequence depends on only a small subset of $\{H_2(k_1, k_2); 0 \le k_1, k_2 \le M - 1\}$ and $\{X(k_1)X(k_2); 0 \le k_1, k_2 \le M - 1\}$. Figure 2.19 and the discussion surrounding it explain this idea. For any k, (6.64) can be written as

$$D(k) = \vec{\eta}(k) + H_1(k)X(k) + \frac{1}{M}\sum_{k_1=0}^{k} H_2(k_1, k - k_1)X(k_1)X(k - k_1)$$

$$+ \frac{1}{M}\sum_{k_1=k+1}^{M-1} H_2(k_1, M + k - k_1)X(k_1)X(M + k - k_1). \quad (6.65)$$

Let $\mathbf{H}_{2,k}$ denote a vector defined as

$$\mathbf{H}_{2,k} = [H_1(k) \quad H_2(0, k) \quad H_2(1, k-1) \cdots H_2(k, 0)$$

$$H_2(k+1, M-1) \quad H_2(k+2, M-2) \cdots H_2(M-1, k+1)]^T \quad (6.66)$$

to indicate the DFT coefficients of first- and second-order Volterra kernels involved in the computation of $Y(k)$. Similarly, define the input DFT vector $\mathbf{X}_{2,k}$ as

$$\mathbf{X}_{2,k} = \frac{1}{M}[MX(k) \quad X(0)X(k) \quad X(1)X(k-1) \cdots X(k)X(0)$$

$$X(k+1)X(M-1) \quad X(k+2)X(M-2) \cdots X(M-1)X(k+1)]^T. \quad (6.67)$$

The output DFT $D(k)$ can be compactly written using these definitions as

$$D(k) = \vec{\eta}(k) + \mathbf{X}_{2,k}^T \mathbf{H}_{2,k}. \quad (6.68)$$

Let us premultiply both sides of this equation by $\mathbf{X}_{2,k}^*$ and take the statistical expectations. This operation results in

$$E\{\mathbf{X}_{2,k}^* D(k)\} = E\{\mathbf{X}_{2,k}^* \mathbf{X}_{2,k}^T\}\mathbf{H}_{2,k}. \quad (6.69)$$

The above derivation utilizes the independence of $\eta(n)$ from the input signal as well as the fact that $\eta(n)$ has zero mean value. Assuming that the autocorrelation matrix $E\{\mathbf{X}_{2,k}^* \mathbf{X}_{2,k}^T\}$ is invertible, we can solve for the coefficient vector $\mathbf{H}_{2,k}$ from (6.69), which gives

$$\mathbf{H}_{2,k} = (E\{\mathbf{X}_{2,k}^* \mathbf{X}_{2,k}^T\})^{-1} E\{\mathbf{X}_{2,k}^* D(k)\}. \quad (6.70)$$

By evaluating (6.70) for k in the range $0, 1, \ldots, (M/2) - 1$ for even values of M or in the range $0, 1, \ldots, (M-1)/2$ for odd values of M, we can solve for all the unknown DFT coefficients of the Volterra kernels. The main advantage of this method is that the matrix inversions involve $O(N) \times O(N)$-element matrices instead of the inversion of $O(N^2) \times O(N^2)$-element matrices required for the direct estimation procedure using time-domain techniques.

6.3.3 Computational Issues

We discuss two important issues here: (1) further reduction in the computational complexity and (2) estimation of the statistical expectations required in the calculations.

6.3.3.1 *Further Reduction in the Computational Complexity* Additional simplifications in the definition of the vectors $\mathbf{X}_{2,k}$ and $\mathbf{H}_{2,k}$ are possible

because of the symmetry of the quadratic kernel. For simplicity of discussion, let us assume that M is even. Since $H_2(k_1, k_2) = H_2(k_2, k_1)$ for symmetric quadratic kernels, we can find a reduced-order coefficient vector of the form

$$\mathbf{H}_{2,k,r} = \left[H_1(k) \quad H_2(0, k) \quad H_2(1, k-1) \cdots H_2\left(\frac{k}{2}, \frac{k}{2}\right) \right.$$

$$\left. H_2(k+1, M-1) \cdots H_2\left(\frac{M+k}{2}, \frac{M+k}{2}\right) \right]^T, \qquad (6.71)$$

with corresponding input DFT vector

$$\mathbf{X}_{2,k,r} = \frac{1}{M} \left[MX(k) \quad 2X(0)X(k) \quad 2X(1)X(k-1) \cdots \right.$$

$$2X\left(\frac{k}{2}-1\right)X\left(\frac{k}{2}+1\right) \quad X\left(\frac{k}{2}\right)X\left(\frac{k}{2}\right)$$

$$2X(k+1)X(M-1)\cdots 2X\left(\frac{M+k}{2}-1\right)X\left(\frac{M+k}{2}+1\right)$$

$$\left. X\left(\frac{M+k}{2}\right)X\left(\frac{M+k}{2}\right) \right]^T \qquad (6.72)$$

when k is even and then perform the calculations using these reduced-order vectors. This will reduce the size of the matrix to be inverted to approximately half in each dimension. A similar simplification can be made for the case when k is odd also.

6.3.3.2 Estimation of the Statistical Expectations

Assuming that the input and output signals are ergodic, we can estimate the autocorrelation matrix of the input vector $\mathbf{X}_{2,k,r}$ and the cross-correlation vector of $D(k)$ and the input vector by time-averaging as follows. First, partition the input and output signals into segments of length L samples each and compute their M-point DFTs. The segments may be overlapped, and they also may be weighted by a window function such as a Hamming window if necessary. The overlap, if any, depends on the window function employed. No overlap is recommended for rectangular window functions. In the case of Hamming windows, 50% overlap of adjacent segments is typically recommended. In order to avoid any lag-domain aliasing of the correlation functions, the DFT size M should be at least two times the segment length L. Let $\mathbf{X}_{2,k,i}$ and $D(k, i)$ represent the reduced-order input vector for the kth DFT coefficient of the ith segment of the input signal and the kth DFT coefficient of the ith segment of the measured output signal, respectively. Then, the autocorrelation matrix and the cross-correlation vector are estimated as

$$\hat{\mathbf{R}}_{xx} = \frac{1}{K} \sum_{i=1}^{K} \mathbf{X}_{2,k,i}^* \mathbf{X}_{2,k,i}^T \qquad (6.73)$$

and

$$\hat{\mathbf{P}}_{xd} = \frac{1}{K} \sum_{l=1}^{K} \mathbf{X}_{2,k,i} D(k, i), \tag{6.74}$$

respectively. These estimates can then be substituted in (6.70) to estimate the coefficients as

$$\hat{\mathbf{H}}_{2,k,r} = \hat{\mathbf{R}}_{xx}^{-1} \hat{\mathbf{P}}_{xd}. \tag{6.75}$$

Example 6.10   We consider the identification of the same system as in Example 6.8. The input signal was obtained by processing a white Gaussian signal $\xi(n)$ with zero mean value and unit variance through a linear system with input–output relationship given by

$$x(n) = 0.6x(n-1) + 0.8\xi(n).$$

The output signal was obtained by processing the input signal $x(n)$ with the nonlinear system and then corrupting its output with a zero-mean and white Gaussian noise sequence with variance 0.001. The noise sequence was independent of the input signals. Fifty independent experiments using 2048 samples each were conducted and the results presented were obtained by averaging over the 50 experiments. Each estimate employed segments of length 32 samples with no overlap and 64-point DFTs.

Figure 6.7 shows the average estimates of the amplitude and phase functions of the linear component obtained using the method presented in this section. The corresponding results for the quadratic components are given in Figure 6.8. Comparing the estimates obtained using the frequency-domain method to those obtained in Example 6.8, we see that the frequency-domain estimates are significantly more noisy than the estimates in Example 6.8. This is because the frequency-domain method estimated the linear and quadratic components in each bin separately, whereas the method in Example 6.8 required the estimation of only four parameters from the same number of samples.

6.3.4   Extension to Higher-Order Volterra Systems

The extension of the method described above to the identification of truncated Volterra systems with higher orders of nonlinearity is straightforward, but somewhat cumbersome. We leave the derivation as an exercise for the reader. The reduction in computational complexity for higher-order systems is not as dramatic as for the quadratic systems. The matrix inversion necessary to implement the method for a Pth-order nonlinear system involves matrices of order $O(N^{P-1}) \times O(N^{P-1})$ elements while the direct time-domain methods require the inversion of an $O(N^P) \times O(N^P)$-element matrix.

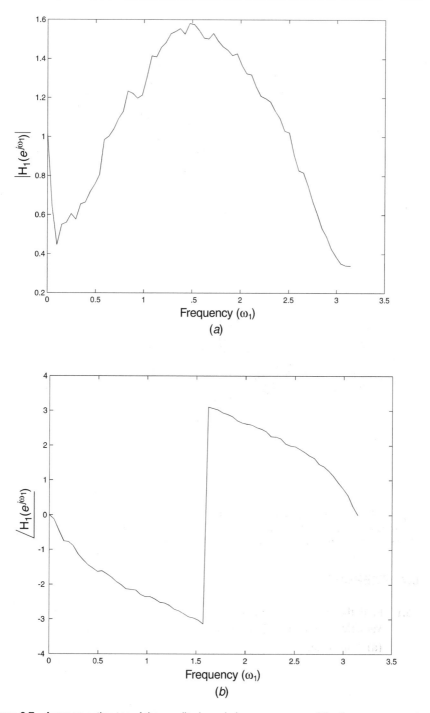

Figure 6.7 Average estimates of the amplitude and phase responses of the linear component of the system in Example 6.10 obtained using the method of this section for Gaussian input signals: (*a*) amplitude response; (*b*) phase response.

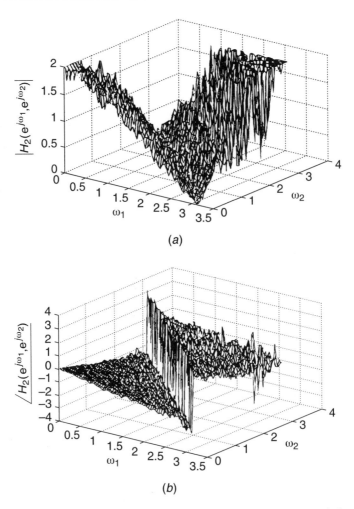

Figure 6.8 Average estimates of the amplitude and phase responses of the quadratic component of the system in Example 6.10 obtained using the method of this section for Gaussian input signals: (a) amplitude response; (b) phase response.

6.4 EXERCISES

6.1 Find the cumulants and moments of orders 1, 2, 3, and 4 for a random variable x with the following probability density functions:

(a) Laplacian:

$$f_x(x) = \frac{a}{2} e^{-a|x|}; \qquad a > 0.$$

(b) Uniform:

$$f_x(x) = \begin{cases} \dfrac{1}{\Delta}; & |x| < \dfrac{\Delta}{2} \\ 0: & \text{otherwise.} \end{cases}$$

6.2 Show that the cumulant generating function $\tilde{\phi}(\omega) = \ln \phi(\omega)$ is zero for $\omega = 0$ for any valid probability density function.

6.3 Suppose that x is a random variable with symmetric probability density function: $f_x(x) = f_x(-x)$. Show that all the odd-ordered moments and cumulants of x are zero.

6.4 Find an expression using (6.11) for $E\{x_1 x_2 x_3 x_4\}$ as a function of the joint cumulants of x_1, x_2, x_3, and x_4 of order up to four.

6.5 Let x be a zero-mean, Gaussian random variable with variance σ_x^2. Show that $E\{x^3\} = 0$ and that $E\{x^4\} = 3\sigma_x^4$ using (6.11) and the fact that all cumulants of Gaussian variables of order larger than two are zero. Extend the results to obtain an expression for $E\{x^p\}$ where $p > 2$.

6.6 Prove that $\text{cum}(x_1, x_2, \ldots, x_n) = 0$ whenever a subset of $x_1, x_2, \ldots, x_n$ is independent of the remaining elements of the set.

6.7 Show that

$$\text{cum}(x_1 + y_1, x_2, \ldots, x_n) = \text{cum}(x_1, x_2, \ldots, x_n) + \text{cum}(y_1, x_2, \ldots, x_n).$$

Use this result to obtain a general expression for $\text{cum}(x_1 + y_1, x_2 + y_2, \ldots, x_n + y_n)$.

6.8 Let $\{x_1, x_2, \ldots, x_n\}$ and $\{y_1, y_2, \ldots, y_n\}$ be two independent sets of random variables. By evaluating the cumulant generating function, show that

$$\text{cum}(x_1 + y_1, x_2 + y_2, \ldots, x_n + y_n)$$
$$= \text{cum}(x_1, x_2, \ldots, x_n) + \text{cum}(y_1, y_2, \ldots, y_n).$$

6.9 Show that the power spectrum of a stationary random process $x(n)$ is real and non-negative. [*Hint:* Show that $c_2^x(0)$ is negative if $c_2^{(x)}(\omega_1)$ is negative for all ω_1. If a portion of $c_2^{(x)}(\omega_1)$ is negative, you can create another process whose power spectrum is zero or negative at all frequencies by appropriately filtering the original signal.]

6.10 Show that all possible symmetries of the bispectrum of a real signal are as shown in Figure 6.2.

6.11 Find all the regions of symmetry of the trispectrum of a real signal.

6.12 Show that the results of Theorem 6.2 are valid for arbitrary, non-Gaussian probability density functions for the noise process $\eta(n)$ as long as it is independent of the input signal $x(n)$.

6.13 Suppose that the input signal $x(n)$ of Theorem 6.2 is a zero-mean and white Gaussian process with variance σ_x^2. The results simplify considerably in this case. Show that

$$\text{cum}\,(y(n), x(n-l_1), \ldots, x(n-l_K)) = K!\sigma_x^{2K}h_K(m_1, m_2, \ldots, m_K)$$

$$+ \sum_{v=1}^{\lfloor \frac{p-K}{2} \rfloor} \frac{(K+2v)!}{v!2^v}\sigma_x^{2(K+v)} \sum_{j_1=0}^{N-1}\sum_{j_2=0}^{N-1}\cdots\sum_{j_v=0}^{N-1}$$

$$h_{K+2v}(m_1, m_2, \ldots, m_K, j_1, j_1, j_2, j_2, \ldots, j_v, j_v)$$

6.14 Derive (6.56) from (6.55).

6.15 Explain how you would use the results of Theorem 6.2 to identify a fifth-order Volterra system.

6.16 Explain how you would modify the parameter estimation algorithm of Section 6.3.2 when the zeroth-order kernel in the system model is not zero.

6.17 *Computing Assignment:* Use the results of Theorem 6.2 to develop a system identification technique for a truncated third-order Volterra system. Generate a zero-mean and white Gaussian random sequence with variance $\sigma_x^2 = 1$ and of length 4096 samples. Process this signal using a third-order nonlinear system with input–output relationship

$$y(n) = x^3(n) - x(n)x^2(n-1) + 0.5x^2(n) + x(n)x(n-1) + 0.2x(n) + x(n-2),$$

and corrupt the output sequence with an additive, zero-mean Gaussian noise process with variance $\sigma_\eta^2 = 0.01$ that is independent of the input sequence $x(n)$.

(a) Find analytical expressions for the first-, second-, and third-order transfer functions of the nonlinear system.

(b) Using 64-point segments with no overlap and 128-point DFTs, implement the system identification method you developed. The following measures may be used to measure the accuracy of the estimate of each transfer function:

$$\|H_1(k_1) - \hat{H}_1(k_1)\|^2 = \frac{\displaystyle\sum_{k_1=0}^{127} |H_1(k_1) - \hat{H}_1(k_1)|^2}{\displaystyle\sum_{k_1=0}^{127} |H_1(k_1)|^2},$$

$$\|H_2(k_1, k_2) - \hat{H}_2(k_1, k_2)\|^2 = \frac{\displaystyle\sum_{k_1=0}^{127}\sum_{k_2=0}^{127} |H_2(k_1, k_2) - \hat{H}_2(k_1, k_2)|^2}{\displaystyle\sum_{k_1=0}^{127}\sum_{k_2=0}^{127} |H_2(k_1, k_2)|^2}$$

and

$$\|H_3(k_1, k_2, k_3) - \hat{H}_3(k_1, k_2, k_3)\|^2$$

$$= \frac{\sum_{k_1=0}^{127} \sum_{k_2=0}^{127} \sum_{k_3=0}^{127} |H_3(k_1, k_2, k_3) - \hat{H}_3(k_1, k_2, k_3)|^2}{\sum_{k_1=0}^{127} \sum_{k_2=0}^{127} \sum_{k_3=0}^{127} |H_3(k_1, k_2, k_3)|^2},$$

where the hats denote estimated quantities.

(c) Repeat this experiment using 50 independent sets of signals and evaluate the quality measures by averaging the results of the 50 experiments.

6.18 Extend the method of Section 6.3.2 to third-order Volterra systems. Repeat the previous experiments using your technique when the input signal is uniformly distributed in the range $[-2,2]$.

7

ADAPTIVE TRUNCATED
VOLTERRA FILTERS

The nonlinear systems we discussed up to this point were assumed to be time- or space-invariant. However, there are many situations in which a system of interest is both nonlinear and time-varying. One example arises in echo cancellation problems in high-speed communication systems. The generation and characteristics of the echoes depend on several factors, including the properties of the hybrid circuit at the far end, load on the circuit, and environmental conditions that might affect the behavior of the amplifiers. In other situations the underlying nonlinear system can be modeled as time-invariant, but the properties of the system itself or the statistics of the signals involved are not known a priori. Adaptive filters are useful in applications characterized by both these situations. This chapter is concerned with adaptive filters that employ truncated Volterra system models.

In order to develop a basic understanding of the theory of adaptive filters, we start our discussion with that of adaptive filters equipped with linear, finite-memory system models. These ideas are then generalized to systems employing truncated Volterra system models.

7.1 ADAPTIVE FILTERS USING LINEAR MODELS

Figure 7.1 depicts a typical adaptive filtering problem. The objective of the adaptive filter is to process its input signal $x(n)$ so that its output $\hat{d}(n)$ is close to the *desired response signal* $d(n)$ in some sense. The adaptive filter uses a system model that is selected primarily on the basis of the user's knowledge of the underlying relationship between the input and the desired response signals. The system adapts the parameters of the model continuously using the *estimation error* $e(n)$, which is the difference between the desired response signal and the output of the adaptive filter, as a basis for making the parameter updates. The adaptive filter attempts to

239

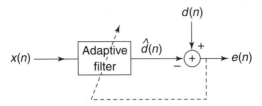

Figure 7.1   Block diagram of an adaptive filter.

adjust the parameters so that some convex function of the error is reduced in the process.

For causal, linear, and finite-memory system models with N-sample memory, the adaptive filter output $\hat{d}(n)$ is computed as

$$\hat{d}(n) = \sum_{m=0}^{N-1} h_1(m; n)x(n - m), \tag{7.1}$$

where $h_1(m; n)$; $m = 0, 1, \ldots, N - 1$ are the coefficients of the adaptive filter at time n. There are several approaches for updating the filter coefficients. We consider two of the most popular classes of algorithms next.

7.1.1 Stochastic Gradient Adaptive Filters

Stochastic gradient adaptive filters [90,172] attempt to iteratively minimize the statistical expectation of some convex function of the error signal using an approximation to the steepest-descent method. Let the cost function that we desire to minimize be

$$J(n) = E\{\phi(e(n))\}, \tag{7.2}$$

where $\phi(\cdot)$ is an even, convex function of $(\cdot)$. In the steepest descent method, the coefficient update is performed using the following strategy:

$$h_1(m; n + 1) = h_1(m; n) - \alpha \frac{\partial}{\partial h_1(m; n)} J(n); \quad m = 0, 1, \ldots, N - 1, \tag{7.3}$$

where α is a small, positive step size parameter that controls the convergence, tracking and steady-state properties of the system.

7.1.1.1 Rationale for the Steepest-Descent Algorithm   Figure 7.2 shows a convex cost function associated with a single coefficient system. At the minimum point of the cost function, which corresponds to the optimal coefficient value, the slope of the function with respect to the coefficient is zero. For coefficient values to the right of the optimal value, the slope is always positive. On the other hand, the slope is negative for all coefficient values to the left of the optimal value. Furthermore, the magnitude of the slope increases with increasing distance of the coefficient value from the optimal value. It should be obvious from this discussion that a reasonable coefficient update strategy is to move the coefficient so that the change in its value is proportional to the negative of the slope of the error surface at

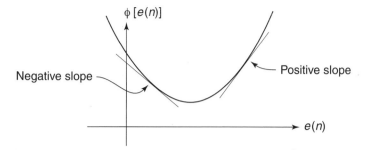

Figure 7.2 A convex cost function explaining the rationale for the steepest-descent algorithm.

the current location of the coefficient. If the constant of proportionality is adequately small, the cost function is reduced after each iteration. Furthermore, the changes are large when the coefficient is far away from its optimal value and small when the coefficient is in the vicinity of its optimal value. The algorithm in (7.3) is an extension of the strategy described above to the case involving N coefficients.

7.1.1.2 The Stochastic Gradient Approximation
While the algorithm described above appears to be sound in its approach, it has one major drawback. To implement (7.3), we must know the partial derivatives of the statistical expectation of the error signal—more specifically $E\{\phi(e(n))\}$—in (7.2). It is only rarely that this information is available to the user. Consequently, the steepest-descent algorithm of (7.3) is not realizable in practice. We must seek more practical algorithms than the steepest-descent approach to update the adaptive filter coefficients. Stochastic gradient adaptive filters use a very simple approximation in the calculation of the partial derivatives in (7.3) to derive a class of realizable algorithms. The stochastic gradient approximation to the steepest-descent algorithm is

$$h_1(m; n+1) = h_1(m; n) - \alpha \frac{\partial}{\partial h_1(m; n)} \phi(e(n)); \qquad m = 0, 1, \ldots, N-1. \quad (7.4)$$

In deriving this equation, we replaced $\partial E\{\phi(e(n))\}/\partial h_1(m; n)$ in the update equation (7.3) for the steepest-descent algorithm with $\partial \phi(e(n))/\partial h_1(m; n)$. Even though this approximation is rather crude, the algorithms that result work reasonably well for small values of the proportionality constant. Furthermore, the mean value of $\partial \phi(e(n))/\partial h_1(m; n)$ is $\partial J(n)/\partial h_1(m, n) = \partial E\{\phi(e(n))\}/\partial h_1(m; n)$, and consequently, we can expect the stochastic gradient adaptive filters to perform like the steepest descent algorithms on average. We now consider a specific example of a stochastic gradient adaptive filter.

7.1.2 Least-Mean-Square Adaptive Filters

The least-mean-square (LMS) adaptive filter uses a cost function defined by the mean-square estimation error

$$J(n) = E\{e^2(n)\}. \qquad (7.5)$$

Since the estimation error is by definition

$$e(n) = d(n) - \sum_{m=0}^{N-1} h_1(m; n)x(n - m), \qquad (7.6)$$

we can evaluate the partial derivatives required to update the coefficients as

$$\frac{\partial}{\partial h_1(m; n)} e^2(n) = -2e(n)x(n - m); \qquad m = 0, 1, \ldots, N - 1. \qquad (7.7)$$

Substituting (7.7) in (7.4) results in the LMS adaptive filter. The complete algorithm consists of the error calculation (7.6) and the coefficient update equation

$$h_1(m; n + 1) = h_1(m; n) + \mu e(n)x(n - m); \qquad m = 0, 1, \ldots, N - 1, \qquad (7.8)$$

where we have defined $\mu = 2\alpha$ for convenience. It is common to employ vector notation to describe adaptive filters. Let us define N-dimensional input and coefficient vectors as

$$\mathbf{X}_1(n) = [x(n), \ x(n - 1), \ldots, x(n - N + 1)]^T \qquad (7.9)$$

and

$$\vec{\mathbf{H}}_1(n) = [h_1(0; n), \ h_1(1; n), \ldots, h_1(N - 1; n)]^T, \qquad (7.10)$$

respectively. Substituting (7.9) and (7.10) in (7.6) and (7.8) results in an equivalent description of the LMS adaptive filter as follows:

$$e(n) = d(n) - \mathbf{X}_1^T(n)\vec{\mathbf{H}}_1(n) \qquad (7.11)$$

$$\vec{\mathbf{H}}_1(n + 1) = \vec{\mathbf{H}}_1(n) + \mu \mathbf{X}_1(n)e(n). \qquad (7.12)$$

7.1.2.1 *Analysis of the Mean Coefficient Behavior of the LMS Adaptive Filter*   Exact analysis of the behavior of most adaptive filters is extremely difficult. Consequently, we resort to several simplifying assumptions to make the analysis feasible [90,172].

Stationarity Assumption   The input vector $\mathbf{X}_1(n)$ and the desired response signal $d(n)$ belong to jointly wide-sense stationary processes with input autocorrelation matrix given by

$$\mathbf{R}_{\mathbf{xx}} = E\{\mathbf{X}_1(n)\mathbf{X}_1^T(n)\} \qquad (7.13)$$

and the cross-correlation vector of $\mathbf{X}_1(n)$ and $d(n)$ given by

$$\mathbf{P}_{d\mathbf{x}} = E\{d(n)\mathbf{X}_1(n)\}. \qquad (7.14)$$

The Independence Assumption   The input signal pair consisting of $\mathbf{X}_1(n)$ and $d(n)$ at time n is independent of the pair $\{\mathbf{X}_1(k), d(k)\}$ whenever $n \neq k$.

The independence assumption is rarely true in practice. Note that $\mathbf{X}_1(n)$ and $\mathbf{X}_1(n-1)$ have $N-1$ common elements in the model of (7.1). In spite of this problem with the independence assumption, it is commonly employed in the analysis of adaptive filters since (1) it considerably simplifies the analysis, and (2) the analyses using the assumption give fairly accurate results for practical values of μ, and consequently, such analyses provide reliable design rules. Mazo [173] has shown that the analyses using the independence assumption results in first-order approximations of the exact analyses.

The Analysis   We can obtain an alternate expression for the coefficient update equation by substituting (7.11) in (7.12). This operation gives

$$\vec{\mathbf{H}}_1(n+1) = \vec{\mathbf{H}}_1(n) + \mu \mathbf{X}_1(n)(d(n) - \mathbf{X}_1^T(n)\vec{\mathbf{H}}_1(n))$$

$$= (\mathbf{I} - \mu \mathbf{X}_1(n)\mathbf{X}_1^T(n))\vec{\mathbf{H}}_1(n) + \mu d(n)\mathbf{X}_1(n). \qquad (7.15)$$

The coefficient vector $\vec{\mathbf{H}}_1(n)$ depends on the input vector $\mathbf{X}_1(k)$ and desired response signal $d(k)$ only for times k prior to n. Consequently, $\vec{\mathbf{H}}_1(n)$ is independent of $\mathbf{X}_1(n)$ because of the independence assumption. Let us now take the statistical expectation of both sides of the preceding equation. This operation results in

$$E\{\vec{\mathbf{H}}_1(n+1)\} = (\mathbf{I} - \mu \mathbf{R}_{\mathbf{xx}})E\{\vec{\mathbf{H}}_1(n)\} + \mu \mathbf{P}_{d\mathbf{x}}. \qquad (7.16)$$

It is convenient at this time to define the coefficient error vector $\mathbf{V}_1(n)$ as

$$\mathbf{V}_1(n) = \vec{\mathbf{H}}_1(n) - \vec{\mathbf{H}}_{\text{opt}}, \qquad (7.17)$$

where

$$\vec{\mathbf{H}}_{opt} = \mathbf{R}_{\mathbf{xx}}^{-1} \mathbf{P}_{d\mathbf{x}} \qquad (7.18)$$

is the optimal solution to the MMSE estimation problem. Substituting (7.17) and (7.18) in (7.16) and simplifying gives

$$E\{\mathbf{V}_1(n+1)\} = (\mathbf{I} - \mu \mathbf{R}_{\mathbf{xx}})E\{\mathbf{V}_1(n)\}. \qquad (7.19)$$

7.1.2.2   *Convergence of the Mean Coefficient Values*   Equation (7.19) indicates that the mean of the coefficient error vector converges to a zero vector if all the eigenvalues of the matrix $(\mathbf{I} - \mu \mathbf{R}_{\mathbf{xx}})$ are bounded by one. Let $\{\lambda_i, i = 0, 1, \ldots, N-1\}$ denote the set of eigenvalues of $\mathbf{R}_{\mathbf{xx}}$. Then, the condition

for convergence of the evolution equation (7.19) becomes

$$-1 < 1 - \mu\lambda_i < 1; \qquad i = 0, 1, \ldots, N - 1, \tag{7.20}$$

which reduces to

$$0 < \mu < \frac{2}{\lambda_i}; \qquad i = 0, 1, \ldots, N - 1. \tag{7.21}$$

The most stringent of these conditions is

$$0 < \mu < \frac{2}{\lambda_{\max}}, \tag{7.22}$$

where $\lambda_{\max}$ represents the largest eigenvalue of $\mathbf{R}_{xx}$. This is the necessary and sufficient condition for the convergence of the mean values of the coefficient of the adaptive filter under the independence assumption.

A Simpler Sufficient Condition for Convergence   Computation of the eigenvalues of the autocorrelation matrix is in general a complex problem, and consequently, it is common to seek easily computable sufficient conditions for the convergence of the adaptive filter. An easy condition to check can be derived in the following manner. Since

$$\lambda_{\max} \le \sum_{i=1}^{N} \lambda_i = \mathrm{Tr}\{\mathbf{R}_{xx}\}, \tag{7.23}$$

we see that a sufficient convergence condition for the mean values is given by

$$0 < \mu < \frac{2}{\mathrm{Tr}\{\mathbf{R}_{xx}\}}, \tag{7.24}$$

where $\mathrm{Tr}\{\cdot\}$ denotes the sum of the diagonal elements of $\{\cdot\}$. For stationary inputs, this condition is equivalent to

$$0 < \mu < \frac{2}{N\sigma_x^2}, \tag{7.25}$$

where σ_x^2 is the mean-squared value of the input signal.

Condition for Mean-Square Convergence   The condition described above only guarantees the convergence of the mean values of the coefficients of the adaptive filter. It does not ensure the convergence of the mean-square or other higher-order moments of the coefficient vector sequence. Ideally, it is desirable to show that the coefficient sequence converges in probability. However, proving that the probability distributions of the coefficients converge to some limiting distributions is extremely difficult. As a result, most analyses of adaptive filters attempt only to show mean-square convergence of the coefficients. We will not present such an analysis here. A useful result to keep in mind is that of a bound for μ that guarantees mean-square convergence of the adaptive filter. It has been shown [90,172] that a sufficient condition for mean-square convergence of the LMS adaptive filter is

$$0 < \mu < \frac{2}{3\mathrm{Tr}\{\mathbf{R}_{xx}\}}. \tag{7.26}$$

Modes of Convergence of the LMS Adaptive Filter Equation (7.19) shows that the behavior of the LMS adaptive filter depends heavily on the input signal statistics. To see this dependence more clearly, let us make a transformation of the coefficient error vector in the following manner. Since $\mathbf{R}_{xx}$ is an autocorrelation matrix, it is at least positive semidefinite. Furthermore, it is a symmetric matrix. Consequently, we can express $\mathbf{R}_{xx}$ as

$$\mathbf{R}_{xx} = \mathbf{Q}^T \mathbf{\Lambda} \mathbf{Q}, \tag{7.27}$$

where $\mathbf{\Lambda}$ is a diagonal matrix containing the eigenvalues of $\mathbf{R}_{xx}$:

$$\mathbf{\Lambda} = \mathrm{diag}\{\lambda_0, \lambda_1, \ldots, \lambda_{N-1}\}, \tag{7.28}$$

and $\mathbf{Q}$ is an orthonormal matrix formed by the eigenvectors of $\mathbf{R}_{xx}$. Now, let

$$\tilde{\mathbf{V}}_1(n) = \mathbf{Q}\mathbf{V}_1(n). \tag{7.29}$$

Substituting this definition in (7.19) results in an evolution equation for a transformed set of coordinates:

$$E\{\tilde{\mathbf{V}}_1(n+1)\} = (\mathbf{I} - \mu\mathbf{\Lambda})E\{\tilde{\mathbf{V}}_1(n)\}. \tag{7.30}$$

The vector equation (7.30) contains a set of N uncoupled scalar equations that describe the evolution of each element of the transformed error vector. The mean value of the mth element of the transformed error vector evolves as

$$E\{\tilde{v}_{1,m}(n+1)\} = (1 - \mu\lambda_m)E\{\tilde{v}_{1,m}(n)\}. \tag{7.31}$$

Iterating this equation n times results in a closed-form expression for $E\{\tilde{v}_{1,m}(n)\}$:

$$E\{\tilde{v}_{1,m}(n)\} = (1 - \mu\lambda_m)^n E\{\tilde{v}_{1,m}(0)\}. \tag{7.32}$$

The preceding analysis clearly shows that the LMS adaptive filter has N modes of convergence. When the condition for convergence given by (7.22) is satisfied, the mean value of the transformed coefficient error for each mode decays exponentially to zero. However, the rate of decay depends on the step size μ as well as the magnitude of the corresponding eigenvalue. In particular, the mode corresponding to the smallest eigenvalue decays at the slowest rate.[1] The mean values of the actual coefficient errors evolve with time as a linear combination of N exponentially decaying terms. The corresponding evolution equation is given by

$$E\{\tilde{v}_{1,m}(n)\} = \sum_{i=0}^{N-1} q_{m,i}(1 - \mu\lambda_i)^n E\{\tilde{v}_{1,i}(0)\}, \tag{7.33}$$

where $q_{m,i}$ is the (m, i)th element of $\mathbf{Q}$. Recall that $\mathbf{Q}$ is an orthonormal matrix and therefore $\mathbf{Q}^{-1} = \mathbf{Q}^T$ when $\mathbf{Q}$ is real. In the worst-case scenario, the mode that decays the slowest may dominate the convergence behavior of the adaptive filter. This can be a significant problem with the LMS adaptive filter—the speed of convergence of the mean values of the coefficients of the adaptive filter is input signal–dependent. Furthermore, in cases where the eigenvalues are highly disparate, the slower modes of convergence may dominate the behavior of the filter. In spite of this drawback, the LMS adaptive filter and its variants form the most widely used class of adaptive filters, mainly because they are simple to implement and work reasonably well in a large number of applications.

7.1.3 Recursive Least-Squares Adaptive Filters

The LMS adaptive filter can be considered as an approximate solution to an optimization problem that attempts to minimize the mean-squared estimation error. Recursive least-squares (RLS) adaptive filters, on the other hand, yield the exact solution to an optimization problem formulated in a deterministic manner. Several variations of the RLS adaptive filters are available. However, we consider only what is known as the exponentially weighted RLS adaptive filter in this book. The objective of the exponentially weighted RLS adaptive filter is to select the coefficients in such a manner that the cost function defined by

$$J(n) = \sum_{k=1}^{n} \lambda^{n-k}(d(k) - \vec{\mathbf{H}}_1^T(n)\mathbf{X}_1(k))^2 \tag{7.34}$$

is minimized at each time. In this formulation, λ is a constant that controls the speed of convergence of the adaptive filter and $0 < \lambda \leq 1$.

The RLS adaptive filter attempts to choose the coefficients $\vec{\mathbf{H}}_1(n)$ so that a weighted average of the squared estimation errors due to this coefficient vector from time $k = 1$ to time $k = n$ is minimized. The exponential weighting factor λ gives the adaptive filter the capability for tracking potential changes in the operating environment. When λ is small, the weight λ^{n-k} decays rapidly as $n - k$ increases and consequently, the solution $\vec{\mathbf{H}}_1(n)$ depends more heavily on the input signals

---

[1] This statement assumes that $1 - \mu\lambda_i > 0$, which is true in almost all practical situations.

closer to time n. On the other hand, when λ is close to 1, λ^{n-k} decays slowly and the solution $\vec{\mathbf{H}}_1(n)$ depends on input signals that span over a longer duration. Consequently, smaller choices of λ enable the adaptive filter to track fast changes in the operating environments. Values of λ close to one can track slow changes. Similar to the LMS adaptive filter, one can expect that the variations of the coefficients about their mean values are small when λ is closer to one and large when λ is away from one.

The solution to the minimization problem in (7.34) can be found by differentiating $J(n)$ with respect to $\vec{\mathbf{H}}_1(n)$, setting the derivatives to zero, and solving the resulting set of simultaneous equations. The optimal coefficient vector at time n is given by

$$\vec{\mathbf{H}}_1(n) = \mathbf{C}^{-1}(n)\mathbf{P}(n), \tag{7.35}$$

where

$$\mathbf{C}(n) = \sum_{k=1}^{n} \lambda^{n-k}\mathbf{X}_1(k)\mathbf{X}_1^T(k) \tag{7.36}$$

is the exponentially weighted least-squares autocorrelation matrix of the input vector and

$$\mathbf{P}(n) = \sum_{k=1}^{n} \lambda^{n-k}\mathbf{X}_1(k)d(k) \tag{7.37}$$

is the exponentially weighted least-squares cross-correlation vector of the input vector $\mathbf{X}_1(n)$ and desired response signal $d(n)$.

7.1.3.1 *Iterative Realization of the Least-Squares Solution* We now develop an iterative realization of the solution in (7.35). We can iteratively update $\mathbf{C}(n)$ and $\mathbf{P}(n)$ using (7.36) and (7.37) as

$$\mathbf{C}(n) = \lambda\mathbf{C}(n-1) + \mathbf{X}_1(n)\mathbf{X}_1^T(n) \tag{7.38}$$

and

$$\mathbf{P}(n) = \lambda\mathbf{P}(n-1) + \mathbf{X}_1(n)d(n), \tag{7.39}$$

respectively. We employ the matrix inversion lemma[2] and the expression for $\mathbf{C}(n)$ in (7.38) to obtain a recursive expression for $\mathbf{C}^{-1}(n)$ as

$$\mathbf{C}^{-1}(n) = \frac{1}{\lambda}\mathbf{C}^{-1}(n-1) - \frac{(1/\lambda)\mathbf{C}^{-1}(n-1)\mathbf{X}_1(n)\mathbf{X}_1^T(n)(1/\lambda)\mathbf{C}^{-1}(n-1)}{1 + (1/\lambda)\mathbf{X}_1^T(n)\mathbf{C}^{-1}(n-1)\mathbf{X}_1(n)}. \tag{7.40}$$

---

[2] If $\mathbf{A}$ is a nonsingular matrix and $\mathbf{U}$ and $\mathbf{V}$ are column vectors of appropriate dimensions,

$$(\mathbf{A} + \mathbf{U}\mathbf{V}^T)^{-1} = \mathbf{A}^{-1} - \frac{\mathbf{A}^{-1}\mathbf{U}\mathbf{V}^T\mathbf{A}^{-1}}{1 + \mathbf{V}^T\mathbf{A}^{-1}\mathbf{U}},$$

provided $(\mathbf{A} + \mathbf{U}\mathbf{V}^T)$ is nonsingular.

Let us define a vector $\mathbf{k}(n)$ as

$$\mathbf{k}(n) = \frac{\mathbf{C}^{-1}(n-1)\mathbf{X}_1(n)}{\lambda + \mathbf{X}_1^T(n)\mathbf{C}^{-1}(n-1)\mathbf{X}_1(n)}. \tag{7.41}$$

This vector is known as the *gain vector* in the literature. Substituting (7.41) in (7.40), the update equation for the inverse of the least-squares autocorrelation matrix becomes

$$\mathbf{C}^{-1}(n) = \frac{1}{\lambda}\mathbf{C}^{-1}(n-1) - \frac{1}{\lambda}\mathbf{k}(n)\mathbf{X}_1^T(n)\mathbf{C}^{-1}(n-1). \tag{7.42}$$

An interesting relationship that the gain vector $\mathbf{k}(n)$ satisfies is that

$$\mathbf{k}(n) = \mathbf{C}^{-1}(n)\mathbf{X}_1(n). \tag{7.43}$$

This can be seen in the following manner. From (7.41), we have

$$\lambda\mathbf{k}(n) + \mathbf{k}(n)\mathbf{X}_1^T(n)\mathbf{C}^{-1}(n-1)\mathbf{X}_1(n) = \mathbf{C}^{-1}(n-1)\mathbf{X}_1(n). \tag{7.44}$$

We can solve for $\mathbf{k}(n)$ from (7.44) as

$$\mathbf{k}(n) = \left[\frac{1}{\lambda}\mathbf{C}^{-1}(n-1) - \frac{1}{\lambda}\mathbf{k}(n)\mathbf{X}_1^T(n)\mathbf{C}^{-1}(n-1)\right]\mathbf{X}_1(n). \tag{7.45}$$

We verify (7.43) by recognizing that the quantity within the square brackets above is identical to the right-hand side of (7.42).

An update equation for the optimal coefficient vector at time n can be obtained by substituting (7.42) and (7.39) in (7.35). This operation results in

$$\vec{\mathbf{H}}_1(n) = \mathbf{C}^{-1}(n-1)\mathbf{P}(n-1) - \mathbf{k}(n)\mathbf{X}_1^T(n)\mathbf{C}^{-1}(n-1)\mathbf{P}(n-1)$$
$$+ \mathbf{C}^{-1}(n)\mathbf{X}_1(n)d(n). \tag{7.46}$$

Substituting (7.43) in the third term on the right-hand-side of (7.46) and recognizing that $\vec{\mathbf{H}}_1(n-1) = \mathbf{C}^{-1}(n-1)\mathbf{P}(n-1)$ enable us to obtain the following simplified expression for $\vec{\mathbf{H}}_1(n)$:

$$\vec{\mathbf{H}}_1(n) = \vec{\mathbf{H}}_1(n-1) + \mathbf{k}(n)\left[d(n) - \mathbf{X}_1^T(n)\vec{\mathbf{H}}_1(n-1)\right]. \tag{7.47}$$

Let us define the a priori estimation error as

$$\epsilon(n) = d(n) - \mathbf{X}_1^T(n)\vec{\mathbf{H}}_1(n-1). \tag{7.48}$$

Then, the desired update equation is

$$\vec{\mathbf{H}}_1(n) = \vec{\mathbf{H}}_1(n-1) + \mathbf{k}(n)\epsilon(n). \tag{7.49}$$

This iterative approach for finding the optimal solution is known as the conventional recursive least-squares (CRLS) adaptive filter. The complete algorithm is summarized in Table 7.1. The algorithm is usually initialized using $\mathbf{C}^{-1}(0) = (1/\delta)\mathbf{I}$, where δ is a small positive constant.

Numerical Properties of the Conventional RLS Adaptive Filter We see from (7.42) that the inverse of the least-squares autocorrelation matrix $\mathbf{C}^{-1}(n)$ is iteratively updated as a difference of two matrices. Ideally, the least-squares autocorrelation matrix and its inverse are positive definite matrices in most situations. However, the differencing operation has the potential to drive some of the eigenvalues of $\mathbf{C}^{-1}(n)$ to zero or negative values because of numerical inaccuracies in the computations. In such situations, the adaptive filter may diverge, or produce a sequence of coefficient values that is not meaningful. It has been shown [323] that the occurrence of this problem can be avoided if care is taken to force $\mathbf{C}^{-1}(n)$ to be a symmetric matrix at all times.

7.1.3.2 A Heuristic Analysis of the RLS Adaptive Filter Using the expression for $\mathbf{k}(n)$ from (7.43), we can rewrite the coefficient update equation as

$$\vec{\mathbf{H}}_1(n) = \vec{\mathbf{H}}_1(n-1) + \mathbf{C}^{-1}(n)\mathbf{X}_1(n)\epsilon(n). \tag{7.50}$$

TABLE 7.1 The Conventional RLS Adaptive Filter

Initialization

$$\mathbf{C}^{-1}(0) = \frac{1}{\delta}\mathbf{I} \; ; \; \delta > 0$$

$$\mathbf{P}(0) = 0$$

$$\vec{\mathbf{H}}_1(0) = 0$$

Main Iteration
Do for $n = 1, 2, 3, \ldots$

$$\epsilon(n) = d(n) - \mathbf{X}_1^T(n)\vec{\mathbf{H}}_1(n-1)$$

$$\mathbf{k}(n) = \frac{\mathbf{C}^{-1}(n-1)\mathbf{X}_1(n)}{\lambda + \mathbf{X}_1^T(n)\mathbf{C}^{-1}(n-1)\mathbf{X}_1(n)}$$

$$\mathbf{C}^{-1}(n) = \frac{1}{\lambda}\mathbf{C}^{-1}(n-1) - \frac{1}{\lambda}\mathbf{k}(n)\mathbf{X}_1^T(n)\mathbf{C}^{-1}(n-1)$$

$$\vec{\mathbf{H}}_1(n) = \vec{\mathbf{H}}_1(n-1) + \mathbf{k}(n)\epsilon(n)$$

$$e(n) = d(n) - \mathbf{X}_1^T(n)\vec{\mathbf{H}}_1(n)$$

Comparing this equation with the coefficient update equation for the LMS adaptive filter in (7.12), we see that the two equations are very similar. The difference is that the step size μ of the LMS algorithm is replaced with the matrix $\mathbf{C}^{-1}(n)$. The use of the *matrix step size* gives the RLS adaptive filter considerable advantage over the LMS adaptive filter. We use a highly simplified analysis to illustrate one such advantage.

Let us assume that λ is very close to 1 so that the adaptive filter has a large memory. Since $\mathbf{C}(n)$ is based on a large number of samples in this case, we can assume that it is an *almost deterministic* quantity when the input signals are stationary and ergodic. The statistical expectation of $\mathbf{C}(n)$ is given by

$$E\{\mathbf{C}(n)\} = \lambda E\{\mathbf{C}(n-1)\} + E\{\mathbf{X}_1(n)\mathbf{X}_1^T(n)\}$$

$$= \lambda E\{\mathbf{C}(n-1)\} + \mathbf{R}_{\mathbf{xx}}. \tag{7.51}$$

The steady-state value of $E\{\mathbf{C}(n)\}$ can be easily found to be

$$\lim_{n \to \infty} E\{\mathbf{C}(n)\} = \mathbf{C}_{ss} = \frac{1}{1-\lambda} \mathbf{R}_{\mathbf{xx}}. \tag{7.52}$$

An approximate analysis for the mean coefficient behavior of the RLS adaptive filter can be made by analyzing the behavior of the update equation

$$\vec{\mathbf{H}}_1(n) = \vec{\mathbf{H}}_1(n-1) + (1-\lambda)\mathbf{R}_{\mathbf{xx}}^{-1}\mathbf{X}_1(n)\epsilon(n). \tag{7.53}$$

Using the analysis technique and assumptions employed for the LMS adaptive filter, we can show that

$$E\{\vec{\mathbf{H}}_1(n)\} = (\mathbf{I} - (1-\lambda)\mathbf{R}_{\mathbf{xx}}^{-1}\mathbf{R}_{\mathbf{xx}})E\{\vec{\mathbf{H}}_1(n-1)\} + (1-\lambda)\mathbf{R}_{\mathbf{xx}}^{-1}\mathbf{P}_{d\mathbf{x}}$$

$$= \lambda E\{\vec{\mathbf{H}}_1(n-1)\} + (1-\lambda)\vec{\mathbf{H}}_{\text{opt}}. \tag{7.54}$$

This result was obtained by replacing μ with $(1-\lambda)\mathbf{R}_{\mathbf{xx}}^{-1}$ in (7.16). Equation (7.54) implies that for all values of λ between 0 and 1, the mean value of the coefficient vector converges to the optimum value at an exponential rate. Furthermore, this convergence does not depend on the statistics of the input signal.

The price we pay for the independence of the adaptive filter's behavior from the input signal statistics is the computational complexity of the algorithm. An operations count shows that the CRLS adaptive filter requires $O(N^2)$ arithmetical operations at each time instant whereas the LMS adaptive filter needs only $O(N)$ arithmetical operations for its realization.

7.2 STOCHASTIC GRADIENT TRUNCATED VOLTERRA FILTERS

For simplicity of presentation, we restrict our discussion in this section to truncated second-order Volterra system models. However, the concepts discussed in this section are valid for higher-order Volterra systems with finite memory. Figure 7.3 describes the basic adaptive filtering problem considered here. Our objective is to adaptively model the desired response signal $d(n)$ as a Volterra series expansion in the most recent N samples as

$$\hat{d}(n) = h_0(n) + \sum_{m_1=0}^{N-1} h_1(m_1; n)x(n - m_1)$$

$$+ \sum_{m_1=0}^{N-1} \sum_{m_2=m_1}^{N-1} h_2(m_1, m_2; n)x(n - m_1)x(n - m_2), \qquad (7.55)$$

where $h_0(n)$ is a bias term, and $h_1(m_1; n)$ and $h_2(m_1, m_2; n)$ are the linear and quadratic coefficients respectively, of the adaptive filter at time n. The objective of the adaptive filter is to iteratively update its coefficients at each time so as to minimize some convex function of the error signal defined as

$$e(n) = d(n) - \hat{d}(n). \qquad (7.56)$$

7.2.1 The Least-Mean-Square Adaptive Second-Order Volterra Filter

Extension of the LMS adaptive filter to the case of truncated Volterra filters is straightforward. The LMS adaptive Volterra filter updates its coefficients at each time using a steepest-descent algorithm that attempts to minimize $e^2(n)$ at each time. The update equations can be easily shown to be [117,165]

$$h_0(n+1) = h_0(n) - \frac{\mu_0}{2}\frac{\partial e^2(n)}{\partial h_0(n)}$$

$$= h_0(n) + \mu_0 e(n), \qquad (7.57)$$

$$h_1(m_1; n+1) = h_1(m_1; n) - \frac{\mu_1(m_1)}{2}\frac{\partial e^2(n)}{\partial h_1(m_1; n)}$$

$$= h_1(m_1; n) + \mu_1(m_1)e(n)x(n - m_1); \quad m_1 = 0, 1, \ldots, N-1 \quad (7.58)$$

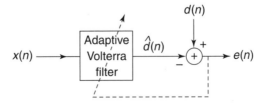

Figure 7.3 Problem formulation for adaptive Volterra filters.

and

$$h_2(m_1, m_2; n+1) = h_2(m_1, m_2; n) - \frac{\mu_2(m_1, m_2)}{2} \frac{\partial e^2(n)}{\partial h_2(m_1, m_2; n)}$$

$$= h_2(m_1, m_2; n) + \mu_2(m_1, m_2)e(n)x(n - m_1)x(n - m_2);$$

$$m_1 = 0, 1, \ldots, N - 1, \quad m_2 = m_1, m_1 + 1, \ldots, N - 1, \quad (7.59)$$

where μ_0, $\mu_1(m_1)$ and $\mu_2(m_1, m_2)$ are small positive constants that control the speed of convergence and the steady-state and/or tracking properties of the filter.

In (7.57)–(7.59), we have used a relatively general framework in using a different step size for each coefficient. This approach to the derivation makes use of the fact that the various signal samples of the form $1, x(n - m_1)$ or $x(n - m_1)x(n - m_2)$ that multiplies the different coefficients of the adaptive filter do not possess identical statistics. For stationary or slowly varying statistics of the input signals, we may choose $\mu_1(m_1)$ to be identical to each other for all values of m_1. Since $x(n - m_1)x(n - m_2)$ do not have identical statistics for different choices of m_1 and m_2, it is probably not optimal to choose the step sizes associated with the coefficients $h_2(m_1, m_2; n)$ to be all the same. However, it is common practice to use identical step sizes for all the higher-order coefficients also in LMS adaptive Volterra filters.

As we did for linear system models, we use matrix notation to rewrite the iterations of the adaptive second-order Volterra filter in a more compact manner. The notation as well as the algorithm itself is summarized in Table 7.2. The structure of this adaptive filter is different from that of the linear case only in the way in which the input and the coefficient vectors are defined. Consequently, extending the performance analysis of Section 7.1.2 to the adaptive Volterra filtering case is

---

TABLE 7.2   The LMS Adaptive Second-Order Volterra Filter

Coefficient vector
$$\vec{H}(n) = [h_0(n), \quad h_1(0; n), \quad h_1(1; n), \ldots, h_1(N - 1; n), \quad h_2(0, 0; n), \quad h_2(0, 1; n),$$

$$\ldots, h_2(0, N - 1; n), \quad h_2(1, 1; n), \ldots, h_2(N - 1, N - 1; n)]^T$$

Input vector
$$\mathbf{X}(n) = [1, \quad x(n), \quad x(n - 1), \ldots, x(n - N + 1), \quad x^2(n), \quad x(n)x(n - 1), \ldots,$$

$$x(n)x(n - N + 1), \quad x^2(n - 1), \ldots, x^2(n - N + 1)]^T$$

Initialization
   $\vec{H}(0)$ can be arbitrarily chosen

Main iteration
$$e(n) = d(n) - \vec{H}^T(n)\mathbf{X}(n)$$

$$\vec{H}(n + 1) = \vec{H}(n) + \boldsymbol{\mu}\mathbf{X}(n)e(n)^a$$

---

[a] $\boldsymbol{\mu}$ is a diagonal matrix whose diagonal entries are not necessarily identical.

relatively easy. For example, one can show that the mean values of the coefficients converge to their optimal values when the convergence parameters μ_0, $\mu_1(m_1)$, and $\mu_2(m_1, m_2)$ are chosen such that

$$0 < \mu_0, \mu_1(m_1), \mu_2(m_1, m_2) < \frac{2}{\lambda_{\max}}, \qquad (7.60)$$

where $\lambda_{\max}$ is the maximum eigenvalue of the statistical autocorrelation matrix of the input vector $\mathbf{X}(n)$. The difficulties with LMS adaptive Volterra filters are similar to those of LMS adaptive linear filters. The most significant problem is that the eigenvalues of the autocorrelation matrix of the input vector control the speed of convergence of the coefficients. In general, the larger the eigenvalue spread (the ratio of the maximum and minimum eigenvalues) is, the slower the convergence speed is. This is particularly troublesome for nonlinear system models, since the eigenvalue spreads are in general large. Even when the input signal is white, the presence of the nonlinear entries in the input vector causes the eigenvalue spread to be more than one.

7.2.1.1 *Removal of Mean Values from the Input Vector* One significant difference between the input vectors for the linear and nonlinear system models is that even when the input $x(n)$ to the adaptive filter has zero mean values, the input vector tends to have nonzero mean values for the nonlinear case. It is well-known that the autocorrelation matrices of nonzero mean vectors tend to be less well conditioned than their covariance matrices. Consequently, we can infer that the convergence properties of the adaptive Volterra filter can be improved by removing the mean values from the input vectors. We can develop an adaptive filter that employs the mean-removed input vectors in the following manner. Let

$$\tilde{d}(n) = d(n) - \overline{d(n)}, \qquad (7.61)$$

where $\overline{d(n)}$ is an estimate of the statistical expected value of $d(n)$. An easy way to estimate the mean value is to employ a first-order lowpass filter of the form

$$\overline{d(n)} = \lambda \overline{d(n-1)} + (1 - \lambda)d(n). \qquad (7.62)$$

Here, λ is a constant between zero and one and is usually very close to one. All the mean values required in the procedure can be estimated in a similar manner. Let us also define the mean-removed input vector as

$$\tilde{\mathbf{X}}(n) = \mathbf{X}(n) - \overline{\mathbf{X}(n)}, \qquad (7.63)$$

where $\overline{\mathbf{X}(n)}$ is an estimate of the mean value of the input vector. This estimate may be obtained using a single-pole lowpass filter as

$$\overline{\mathbf{X}(n)} = \lambda\overline{\mathbf{X}(n-1)} + (1-\lambda)\mathbf{X}(n). \tag{7.64}$$

Then, the corresponding adaptive filter is given by

$$e(n) = \tilde{d}(n) - \tilde{\mathbf{X}}^T(n)\vec{\mathbf{H}}(n), \tag{7.65}$$

and

$$\vec{\mathbf{H}}(n+1) = \vec{\mathbf{H}}(n) + \mu\tilde{\mathbf{X}}(n)e(n). \tag{7.66}$$

The estimate $\hat{d}(n)$ of the desired response signal $d(n)$ is given by

$$\hat{d}(n) = \tilde{\mathbf{X}}^T(n)\vec{\mathbf{H}}(n) + \overline{d(n)}. \tag{7.67}$$

Example 7.1 In this example, we consider the identification of a simple second-order system with input–output relationship

$$y(n) = 1 + x(n) + x^2(n)$$

using measured values of the input signal $x(n)$ and the desired output signal $d(n)$ obtained as the sum of $y(n)$ and measurement noise that is uncorrelated with $x(n)$. The input signal $x(n)$ is a uniformly distributed random variable with variance one. The mean value of the signal changed from experiment to experiment. The measurement noise is a zero-mean, Gaussian signal with unit variance. We now compare the performances of the LMS adaptive Volterra filters with and without mean removal in this identification problem. Plots of the ensemble-averaged, squared estimation error obtained over 50 independent runs spanning 2000 samples each are plotted in Figure 7.4 for the two approaches for three different mean values for the input signal given by 0, 1, and 10. Both the algorithms employed a step size of $\mu = 0.01$ for all the coefficients. The algorithm employing mean removal used $\lambda = 0.99$ for the lowpass filters. We can see from the figures that the performance of the algorithm that removes the mean values from the input vector does not change significantly with the input mean value. On the other hand, the behavior of the algorithm that does not remove the mean values from the input vector entries suffers from significant performance degradation, especially when the mean value of the input signal is large.

7.2.1.2 *Normalized LMS Adaptive Volterra Filters* One observation that we can make from the analysis of the LMS adaptive filters in Section 7.1.2 is that the performance of the algorithm varies with the input signal strength. If the input signal

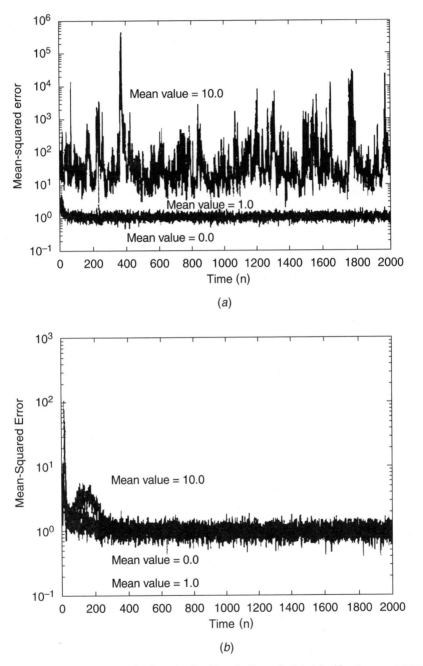

Figure 7.4 Learning curves for the adaptive filters in Example 7.1: (*a*) without mean removal; (*b*) with mean removal.

is scaled by a factor α, the eigenvalues of the autocorrelation matrix of the input vector $\mathbf{X}_1(n)$ to the adaptive linear filter are scaled by α^2. Consequently, the behavior of each mode of convergence of the adaptive filter changes when the input signal power changes. A common way of mitigating the difficulties that arise due to the dependence of the LMS adaptive filters on the input signal power is to use a normalized LMS adaptive filter. For the case of linear FIR system models, the coefficient update equations for the normalized LMS (NLMS) filter is given by

$$\vec{\mathbf{H}}_1(n+1) = \vec{\mathbf{H}}_1(n) + \frac{\mu}{\|\mathbf{X}_1(n)\|^2} \mathbf{X}_1(n)e(n), \tag{7.68}$$

where $\|\mathbf{X}_1(n)\|^2 = \mathbf{X}_1^T(n)\mathbf{X}_1(n)$ is the squared Euclidean norm of the input vector. An extension of this method can be made for the adaptive Volterra filters also. However, since the powers associated with various entries of the input vector may differ significantly in the nonlinear filtering case, the NLMS adaptive Volterra filters with update equations similar to that in (7.68) are always not as effective as the NLMS adaptive linear filters.

7.2.1.3 *Variations of the LMS Adaptive Volterra Filter*   It appears that Coker and Simkins were the first to publish an adaptive LMS Volterra filter in the literature [58]. Koh and Powers [117] analyzed the performance of second-order LMS adaptive filters for a stationary operating environment and Gaussian input signals. Stochastic gradient adaptive Volterra filters found several applications even in the early stages of their development. Such applications included echo cancellation [45,283], channel equalization [19], prediction of television images for coding [291], and noise cancellation [297]. Several variations of the standard LMS adaptation scheme have been applied to obtain improved convergence and tracking performance. One such method involves using time-varying step size sequences for the coefficients [165,289].

7.3 RLS ADAPTIVE VOLTERRA FILTERS

Development of the conventional RLS adaptive Volterra filter can be achieved easily using the same procedure employed for deriving the RLS adaptive filters employing linear system models. The algorithm for truncated Volterra system models is identical to the one presented in Table 7.1 with the exception that the coefficient and input vectors must be defined appropriately for the problem. For example, using the definitions of $\vec{\mathbf{H}}(n)$ and $\mathbf{X}(n)$ given in Table 7.2 in place of $\vec{\mathbf{H}}_1(n)$ and $\mathbf{X}_1(n)$ in Table 7.1 will result in the adaptive RLS quadratic filter.

Examples presented later in this section show that the adaptive quadratic filters retain the fast convergence property of the RLS adaptive linear filters. However, the computational complexity of this algorithm corresponds to that of $O(N^4)$ arithmetical operations per iteration for quadratic system models. Recall that the LMS adaptive quadratic filter and its variants can be implemented with a complexity proportional to $O(N^2)$ arithmetical operations per iteration. More efficient RLS algorithms requiring $O(N^3)$ arithmetical operations per iteration have been devised [136,306,307] and are discussed later in this chapter. However, even this complexity is an order of magnitude larger than that of stochastic gradient adaptive filters.

7.3.1 Stability of RLS Adaptive Filters

Exponentially windowed RLS adaptive filters are stable when the parameter λ is chosen such that $0 < \lambda < 1$. However, in order to get reasonably good estimates for a pth order Volterra filter, one must choose λ such that

$$1 - \lambda \ll \frac{1}{N^p}. \tag{7.69}$$

The rationale for this suggestion is that the time constant of the exponential window can be shown to be approximately $1/(1 - \lambda)$ time samples. We can make the heuristic argument that the RLS adaptive filter estimates the parameters using the data that occur during one time constant. Since a pth-order Volterra filter has $O(N^p)$ coefficients, it should be apparent that a reliable estimation procedure should be based on a much larger number of samples than the number of parameters. Selection of λ as given in (7.69) ensures this.

7.3.2 Performance Evaluation of RLS Adaptive Volterra Filters

Lee and Mathews presented a performance analysis of an RLS adaptive second-order Volterra filter in [136]. The operating environment considered in this analysis involved that of the identification of a time-varying second-order truncated Volterra system using stationary, Gaussian input signals with zero mean value. The unknown system changed according to a random-walk model described by

$$\vec{\mathbf{H}}_{\mathrm{opt}}(n) = \vec{\mathbf{H}}_{\mathrm{opt}}(n - 1) + \mathbf{M}(n), \tag{7.70}$$

where $\mathbf{M}(n)$ is an IID Gaussian process with zero mean value, variance $\sigma_m^2 \mathbf{I}$, and independent of the input signal or the measurement noise. The measurement noise was assumed to be an IID sequence with zero mean value and variance σ_η^2, and was independent of the input signal. Using a set of commonly employed approximations, it was shown in [133] that the steady-state value of the excess mean-squared error is given by

$$e_{ex}(\infty) = \frac{(1 - \lambda)}{2} \sigma_\eta^2 \left(\frac{N^2 + 3N}{2} \right) + \frac{\sigma_m^2 \mathrm{tr}\{\mathbf{R_{xx}}\}}{2(1 - \lambda)} \tag{7.71}$$

when $1 - \lambda$ is very small. The first term in the expression is the *adaptation noise* contributed by the measurement noise, and the second term is the *lag noise* contributed by the time variations in the operating environment. We can see from

(7.71) that the adaptation noise does not depend on the input signal statistics for the RLS adaptive Volterra filter. The derivations are lengthy, but straightforward. The interested reader can obtain the details from Lee and Mathews [136].

7.3.3 Approximate RLS Volterra Filters

The early attempts to reduce the computational complexity of recursive least-squares adaptive Volterra filters involved developing approximations based on the relationship of the higher-order statistics of Gaussian signals to their second-order statistics [63,335]. Even though they were successful in reducing the computational burden, they either suffered from slow convergence or poor numerical properties. Consequently, we do not discuss such methods in this chapter.

Example 7.2 This example compares the performance of the RLS Volterra filters with that of the LMS Volterra filters. We consider a system identification problem in which the unknown system to be identified is a second-order Volterra system with four-sample memory. At the start of the experiments, its input–output relationship is described by the difference equation

$$y(n) = -0.78x(n) - 1.48x(n-1) + 1.39x(n-2) + 0.04x(n-3) + 0.54x^2(n)$$
$$+ 3.72x(n)x(n-1) + 1.86x(n)x(n-2) - 0.76x(n)x(n-3) - 1.62x^2(n-1)$$
$$+ 0.76x(n-1)x(n-2) - 0.12x(n-1)x(n-3)$$
$$+ 1.41x^2(n-2) - 1.52x(n-2)x(n-3) - 0.13x^2(n-3).$$

The input signal $x(n)$ was obtained by processing an IID Gaussian process with zero mean value and variance 0.05 with a linear time-invariant filter with impulse response function

$$h(n) = \begin{cases} 0.9045; & n=0 \\ 1.0; & n=1 \\ 0.9045; & n=2 \\ 0.0; & \text{otherwise.} \end{cases}$$

With this setup, the power of $y(n)$ was approximately one during the initial stages of the experiments.

Figure 7.5 displays ensemble averages of 50 independent experiments conducted in a nonstationary operating environment. This figure compares the convergence speeds of the LMS and RLS adaptive Volterra filters when the adaptive filters used the same system model as the unknown system. In particular, the experiments did not employ a bias term, since the rest of the model was adequate to exactly match the unknown system. The RLS adaptive filter employed a forgetting factor of 0.9966 and the LMS adaptive filter used a step-size value of 0.055. These values were chosen so that the steady-state excess mean-square errors produced by both algorithms were

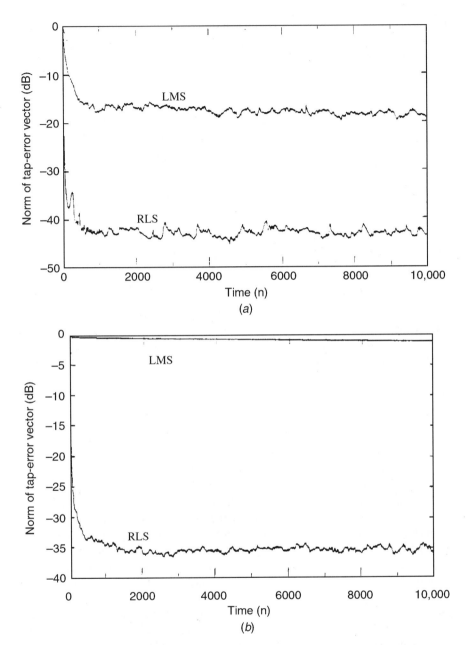

Figure 7.5 Evolution of error norms in Example 7.2 for a time-varying unknown system: *(a)* error norm for the linear coefficients; *(b)* error norm for the quadratic coefficients. (Courtesy of J. Lee. Copyright © 1993 IEEE.)

approximately the same when the coefficient of the unknown system did not change with time. The performance measure employed in the figures is the squared norm of the coefficient errors averaged over the 50 runs. The error norms were tabulated separately for the linear and quadratic parts, and are defined as

$$\|V_L(n)\| = 10 \log \left\{ \frac{\sum\limits_{m_1=0}^{N-1} |h_1(m_1; n) - h_{1,\text{opt}}(m_1; n)|^2}{\sum\limits_{m_1=0}^{N-1} |h_{1,\text{opt}}(m_1; n)|^2} \right\}$$

and

$$\|V_Q(n)\| = 10 \log \left\{ \frac{\sum\limits_{m_1=0}^{N-1} \sum\limits_{m_2=m_1}^{N-1} |h_2(m_1, m_2; n) - h_{2,\text{opt}}(m_1, m_2; n)|^2}{\sum\limits_{m_1=0}^{N-1} \sum\limits_{m_2=m_1}^{N-1} |h_{2,\text{opt}}(m_1, m_2; n)|^2} \right\},$$

respectively.

The coefficients changed their values continuously according to a random-walk model in the experiments. They were allowed to evolve in time in accordance with the relationship

$$\vec{H}_{\text{opt}}(n) = \vec{H}_{\text{opt}}(n-1) + \mathbf{M}(n),$$

where $\mathbf{M}(n)$ is an IID vector Gaussian process with zero mean value and covariance $\sigma_m^2 \mathbf{I}$. In our experiments, σ_m^2 was selected to be 2.0×10^{-7}.

We observe from the plots in Figure 7.5 that the RLS adaptive filter converges much faster than the LMS adaptive filter. This is an advantage that the RLS adaptive filters possess over the LMS adaptive filters in general. A second observation that we can make at this time is that the error norm for the quadratic coefficients is larger than that for the linear coefficients in all our experiments. This phenomenon can be attributed at least in part to the dependence of the quadratic filter coefficients on the higher-order statistics of the input signal which are more difficult to estimate.

Example 7.3 The purpose of this example is to investigate the performance of an adaptive second-order Volterra filter when the underlying system is different from the selected system model. For this purpose, we consider the identification of a nonlinear channel using the adaptive filter as described in Figure 7.6. The nonlinear channel model shown in Figure 7.6 is a highly simplified model of a digital satellite transmission system [17]. Satellite digital transmission represents one of the most important cases of a digital communication system employing a nonlinear channel. The memoryless nonlinear device is an AM/AM converter whose characteristics are as shown in Figure 7.7. The transfer function of the fourth-order lowpass Butter-

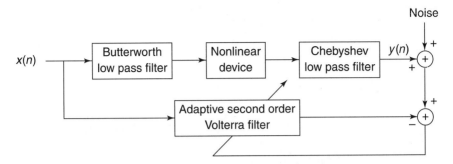

Figure 7.6 Application of the second-order Volterra filter to identify a nonlinear transmission system. (Courtesy of J. Lee. Copyright © 1993 IEEE.)

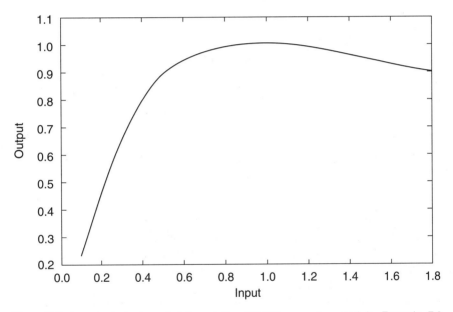

Figure 7.7 Input–output characteristics of the AM/AM converter used in Example 7.3. (Courtesy of J. Lee. Copyright © 1993 IEEE.)

worth and Chebyshev filters, denoted as $H_B(z)$ and $H_C(z)$, respectively, are given by

$$H_B(z) = \frac{(0.078 + 0.1559z^{-1} + 0.078z^{-2})(0.0619 + 0.1238z^{-1} + 0.0619z^{-2})}{(1.0 - 1.3209z^{-1} + 0.6327z^{-2})(1.0 - 1.048z^{-1} + 0.2961z^{-2})}$$

and

$$H_C(z) = \frac{(0.4638 - 0.4922z^{-1} + 0.4638z^{-2})(0.183 + 0.1024z^{-1} + 0.183z^{-2})}{(1.0 - 1.2556z^{-1} + 0.6891z^{-2})(1.0 - 0.7204z^{-1} + 0.1888z^{-2})}$$

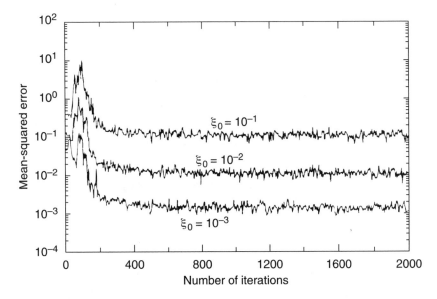

Figure 7.8 Learning curves associated with the a priori MSE over the first 2000 samples in Example 7.3. (Courtesy of J. Lee. Copyright © 1993 IEEE.)

Both filters have a cutoff frequency of 0.1 cycle per sample. The input signal $x(n)$ is uniformly distributed on the interval [0.12, 1.78] so that the AM/AM converter was operating at the saturation region most of the time. With the setup described above, the power of $y(n)$ was approximately one. We consider the performance of an RLS adaptive filter that employs a second-order Volterra system model with only linear and quadratic terms, $N = 12$ and $\lambda = 0.9985$. Figure 7.8 shows the a priori MSE over the first 2000 samples for three different measurement noise levels. The measurement noise was an IID Gaussian sequence with zero mean value and variance as shown for each case. The results are averages calculated over 50 independent experiments. The steady-state MSEs obtained by time-averaging the ensemble averages in the range [9000, 10000] are given in Table 7.3. It appears from the results of this example that the second-order Volterra system model works well in this problem despite the fact that the structure of the adaptive filter is completely different from that of the system model.

TABLE 7.3 Steady-state MSE for Example 7.3

	$\zeta_0 = 10^{-3}$	$\zeta_0 = 10^{-2}$	$\zeta_0 = 10^{-1}$
$N = 12$	0.00131	0.01099	0.10623

Source: Courtesy of J. Lee. Copyright © 1993 IEEE.

7.4 FAST RLS TRUNCATED VOLTERRA FILTERS

The recursive least-squares truncated Volterra filters presented in the previous section employed the matrix inversion lemma to obtain an iterative solution to the adaptive filtering problem. Even though successive input vectors to the adaptive filter have a large number of common elements, the derivations did not make use of any relationships that exist between successive input vectors. As a result, the algorithm we developed in Table 7.1, and its adaptations to the truncated Volterra filtering problem can be considerably simplified. In this section, we derive a fast algorithm for recursive least-squares second-order truncated Volterra filtering. Understanding fast RLS Volterra filters requires a good understanding of the theory of fast algorithms for updating RLS adaptive linear filters. This section may be skipped without loss of continuity by readers not interested in this topic. Detailed derivations of the fast RLS adaptive filters for the linear case may be found in Mathews and Douglas [172]. Some of the most important contributions to the development of fast RLS linear filters were made by Ljung et al. [147], Carayannis et al. [38], Cioffi and Kailath [54], and Slock and Kailath [294]. A fast RLS adaptive second-order Volterra filter was developed by Lee and Mathews [136]. Our description follows that of [136].

7.4.1 The Basic Approach to Computational Simplifications

The approach employed in the derivation of the fast RLS adaptive Volterra filters is to consider the nonlinear filtering problem as a multichannel but linear filtering problem as shown in Section 5.4.1 during the derivation of the lattice realizations of Volterra filters. From the definition of the input signal vector $\mathbf{X}(n)$ for the second-order Volterra filtering problem, we can see that the $N + 1$ elements contained in the vector

$$\mathbf{r}^T(n) = [x(n - N) \ x(n - N)x(n - 1) \ x(n - N)x(n - 2) \cdots x^2(n - N)] \qquad (7.72)$$

are discarded from the input vector $\mathbf{X}(n - 1)$, and another set of elements contained in

$$\mathbf{v}^T(n) = [x(n) \ x^2(n) \ x(n)x(n - 1) \cdots x(n)x(n - N + 1)] \qquad (7.73)$$

are added to the remaining elements to form the input data vector $\mathbf{X}(n)$.

The "trick" used in all fast RLS adaptive filters is to exploit the relationships among the forward predictor, the backward predictor, and the gain vector to obtain the relevant update equations. The forward predictor estimates the vector $\mathbf{v}(n)$ using the input vector $\mathbf{X}(n - 1)$ and the backward predictor estimates $\mathbf{r}(n)$ using $\mathbf{X}(n)$. The gain vector $\mathbf{k}(n)$ can be shown to be the coefficient vector for the filter that estimates the special sequence known as the pinning sequence and defined as

$$
\pi_k(n) = \begin{cases} 1; & n = k \\ 0; & \text{otherwise,} \end{cases}
\tag{7.74}
$$

using $\mathbf{X}(n)$. The structure of each of the three estimation problems described above is identical to the estimation problem for $d(n)$. Furthermore, the coefficient update in each case is based on the exponentially weighted least-squares criterion. Consequently, it is straightforward to show using the matrix inversion lemma that the update equations for the forward predictor coefficient vector $\mathbf{A}(n)$ and the backward predictor coefficient vector $\mathbf{B}(n)$ are given by

$$
\mathbf{A}(n) = \mathbf{A}(n-1) + \mathbf{k}(n-1)\mathbf{f}_{n-1}^T(n)
\tag{7.75}
$$

and

$$
\mathbf{B}(n) = \mathbf{B}(n-1) + \mathbf{k}(n)\mathbf{b}_{n-1}^T(n),
\tag{7.76}
$$

respectively, where the prediction error vectors $\mathbf{f}_k(n)$ and $\mathbf{b}_k(n)$ are defined as

$$
\mathbf{f}_k(n) = \mathbf{v}(n) - \mathbf{A}^T(k)\mathbf{X}(n-1)
\tag{7.77}
$$

and

$$
\mathbf{b}_k(n) = \mathbf{r}(n) - \mathbf{B}^T(k)\mathbf{X}(n),
\tag{7.78}
$$

respectively. The subscript k in these last two equations indicates the time instant at which the coefficient vectors were computed.

Remark 7.1: Notation for A Priori and A Posteriori Estimation Errors Contrary to our practice during the derivation of the conventional RLS adaptive filter, we do not use distinctly different notation for a priori and a posteriori estimation errors in the derivation of the fast RLS Volterra filters. Instead, we denote the a priori estimation errors using the subscript $n - 1$ as in $e_{n-1}(n)$, $\mathbf{f}_{n-1}(n)$ and $\mathbf{b}_{n-1}(n)$ for the errors associated with the estimation of the desired response signal, forward prediction, and backward prediction, respectively. Similarly, $e_n(n)$, $\mathbf{f}_n(n)$, and $\mathbf{b}_n(n)$ denote the a posteriori estimation errors for the three problems.

Recall from the derivation of the conventional RLS adaptive filter that we update the coefficient vector $\vec{\mathbf{H}}(n)$ in a similar manner using

$$
\vec{\mathbf{H}}(n) = \vec{\mathbf{H}}(n-1) + \mathbf{k}(n)e_{n-1}(n).
\tag{7.79}
$$

In a recursive implementation of the adaptive filter, we would compute $\bar{\mathbf{H}}(n-1)$, $\mathbf{A}(n-1)$, $\mathbf{B}(n-1)$, and $\mathbf{k}(n-1)$ at time $n-1$. Consequently, we can update the forward predictor coefficients at time n using the quantities that were evaluated at the previous time instant. However, before we can update the coefficient vectors $\mathbf{B}(n)$ and $\bar{\mathbf{H}}(n)$, we must update the gain vector $\mathbf{k}(n)$. Recall from Table 7.1 that updating the gain vector was the most computation-intensive component of the conventional RLS adaptive filter. Consequently, the key to the development of computationally efficient RLS adaptive Volterra filters is the derivation of an efficient update structure for the gain vector.

7.4.2 Strategy for Updating the Gain Vector

The strategy used to update $\mathbf{k}(n)$ is as follows. We recognize that the gain vector estimates the pinning sequence using the input vector $\mathbf{X}(n)$ at time n and using the input vector $\mathbf{X}(n-1)$ at time $n-1$. We consider an *augmented input vector* $\mathbf{X}_{N+1}(n)$ defined as

$$\mathbf{X}_{N+1}(n) = \begin{bmatrix} \mathbf{v}(n) \\ \mathbf{X}(n-1) \end{bmatrix}, \tag{7.80}$$

which contains all the elements of $\mathbf{X}(n)$ and $\mathbf{X}(n-1)$. By appropriately rearranging the elements of $\mathbf{X}_{N+1}(n)$ using a permutation matrix $\mathbf{J}$, we can also explicitly show the dependence of $\mathbf{X}_{N+1}(n)$ on $\mathbf{X}(n)$ as

$$\mathbf{J}\mathbf{X}_{N+1}(n) = \begin{bmatrix} \mathbf{X}(n) \\ \mathbf{r}(n) \end{bmatrix}. \tag{7.81}$$

Because the augmented input vector contains the elements of $\mathbf{X}(n)$ as well as those of $\mathbf{X}(n-1)$, a reasonable strategy to update the gain vector is to first find an *augmented gain vector* $\mathbf{k}_{N+1}(n)$ that estimates the pinning sequence $\pi_k(n)$ using $\mathbf{X}_{N+1}(n)$ and then determine any relationship that might exist among $\mathbf{k}(n)$, $\mathbf{k}_{N+1}(n)$, and $\mathbf{k}(n-1)$. Consequently, we attempt to find an efficient iteration that first updates $\mathbf{k}_{N+1}(n)$ from $\mathbf{k}(n-1)$, and then solves for $\mathbf{k}(n)$ from $\mathbf{k}_{N+1}(n)$.

7.4.3 Solving for the Augmented Gain Vector

Both $\mathbf{k}_{N+1}(n)$ and $\mathbf{k}(n-1)$ estimate a sequence that is one at a single time instant and zero everywhere else. Furthermore, $\mathbf{k}_{N+1}(n)$ uses $\mathbf{X}_{N+1}(n)$ to make the estimate. The augmented input vector contains all the elements of $\mathbf{X}(n-1)$ and $\mathbf{v}(n)$. Consequently, we can express the estimate $\mathbf{k}_{N+1}^T(n)\mathbf{X}_{N+1}(n)$ as a sum of the estimate using $\mathbf{k}(n-1)$ given by $\mathbf{k}^T(n-1)\mathbf{X}(n-1)$, and the estimate using the *new information* contained in $\mathbf{v}(n)$. The new or additional information contained in $\mathbf{v}(n)$ is its component that is orthogonal (in the least-squares sense) to $\mathbf{X}(n-1)$. This component is simply the error in estimating $\mathbf{v}(n)$ using $\mathbf{X}(n-1)$, that is, the forward prediction error given by

$$\mathbf{f}_n(n) = \mathbf{v}(n) - \mathbf{A}^T(n)\mathbf{X}(n-1). \tag{7.82}$$

Thus, we express the estimate $\mathbf{k}_{N+1}^T(n)\mathbf{X}_{N+1}(n)$ as

$$\mathbf{k}_{N+1}^T(n)\mathbf{X}_{N+1}(n) = \mathbf{k}^T(n-1)\mathbf{X}(n-1) + \boldsymbol{\rho}_f^T(n)\mathbf{f}_n(n), \tag{7.83}$$

where $\boldsymbol{\rho}_f(n)$ is the coefficient vector for $\mathbf{f}_n(n)$ defined as

$$\boldsymbol{\rho}_f(n) = \boldsymbol{\alpha}^{-1}(n)\mathbf{f}_n(n), \tag{7.84}$$

where $\boldsymbol{\alpha}(n)$ represents the least-squares autocorrelation matrix of the forward prediction error sequence given by

$$\boldsymbol{\alpha}(n) = \sum_{k=1}^{n} \lambda^{n-k}\mathbf{f}_n(k)\mathbf{f}_n^T(k), \tag{7.85}$$

and the forward prediction error vector $\mathbf{f}_n(n)$ can also be interpreted as the cross-correlation vector of the forward prediction error sequence and the pinning sequence as can be seen from

$$\mathbf{f}_n(n) = \sum_{k=1}^{n} \lambda^{n-k}\pi_n(k)\mathbf{f}_n(k). \tag{7.86}$$

We will derive a recursion for $\boldsymbol{\alpha}(n)$ later. We can obtain an update equation for $\mathbf{k}_{N+1}(n)$ by rewriting (7.83) as

$$\mathbf{k}_{N+1}^T(n)\mathbf{X}_{N+1}(n) = \mathbf{k}^T(n-1)\mathbf{X}(n-1) + \boldsymbol{\rho}_f^T(n)(\mathbf{v}_n(n) - \mathbf{A}^T(n)\mathbf{X}(n-1))$$

$$= \begin{bmatrix} \boldsymbol{\rho}_f(n) \\ \mathbf{k}(n-1) - \mathbf{A}(n)\boldsymbol{\rho}_f(n) \end{bmatrix}^T \begin{bmatrix} \mathbf{v}(n) \\ \mathbf{X}(n-1) \end{bmatrix}, \tag{7.87}$$

and then equating the coefficients of $\mathbf{X}_{N+1}(n)$ on both sides of the equation. This operation results in

$$\mathbf{k}_{N+1}(n) = \begin{bmatrix} \boldsymbol{\rho}_f(n) \\ \mathbf{k}(n-1) - \mathbf{A}(n)\boldsymbol{\rho}_f(n) \end{bmatrix}, \tag{7.88}$$

which represents an efficient approach for obtaining $\mathbf{k}_{N+1}(n)$ from $\mathbf{k}(n-1)$.

7.4.4 Another Expression for the Augmented Gain Vector

We now derive an expression similar to (7.88) that involves $\mathbf{k}(n)$ rather than $\mathbf{k}(n-1)$. Given the two expressions for the augmented gain vector, we will then solve for $\mathbf{k}(n)$ from the second expression for $\mathbf{k}_{N+1}(n)$. Recall from (7.81) that we can decompose $\mathbf{X}_{N+1}(n)$ as

$$\mathbf{J}\mathbf{X}_{N+1}(n) = \begin{bmatrix} \mathbf{X}(n) \\ \mathbf{r}(n) \end{bmatrix}. \tag{7.89}$$

Therefore, we can also obtain an expression for $\mathbf{k}_{N+1}^T(n)\mathbf{X}_{N+1}(n)$ using this decomposition as

$$\mathbf{k}_{N+1}^T(n)\mathbf{X}_{N+1}(n) = \mathbf{k}^T(n)\mathbf{X}(n) + \boldsymbol{\rho}_b^T(n)\mathbf{b}_n(n), \tag{7.90}$$

where $\mathbf{b}_n(n)$ is the component of $\mathbf{r}(n)$ that is orthogonal to $\mathbf{X}(n)$, and is the backward prediction error given by

$$\mathbf{b}_n(n) = \mathbf{r}(n) - \mathbf{B}^T(n)\mathbf{X}(n). \tag{7.91}$$

The coefficient vector $\boldsymbol{\rho}_b(n)$ is given by

$$\boldsymbol{\rho}_b(n) = \boldsymbol{\beta}^{-1}(n)\mathbf{b}_n(n), \tag{7.92}$$

where $\boldsymbol{\beta}(n)$ is the least-squares autocorrelation matrix of the backward prediction error sequence, defined as

$$\boldsymbol{\beta}(n) = \sum_{k=1}^{n} \lambda^{n-k}\mathbf{b}_n(k)\mathbf{b}_n^T(k). \tag{7.93}$$

Following the same procedure employed to derive (7.88), we substitute (7.91) in (7.90) and collect similar terms together to get

$$\mathbf{k}_{N+1}^T(n)\mathbf{X}_{N+1}(n) = \begin{bmatrix} \mathbf{k}(n) - \mathbf{B}(n)\boldsymbol{\rho}_b(n) \\ \boldsymbol{\rho}_b(n) \end{bmatrix}^T \begin{bmatrix} \mathbf{X}(n) \\ \mathbf{r}(n) \end{bmatrix}. \tag{7.94}$$

Because $\mathbf{J}$ is a permutation matrix, it is relatively straightforward to show that $\mathbf{J}^T\mathbf{J} = \mathbf{J}\mathbf{J}^T = \mathbf{I}$. We substitute this relationship to rewrite (7.94) as

$$\mathbf{k}_{N+1}^T(n)\mathbf{X}_{N+1}(n) = \begin{bmatrix} \mathbf{k}(n) - \mathbf{B}(n)\boldsymbol{\rho}_b(n) \\ \boldsymbol{\rho}_b(n) \end{bmatrix}^T \mathbf{J}\mathbf{J}^T \begin{bmatrix} \mathbf{X}(n) \\ \mathbf{r}(n) \end{bmatrix}. \tag{7.95}$$

It immediately follows from (7.89) and (7.95) that the augmented gain vector is related to $\mathbf{k}(n)$ as

$$\mathbf{k}_{N+1}(n) = \mathbf{J}^T \begin{bmatrix} \mathbf{k}(n) - \mathbf{B}(n)\boldsymbol{\rho}_b(n) \\ \boldsymbol{\rho}_b(n) \end{bmatrix}. \tag{7.96}$$

7.4.5   Update for the Gain Vector

Let us premultiply both sides of (7.96) with $\mathbf{J}$. This operation results in a partition for the rearranged version of the augmented gain vector that results as given by

$$\mathbf{J}\mathbf{k}_{N+1}(n) = \begin{bmatrix} \mathbf{k}(n) - \mathbf{B}(n)\boldsymbol{\rho}_b(n) \\ \boldsymbol{\rho}_b(n) \end{bmatrix} = \begin{bmatrix} \mathbf{m}(n) \\ \mathbf{s}(n) \end{bmatrix}, \tag{7.97}$$

where $\mathbf{m}(n)$ contains $(N^2 + 3N)/2$ elements and $\mathbf{s}(n)$ contains $N + 1$ elements. From this equation, we can obtain an expression for $\mathbf{k}(n)$ as

$$\mathbf{k}(n) = \mathbf{m}(n) + \mathbf{B}(n)\boldsymbol{\rho}_b(n). \tag{7.98}$$

This expression requires knowledge of $\mathbf{B}(n)$ for its implementation. However, we can see from (7.76) that updating $\mathbf{B}(n)$ requires the gain vector $\mathbf{k}(n)$. We now substitute (7.76) in (7.98) to get

$$\mathbf{k}(n) = \mathbf{m}(n) + (\mathbf{B}(n-1) + \mathbf{k}(n)\mathbf{b}_{n-1}^T(n))\boldsymbol{\rho}_b(n). \tag{7.99}$$

From (7.97) we see that $\boldsymbol{\rho}_b(n) = \mathbf{s}(n)$. Consequently, we do not need to compute $\boldsymbol{\rho}_b(n)$ separately. We can solve for $\mathbf{k}(n)$ from (7.99) after substituting $\mathbf{s}(n)$ for $\boldsymbol{\rho}_b(n)$ to get an implementable and computationally efficient update equation for the gain vector as

$$\mathbf{k}(n) = (1 - \mathbf{b}_{n-1}^T(n)\mathbf{s}(n))^{-1}(\mathbf{m}(n) + \mathbf{B}(n-1)\mathbf{s}(n)). \tag{7.100}$$

7.4.6   The Final Pieces of the Derivation

We now have almost all the information required to successfully implement the fast version of the RLS adaptive second-order Volterra filter. In this subsection, we present some additional relationships to complete our derivation.

7.4.6.1   Updating the Least-Squares Forward Prediction Error Matrix
We leave it as an exercise for the reader to show that

$$\boldsymbol{\alpha}(n) = \mathbf{C}_{vv}(n) - \mathbf{P}_{vX}(n)\mathbf{A}(n), \tag{7.101}$$

where

$$\mathbf{C}_{\mathbf{vv}}(n) = \sum_{k=1}^{n} \lambda^{n-k} \mathbf{v}(k) \mathbf{v}^{T}(k) \tag{7.102}$$

and

$$\mathbf{P}_{\mathbf{vX}}(n) = \sum_{k=1}^{n} \lambda^{n-k} \mathbf{v}(k) \mathbf{X}^{T}(k-1) \tag{7.103}$$

are the exponentially weighted least-squares autocorrelation matrix of $\mathbf{v}(n)$ and the exponentially weighted least-squares cross-correlation matrix of $\mathbf{v}(n)$ and $\mathbf{X}(n-1)$, respectively. A similar expression was derived for the minimum mean-square error estimation problem in (5.19). The correlation matrices can be iteratively updated as

$$\mathbf{C}_{\mathbf{vv}}(n) = \lambda \mathbf{C}_{\mathbf{vv}}(n-1) + \mathbf{v}(n) \mathbf{v}^{T}(n) \tag{7.104}$$

and

$$\mathbf{P}_{\mathbf{vX}}(n) = \lambda \mathbf{P}_{\mathbf{vX}}(n-1) + \mathbf{v}(n) \mathbf{X}^{T}(n-1), \tag{7.105}$$

respectively. Furthermore, the predictor coefficient matrix $\mathbf{A}(n)$ is updated using

$$\mathbf{A}(n) = \mathbf{A}(n-1) + \mathbf{k}(n-1) \mathbf{f}_{n-1}^{T}(n). \tag{7.106}$$

We substitute the three update equations (7.104)–(7.106) in (7.101) to get

$$\begin{aligned}
\boldsymbol{\alpha}(n) &= \lambda \mathbf{C}_{\mathbf{vv}}(n-1) + \mathbf{v}(n) \mathbf{v}^{T}(n) \\
&\quad - (\lambda \mathbf{P}_{\mathbf{vX}}(n-1) + \mathbf{v}(n) \mathbf{X}^{T}(n-1))(\mathbf{A}(n-1) + \mathbf{k}(n-1) \mathbf{f}_{n-1}^{T}(n)) \\
&= \lambda (\mathbf{C}_{\mathbf{vv}}(n-1) - \mathbf{P}_{\mathbf{vX}}(n-1) \mathbf{A}(n-1)) \\
&\quad + \mathbf{v}(n) \mathbf{v}^{T}(n) - \mathbf{P}_{\mathbf{vX}}(n) \mathbf{k}(n-1) \mathbf{f}_{n-1}^{T}(n) - \mathbf{v}(n) \mathbf{X}^{T}(n-1) \mathbf{A}(n-1). \tag{7.107}
\end{aligned}$$

We can show, using the definition of the gain vector, that $\mathbf{P}_{\mathbf{vX}}(n) \mathbf{k}(n-1) = \mathbf{A}^{T}(n) \mathbf{X}(n-1)$. Furthermore, $\mathbf{A}^{T}(n) \mathbf{X}(n-1) = \mathbf{v}(n) - \mathbf{f}_{n}(n)$. Substituting these two relationships in (7.107), and recognizing that $\mathbf{C}_{\mathbf{vv}}(n-1) - \mathbf{P}_{\mathbf{vX}}(n-1) \mathbf{A}(n-1) = \boldsymbol{\alpha}(n-1)$, we get the update equation for the least-squares prediction error matrix to be

$$\begin{aligned}
\boldsymbol{\alpha}(n) &= \lambda \boldsymbol{\alpha}(n-1) + \mathbf{v}(n)(\mathbf{v}(n) - \mathbf{A}^{T}(n-1) \mathbf{X}(n-1))^{T} - (\mathbf{v}(n) - \mathbf{f}_{n}(n)) \mathbf{f}_{n-1}^{T}(n) \\
&= \lambda \boldsymbol{\alpha}(n-1) + \mathbf{f}_{n}(n) \mathbf{f}_{n-1}^{T}(n). \tag{7.108}
\end{aligned}$$

7.4.6.2 Relationship between A Priori and A Posteriori Errors There exist very simple relationships between the a priori estimation errors of the form

$e_{n-1}(n)$ and a posteriori estimation errors of the form $e_n(n)$. To see this, we substitute the update equation for the coefficient vector in the definition of the estimation error to get

$$
\begin{aligned}
e_n(n) &= d(n) - \vec{\mathbf{H}}^T(n)\mathbf{X}(n) \\
&= d(n) - (\vec{\mathbf{H}}(n-1) + \mathbf{k}(n)e_{n-1}(n))^T\mathbf{X}(n) \\
&= (1 - \mathbf{k}^T(n)\mathbf{X}(n))e_{n-1}(n) \\
&= \gamma(n)e_{n-1}(n),
\end{aligned}
\tag{7.109}
$$

where we have defined $\gamma(n)$ to be

$$
\gamma(n) = 1 - \mathbf{k}^T(n)\mathbf{X}(n). \tag{7.110}
$$

The parameter $\gamma(n)$ is the error in estimating the pinning sequence $\pi_k(n)$ at time n, and is commonly referred to as the *likelihood variable*. In a manner similar to the derivation of (7.109), we can also show that

$$
\mathbf{f}_n(n) = \gamma(n-1)\mathbf{f}_{n-1}(n) \tag{7.111}
$$

and

$$
\mathbf{b}_n(n) = \gamma(n)\mathbf{b}_{n-1}(n). \tag{7.112}
$$

We can now substitute (7.111) in (7.108) and apply the matrix inversion lemma to the resulting expression to obtain a recursive update equation for the inverse of the least-squares prediction error matrix as

$$
\boldsymbol{\alpha}^{-1}(n) = \lambda^{-1}\left[\boldsymbol{\alpha}^{-1}(n-1) - \frac{\boldsymbol{\alpha}^{-1}(n-1)\mathbf{f}_{n-1}(n)\mathbf{f}_{n-1}^T(n)\boldsymbol{\alpha}^{-1}(n-1)}{\dfrac{\lambda}{\gamma(n-1)} + \mathbf{f}_{n-1}^T(n)\boldsymbol{\alpha}^{-1}(n-1)\mathbf{f}_{n-1}(n)}\right]. \tag{7.113}
$$

7.4.6.3 *Updating the Likelihood Variable*   Instead of evaluating the likelihood variable using its definition, we derive and use a simple update equation for $\gamma(n)$. The procedure we follow for updating $\gamma(n)$ is to find an expression for $\gamma_{N+1}(n)$ defined as

$$
\gamma_{N+1}(n) = 1 - \mathbf{k}_{N+1}^T(n)\mathbf{X}_{N+1}(n) \tag{7.114}
$$

employing $\gamma(n-1)$ and then derive an expression for $\gamma(n)$ that depends on $\gamma_{N+1}(n)$. For the first of these two tasks, we substitute the update equation (7.88) for the augmented gain vector in (7.114) to get

$$\gamma_{N+1}(n) = 1 - \left[\begin{array}{c} \boldsymbol{\rho}_f(n) \\ \mathbf{k}(n-1) - \mathbf{A}(n)\boldsymbol{\rho}_f(n) \end{array} \right]^T \left[\begin{array}{c} \mathbf{v}(n) \\ \mathbf{X}(n-1) \end{array} \right]$$

$$= 1 - \mathbf{k}^T(n-1)\mathbf{X}(n-1) - \boldsymbol{\rho}_f^T(n)(\mathbf{v}(n) - \mathbf{A}^T(n)\mathbf{X}(n-1))$$

$$= \gamma(n-1) - \boldsymbol{\rho}_f^T(n)\mathbf{f}_n(n). \tag{7.115}$$

In a similar manner, we can show, using (7.96) and the fact that $\boldsymbol{\rho}_b(n) = \mathbf{s}(n)$, that

$$\gamma_{N+1}(n) = \gamma(n) - \mathbf{s}^T(n)\mathbf{b}_n(n). \tag{7.116}$$

Substituting $\gamma(n)\mathbf{b}_{n-1}(n)$ for $\mathbf{b}_n(n)$ in this expression and solving for $\gamma(n)$, we get the desired update equation for $\gamma(n)$ to be

$$\gamma(n) = (1 - \mathbf{s}^T(n)\mathbf{b}_{n-1}(n))^{-1}\gamma_{N+1}(n). \tag{7.117}$$

7.4.7 The Complete Algorithm

The complete algorithm that results from the derivations is tabulated in Table 7.4. The initialization process for the various parameters is also described in the table. The most computationally burdensome parts of the algorithm requires $O(N^3)$ arithmetical operations. Consequently, the overall complexity of the fast RLS adaptive second-order Volterra filter corresponds to $O(N^3)$ operations per iteration. This complexity is an order of magnitude lower than the $O(N^4)$ arithmetical operations required to implement the conventional RLS adaptive Volterra filter, and thus the algorithm of Table 7.4 represents one of the most efficient algorithms available at present for least-squares adaptive second-order Volterra filtering.

The reader should keep in mind that the fast Volterra filter described above implements the exact solution to the least-squares estimation problem in an efficient manner, and therefore this algorithm retains the good convergence and tracking properties of the RLS adaptive filters demonstrated in Examples 7.2 and 7.3. The only differences in the performances of various realizations of the exact RLS solution occur because of the differences in the numerical properties of the implementations. This issue is discussed in a later subsection.

7.4.8 Extension to Higher-Order Volterra Filters

Extension of the fast RLS second-order Volterra filter to the higher-order case is straightforward. In fact, Table 7.4 can also be used to describe a fast RLS pth-order Volterra filter with a memory span of N samples if the input vector and all other matrices and vectors are defined properly. In this case, the input vector has $O(N^p)$

TABLE 7.4 Fast RLS Adaptive Second-Order Volterra Filter

Initialization

$$\boldsymbol{\alpha}^{-1}(0) = \delta^{-1}\mathbf{I}_{N+1}$$

$$\gamma(0) = 1$$

$$\mathbf{A}(0) = \mathbf{0}_{((N^2+3N)/2)\times(N+1)}$$

$$\mathbf{B}(0) = \mathbf{0}_{((N^2+3N)/2)\times(N+1)}$$

$$\mathbf{k}(0) = \mathbf{0}_{((N^2+3N)/2)\times1}$$

$$\vec{\mathbf{H}}(0) = \mathbf{0}_{((N^2+3N)/2)\times1}$$

Main iteration

$$\mathbf{v}^T(n) = [x(n)\ x^2(n)\ x(n)x(n-1)\ \cdots\ x(n)x(n-N+1)]$$

$$\mathbf{X}_{N+1}(n) = \begin{bmatrix} \mathbf{v}(n) \\ \mathbf{X}(n-1) \end{bmatrix}$$

$$\begin{bmatrix} \mathbf{X}(n) \\ \mathbf{r}(n) \end{bmatrix} = \mathbf{J}\mathbf{X}_{N+1}(n)$$

$$\mathbf{f}_{n-1}(n) = \mathbf{v}(n) - \mathbf{A}^T(n-1)\mathbf{X}(n-1)$$

$$\mathbf{f}_n(n) = \gamma(n-1)\mathbf{f}_{n-1}(n)$$

$$\boldsymbol{\alpha}^{-1}(n) = \lambda^{-1}\left[\boldsymbol{\alpha}^{-1}(n-1) - \frac{\boldsymbol{\alpha}^{-1}(n-1)\mathbf{f}_{n-1}(n)\mathbf{f}_{n-1}^T(n)\boldsymbol{\alpha}^{-1}(n-1)}{\lambda/\gamma(n-1) + \mathbf{f}_{n-1}^T(n)\boldsymbol{\alpha}^{-1}(n-1)\mathbf{f}_{n-1}(n)}\right]$$

$$\boldsymbol{\rho}_f(n) = \boldsymbol{\alpha}^{-1}(n)\mathbf{f}_n(n)$$

$$\gamma_{N+1}(n) = \gamma(n-1) - \boldsymbol{\rho}_f^T(n)\mathbf{f}_n(n)$$

$$\mathbf{A}(n) = \mathbf{A}(n-1) + \mathbf{k}(n-1)\mathbf{f}_{n-1}^T(n)$$

$$\mathbf{k}_{N+1}(n) = \begin{bmatrix} \boldsymbol{\rho}_f(n) \\ \mathbf{k}(n-1) - \mathbf{A}(n)\boldsymbol{\rho}_f(n) \end{bmatrix}$$

$$\begin{bmatrix} \mathbf{m}(n) \\ \mathbf{s}(n) \end{bmatrix} = \mathbf{J}\mathbf{k}_{N+1}(n)$$

$$\mathbf{b}_{n-1}(n) = \mathbf{r}(n) - \mathbf{B}^T(n-1)\mathbf{X}(n)$$

$$\mathbf{k}(n) = (1 - \mathbf{b}_{n-1}^T(n)\mathbf{s}(n))^{-1}(\mathbf{m}(n) + \mathbf{B}(n-1)\mathbf{s}(n))$$

$$\gamma(n) = (1 - \mathbf{b}_{n-1}^T(n)\mathbf{s}(n))^{-1}\gamma_{N+1}(n)$$

$$\mathbf{B}(n) = \mathbf{B}(n-1) + \mathbf{k}(n)\mathbf{b}_{n-1}^T(n)$$

$$e_{n-1}(n) = d(n) - \vec{\mathbf{H}}^T(n-1)\mathbf{X}(n)$$

$$e_n(n) = \gamma(n)e_{n-1}(n)$$

$$\vec{\mathbf{H}}(n) = \vec{\mathbf{H}}(n-1) + \mathbf{k}(n)e_{n-1}(n)$$

---

elements, and the computational complexity of the algorithm corresponds to $O(N^{2p-1})$ multiplications during each iteration. Even though this represents a savings over the conventional RLS adaptive filter, this complexity is significant for even moderately large values of N and p. Much additional work is needed in reducing the computational complexity of RLS adaptive Volterra filters.

7.4.9 Numerical Properties of the Fast RLS Volterra Filter

One of the major problems of fast RLS algorithms employing the direct-form structure is their poor numerical properties. Several researchers have investigated the numerical error propagation in fast RLS adaptive linear filters [54,55,144,148,294] and suggested methods to overcome the lack of robustness of these algorithms to finite precision errors in their implementation. One common approach for improving the numerical properties of fast RLS adaptive filters is to use *rescue devices* [144]. It is not difficult to show that the likelihood variable is bounded according to $0 \leq \gamma(n) \leq 1$ if the algorithm is implemented with infinite precision. If $\gamma(n)$ goes out of this range, it is only because of numerical errors. Experiments by several researchers have shown that $\gamma(n)$ usually goes out of its range just before the onset of divergence of the algorithm. The rescue method attempts to stabilize the algorithm by reinitializing the adaptive filter whenever $\gamma(n)$ goes below zero or exceeds one. The reinitialization is achieved by resetting all the variables of the adaptive filter to their initial values as given in Table 7.4 with the exception that $\vec{\mathbf{H}}(n)$ is left unchanged. All the experiments described in the previous section were performed using fast RLS Volterra filters, and we did not observe any difficulties due to numerical instabilities for the duration of the simulations. The rescue method is not guaranteed to stabilize the fast RLS adaptive filters using direct-form system models, and it is almost certain that such algorithms will eventually diverge as a result of poor numerical properties. However, it has often been the case that the rescue device substantially increases the duration for which the adaptive filter operates in a reliable manner.

A method for stabilizing the fast RLS filters by introducing redundancy in the calculations by computing the same variables using several methods and then using feedback of the numerical errors estimated from these multiple calculations has been described in [294]. This method was shown to be exponentially stable if the feedback constants are appropriately chosen. This method can be extended to the case of the fast Volterra filters also. However, no analysis on the selection of the feedback constants has yet been performed for Volterra filters.

Fast RLS adaptive Volterra filters implemented using the lattice structure described in Section 5.4 are numerically stable. Variations of the lattice filters that employ QR-decomposition-based algorithms are also numerically stable. The interested reader is referred to Syed and Mathews [306,307] for additional information.

7.5 ADAPTIVE VOLTERRA FILTERS USING DISTRIBUTED ARITHMETIC

In this section we discuss how LMS adaptive Volterra filters can be implemented using distributed arithmetic. We use the same notation employed in Sections 3.2.4 and 3.4.2. Recall from (3.85) that an estimate of the desired response signal $d(n)$

obtained as the output of a homogeneous pth-order adaptive Volterra filter can be written as

$$\hat{d}(n) = \mathbf{X}_p^T(n)\vec{\mathbf{H}}_p(n) = \mathbf{P}_p^T\mathbf{Q}_p^T(n)\vec{\mathbf{H}}_p(n), \tag{7.118}$$

where the input vector $\mathbf{X}_p(n)$ has been decomposed as $\mathbf{Q}_p(n)\mathbf{P}_p$ using the distributed arithmetic realization. Efficient realizations in distributed arithmetic involves first calculating the matrix product $\mathbf{Q}_p^T(n)\vec{\mathbf{H}}_p(n)$ and then computing the output signal by multiplying this vector with $\mathbf{P}_p^T$. The approach followed in the distributed arithmetic realization of the adaptive filter is to update the product vector $\mathbf{Q}_p^T(n)\vec{\mathbf{H}}_p(n)$ directly rather than to update the coefficient vector $\vec{\mathbf{H}}_p(n)$.

The conventional LMS adaptive filter updates the coefficients $\vec{\mathbf{H}}_p(n)$ using the relationship

$$\vec{\mathbf{H}}_p(n+1) = \vec{\mathbf{H}}_p(n) + \mu_p\mathbf{X}_p(n)e(n) \tag{7.119}$$

where μ_p is the step size of the adaptive filter and $e(n)$ is the estimation error given by

$$e(n) = d(n) - \hat{d}(n). \tag{7.120}$$

The distributed arithmetic realization of the adaptive filter assumes an update equation for $\mathbf{Q}_p^T(n)\mathbf{H}_p(n)$ with a structure similar to that in (7.119) as given by

$$\mathbf{Q}_p^T(n)\vec{\mathbf{H}}_p(n+1) = \mathbf{Q}_p^T(n)\vec{\mathbf{H}}_p(n) + \mu_p\mathbf{Q}_p^T(n)\mathbf{X}_p(n)e(n)$$
$$= \mathbf{Q}_p^T(n)\vec{\mathbf{H}}_p(n) + \mu_p\mathbf{Q}_p^T(n)\mathbf{Q}_p(n)\mathbf{P}_pe(n). \tag{7.121}$$

In the memory-oriented realizations using distributed arithmetic, the rows of $\mathbf{Q}_p^T(n)$ serve as address locations in which the appropriate entries of the product vector $\mathbf{Q}_p^T(n)\vec{\mathbf{H}}_p(n)$ are stored. Consequently, the adaptive filter updates in (7.121) are applied to the memory locations specified by the rows of $\mathbf{Q}_p^T(n)$. At time $(n+1)$, the rows of $\mathbf{Q}_p^T(n+1)$ are employed as the addresses of the memory locations that contain entries of $\mathbf{Q}_p^T(n+1)\vec{\mathbf{H}}_p(n+1)$, which are then employed to find $\hat{d}(n+1)$. The iterations continue by computing the error and updating the memory locations as in (7.120) and (7.121), respectively. If the input bits are uncorrelated with each other, we can show that the adaptive filter will update the entries in all the memory locations within a finite number of iterations.

7.5.1 A Simpler Update Equation

A reasonable approximation that one can employ is to replace the matrix $\mathbf{Q}_p^T(n)\mathbf{Q}_p(n)$ in the update equation with its statistical expectation or an approximation of the statistical expectation. It has been shown in [288] that $E\{\mathbf{Q}_p^T(n)\mathbf{Q}_p(n)\}$ is a sparse matrix when the input signal has zero mean value and their constituent bits are uncorrelated with each other. When the offset binary code is used, the nonzero elements of $E[\mathbf{Q}_p^T(n)\mathbf{Q}_p(n)]$ take their values from the set $N, N^2, \ldots, N^p$. In the linear case, this matrix becomes diagonal, while in the nonlinear case, the elements with value N^p are located on the diagonal of the matrix. Even for moderately large values of N, the diagonal terms dominate the other entries in the matrix. Consequently, it is reasonable to approximate the matrix $E[\mathbf{Q}_p^T(n)\mathbf{Q}_p(n)]$ with $N^p\mathbf{I}$. With this approximation, the adaptation algorithm becomes [288]

$$\mathbf{Q}_p^T(n)\vec{\mathbf{H}}_p(n+1) = \mathbf{Q}_p^T(n)\vec{\mathbf{H}}_p(n) + \mu N^p \mathbf{P}_p e(n), \tag{7.122}$$

which is considerably simpler than the update equation in (7.121). Additional complexity reductions can be obtained by using the address compression and split address techniques described in Section 3.2.4.

Remark 7.2 It is technically possible for multiple rows in $\mathbf{Q}_p^T(n+1)$ to be identical at some time instant. Such a situation will result in more than one update of the entry in the same memory location at that time. This may lead to unstable behavior of the adaptive filter when the step size is chosen to be relatively large. The potential for unstable behavior in such situations can be avoided by updating the entry in a particular memory location no more than once during each iteration.

Example 7.4 This example considers the memory-oriented realization of an adaptive second-order Volterra filter using distributed arithmetic. The adaptive filter was employed in a system identification problem. The unknown system was identical to the system in Example 7.2. The input signals to the adaptive filter were represented using finite wordlengths of $B = 2, B = 4$, and $B = 8$ bits, and coded with the offset binary arithmetic. Furthermore, the input samples belonged to IID processes with equal probability of taking any of the possible 2^B distinct values. The output signal was corrupted with a white Gaussian noise sequence such that the output signal-to-noise ratio (SNR) was equal to 30 dB in each experiment. Figure 7.9 displays the ensemble averages of the squared estimation error associated with 50 independent experiments. For each experiment, the mean-square error was computed as the time average of 50 consecutive samples and the time averages were further averaged over the different experiments. The adaptation algorithm in (7.122) was used for updating the coefficients, with the constraint that the entry in any memory location was updated no more than once during each iteration. The continuous curve in Figure 7.9 refers to the case of an input signal with $B = 2$ bit/sample resolution, the dashed curve corresponds to the case when the input signal is represented using $B = 4$ bits, and the dotted and dashed curve refers to the case in which the input signal is represented using $B = 8$ bits. In the first set of experiments the step size was chosen to be equal to 3×2^{-5}, while in the other two sets it was chosen to be equal to 2^{-5}. The dotted curve in Figure 7.9 shows the mean-square error associated with the conventional LMS adaptation algorithm for $B = 8$ and $\mu = 0.7 \times 2^{-5}$. The step sizes for the different experiments were chosen such that the steady-state estimation error was approximately the same in all the experiments. Figure 7.9 shows that the distributed arithmetic realizations are characterized by slower convergence rates

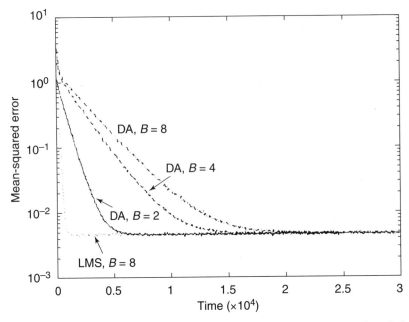

Figure 7.9 Adaptation plots for a second-order Volterra filter using memory-oriented distributed arithmetic realizations and the conventional LMS algorithms. (Courtesy of A. Carini.)

when compared with the conventional LMS algorithm. On the other hand, its complexity is remarkably lower since all the multiplications necessary to compute the products of the input samples as well as the multiplications associated with the coefficients are avoided in the distributed arithmetic realization.

7.6 ADAPTIVE LATTICE VOLTERRA FILTERS FOR GAUSSIAN INPUTS

As was discussed in Section 7.1.2, the performance of the LMS adaptive filter degrades considerably when the elements of the input vector to the adaptive filter are correlated with each other. If the elements of the input vector are mutually orthogonal, we can select the step size associated with each element such that every coefficient converges at about the same rate. Furthermore, the convergence behaviors of the different coefficients are not coupled to each other in this case. We now develop an adaptive lattice Volterra filter [171] that uses the lattice orthogonalization technique described in Section 5.3.5 for Gaussian input signals.

7.6.1 An Adaptive Lattice Linear Predictor

The key component in the lattice realization of the Volterra filter for Gaussian input signals is the linear lattice predictor shown in Figure 7.10. A brief overview of linear

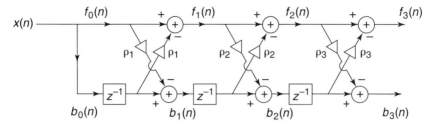

Figure 7.10 A linear lattice predictor.

lattice predictors is given in Appendix C, where it is shown that the reflection coefficient ρ_k can be interpreted in two different ways:

1. ρ_k estimates the forward prediction error signal $f_{k-1}(n)$ using the delayed backward prediction error signal $b_{k-1}(n-1)$.
2. ρ_k estimates $b_{k-1}(n-1)$ using $f_{k-1}(n)$.

Furthermore, the forward and backward prediction errors $f_k(n)$ and $b_k(n)$ can be interpreted as the estimation error signals associated with problems 1 and 2, respectively. Consequently, we can adapt ρ_k independently of other stages. The most common approach to implementing such an adaptive filter is to update ρ_k such that the sum of the squares of the forward and backward prediction errors is reduced at each iteration using a gradient descent algorithm. This procedure results in the following update equations for the reflection coefficients:

$$f_k(n) = f_{k-1}(n) - \rho_k(n)b_{k-1}(n-1), \tag{7.123}$$

$$b_k(n) = b_{k-1}(n-1) - \rho_k(n)f_{k-1}(n) \tag{7.124}$$

and

$$\rho_k(n+1) = \rho_k(n) - \frac{\mu_k}{2}\frac{\partial}{\partial \rho_k(n)}\{f_k^2(n) + b_k^2(n)\}$$

$$= \rho_k(n) + \mu_k\{f_k(n)b_{k-1}(n-1) + b_k(n)f_{k-1}(n)\}. \tag{7.125}$$

Since the input signal power at any stage is in general different from that at other stages, it is common practice to normalize the step sizes with an estimate of the input signal power. Let $\sigma_k^2(n)$ correspond to an estimate of the sum of the signal powers at the kth stage. One way of computing $\sigma_k^2(n)$ is as the output of a first-order lowpass filter whose input is $f_{k-1}^2(n) + b_{k-1}^2(n-1)$. This results in an estimate iteratively computed as

$$\sigma_k^2(n) = \lambda\sigma_k^2(n-1) + (1-\lambda)\{f_{k-1}^2(n) + b_{k-1}^2(n-1)\}. \tag{7.126}$$

The normalized LMS lattice predictor uses $\mu/\sigma_k^2(n)$ in place of μ_k to update the reflection coefficients as

$$\rho_k(n+1) = \rho_k(n) + \frac{\mu}{\sigma_k^2(n)}\{f_k(n)b_{k-1}(n-1) + b_k(n)f_{k-1}(n-1)\}. \qquad (7.127)$$

Equations (7.123), (7.124), (7.126), and (7.127) constitute the kth stage of a normalized adaptive linear lattice predictor. Since zeroth-order prediction involves no prediction, the zeroth-order forward and backward prediction errors are identical to the input signal itself. Therefore, the lattice predictor is initialized with

$$f_0(n) = b_0(n) = x(n). \qquad (7.128)$$

This initialization procedure is shown in the block diagram as a direct connection of the input signal to the lower and upper paths of the lattice predictor.

7.6.2 Extending the Adaptive Lattice Predictor to the Second-Order Volterra Filtering

Recall from Section 5.3.5 that the signal set consisting of

$$\{1, \ b_0(n), \ b_1(n), \ldots, b_{N-1}(n), \ b_0^2(n) - \sigma_0^2(n), \ b_0(n)b_1(n),$$
$$\ldots, b_0(n)b_{N-1}(n), \ b_1^2(n) - \sigma_1^2(n), \ldots, b_{N-1}^2(n) - \sigma_N^2(n)\}, \qquad (7.129)$$

where $\sigma_k^2(n)$ represents the mean-squared value of $b_k(n)$ at time n, comprises an orthogonal basis set for the second-order Volterra estimation problem involving N-sample memory. Furthermore, the coefficients corresponding to each element of the set in (7.129) can be evaluated independently of all others. Consequently, we can develop an adaptive filter whose coefficients are adapted individually. The block diagram for generating the input signals to the nonlinear estimation structure is shown in Figure 7.11. Let $e_b(n)$ represent the residual after removing the mean value of the desired response signal from $d(n)$, and $e_i^{(1)}(n)$ represents the error in estimating the mean-removed desired response signal $e_b(n)$ using the first i backward prediction error signals, i.e.,

$$e_i^{(1)}(n) = e_b(n) - \sum_{k=1}^{i} g_k^{(1)}(n)b_k(n). \qquad (7.130)$$

By defining $e_{-1}^{(1)}(n) = e_b(n)$, we can evaluate $e_i^{(1)}(n)$ sequentially as

$$e_i^{(1)}(n) = e_{i-1}^{(1)}(n) - g_i^{(1)}(n)b_i(n); \quad i = 0, 1, \ldots, N-1. \qquad (7.131)$$

Except for the computations involving the mean values, this estimation procedure makes use of only linear terms. The estimation errors due to the nonlinear

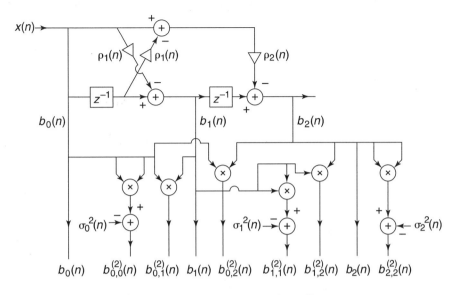

Figure 7.11 Adaptive lattice Volterra filter.

components are also computed sequentially as

$$e_{i,j}^{(2)}(n) = e_{i,j-1}^{(2)}(n) - g_{i,j}^{(2)}(n)b_{i,j}^{(2)}(n), \tag{7.132}$$

where

$$b_{i,j}^{(2)}(n) = b_i(n)b_j(n) - \sigma_{1,i}^2(n)\delta(i-j) \tag{7.133}$$

is obtained by removing an estimate of the mean value from the product signals, $\sigma_{1,i}^2(n)$ is an estimate of the mean-squared value of the ith-order backward prediction error signal, $g_{i,j}^{(2)}(n)$ is the coefficient multiplying $b_{i,j}^{(2)}(n)$ at time n, and $e_{i,-1}^{(2)}(n)$ is defined as

$$e_{i,-1}^{(2)}(n) = \begin{cases} e_{i-1,N-1}^{(2)}(n); & i = 1, 2, \ldots, N-1 \\ e_{N-1}^{(1)}(n); & i = 0 \end{cases} \tag{7.134}$$

The steps involved in the adaptive estimation of the components of the system of Figure 7.11 are described below.

7.6.2.1 Estimation of the Mean Values
The mean value of the desired response signal is estimated as

$$\overline{d(n)} = \lambda\overline{d(n-1)} + (1-\lambda)d(n), \tag{7.135}$$

where λ is a constant in the range $(0,1)$. Then, $e_b(n) = e_{-1}^{(1)}(n)$ is estimated adaptively as

$$e_b(n) = d(n) - \overline{d(n)}. \tag{7.136}$$

For the quadratic estimator, the squared backward prediction error signals have nonzero mean values. These mean values are removed by iteratively estimating the mean-square values of the backward prediction error signals as

$$\sigma_{1,i}^2(n) = \lambda\sigma_{1,i}^2(n-1) + (1-\lambda)b_i^2(n) \tag{7.137}$$

and then evaluating $b_{i,j}^{(2)}(n)$ using (7.133).

7.6.2.2 Coefficient Adaptation
It is important to realize that the elements of the signal set defined using $\{b_i(n), b_{i,j}^{(2)}(n); 0 \le i, j \le N-1\}$ are not necessarily mutually orthogonal. However, we expect that, in time, the adaptive processor will generate a signal set whose elements are almost orthogonal to each other. We develop the adaptation procedure as if the signal set is an orthogonal basis set for our estimation problem at all times. Consequently, each coefficient is updated independently of the others, since each coefficient is assumed to perform a single coefficient estimation problem on its own. For example, the coefficient $g_i^{(1)}(n)$ may be thought of as the single coefficient of the adaptive filter that estimates $e_{i-1}^{(1)}(n)$ using $b_i(n)$. Therefore we can develop a single-coefficient normalized LMS adaptive filter for this problem as

$$e_i^{(1)}(n) = e_{i-1}^{(1)}(n) - g_i^{(1)}(n)b_i(n), \tag{7.138}$$

and

$$g_i^{(1)}(n+1) = g_i^{(1)}(n) + \frac{\mu}{\sigma_{1,i}^2(n)}e_i^{(1)}(n)b_i(n). \tag{7.139}$$

Similar derivations can be made for the nonlinear coefficients also. The single-coefficient adaptive filter for updating $g_{i,j}^{(2)}(n)$ may be derived in the following manner. We first estimate the power in $b_{i,j}^{(2)}(n)$ using a first-order lowpass filter as

$$\sigma_{2,i,j}^2(n) = \lambda\sigma_{2,i,j}^2(n-1) + (1-\lambda)(b_{i,j}^{(2)}(n))^2. \tag{7.140}$$

The coefficients are now updated as

$$g_{i,j}^{(2)}(n+1) = g_{i,j}^{(2)}(n) + \frac{\mu}{\sigma_{2,i,j}^2(n)}e_{i,j}^{(2)}(n)b_{i,j}^{(2)}(n). \tag{7.141}$$

The complete algorithm for the second-order adaptive Volterra filter with lattice orthogonalization for Gaussian input signals is given in Table 7.5. The adaptive lattice second-order Volterra filter was developed by Koh and Powers [115]. Extensions to the higher-order Volterra systems can be found in Mathews [171].

Example 7.5 In this example, we present the results of an experiment that demonstrates the properties of the adaptive lattice Volterra filter when its input signals have narrowband characteristics. The results presented are ensemble averages over 50 independent simulations of a system identification problem. The unknown system was a second-order Volterra filter described by the following input–output relationship:

$$y(n) = -0.78x(n) - 1.48x(n-1) + 1.39x(n-2) + 0.04x(n-3)$$
$$+ 0.54x^2(n) + 3.72x(n)x(n-1) + 1.86x(n)x(n-2) - 0.76x(n)x(n-3)$$
$$- 1.62x^2(n-1) + 0.76x(n-1)x(n-2) - 0.12x(n-1)x(n-3)$$
$$+ 1.41x^2(n-2) - 1.52x(n-2)x(n-3) - 0.13x^2(n-3)$$

This system is identical to the one used in Example 7.2. Four different types of input signals were used in the simulations. Each signal set was generated as the output of a linear system with input–output relationship

$$x(n) = bx(n-1) + \sqrt{1-b^2}\xi(n),$$

where $\xi(n)$ was a zero-mean and white Gaussian noise with unit variance and b was a parameter between 0 and 1 that determined the level of correlation between adjacent samples of the process $x(n)$. Experiments were conducted with the parameter b set to 0.00, 0.50, 0.90, and 0.99. When $b = 0$, the input signal is white. As the parameter b approaches 1, the signal characteristics become highly lowpass in nature. The desired response signals were generated by passing the input signals described above through the unknown system, and corrupting the output signals with additive zero-mean and Gaussian noise with variance 0.1. The measurement noise sequence and the input signal $x(n)$ were mutually uncorrelated. In all the experiments, μ and β were chosen to be 0.001 and 0.999, respectively. Figure 7.12 displays overlaid plots of the squared estimation error signal, averaged over the 50 runs. These error curves were further smoothed by time-averaging over 10 consecutive samples. It can be seen from the figure that the rate of convergence of the adaptive filter is similar in all cases, in spite of the fairly large disparity in the spectra of the input signals. It appears from the results of this experiment that the objective of designing an adaptive filter that is relatively insensitive to the statistics of the input signals has been achieved.

7.7 ADAPTIVE FILTERS FOR PARALLEL-CASCADE STRUCTURES

We saw in Sections 3.2.3 and 3.4.1 that parallel-cascade realizations of truncated Volterra filters provide the advantages of modularity as well as the ability to simplify the filter through approximate representations. An adaptive Volterra filter in which

TABLE 7.5 Adaptive Second-Order Volterra Filter with Lattice Orthogonalization for Gaussian Inputs

Mean removal

$$\overline{x(n)} = \lambda \overline{x(n-1)} + (1-\lambda)x(n)$$

$$\overline{d(n)} = \lambda \overline{d(n-1)} + (1-\lambda)d(n)$$

$$x_b(n) = x(n) - \overline{x(n)}$$

$$e_b(n) = d(n) - \overline{d(n)}$$

Linear lattice predictor

$$f_0(n) = b_0(n) = x_b(n)$$

For $k = 1, 2, \ldots, N-1$

$$f_k(n) = f_{k-1}(n) - \rho_k(n)b_{k-1}(n-1)$$

$$b_k(n) = b_{k-1}(n-1) - \rho_k(n)f_{k-1}(n)$$

$$\sigma_k^2(n) = \lambda\sigma_k^2(n-1) + (1-\lambda)\{f_{k-1}^2(n) + b_{k-1}^2(n-1)\}$$

$$\rho_k(n+1) = \rho_k(n) + \frac{\mu}{\sigma_k^2(n)}\{f_k(n)b_{k-1}(n-1) + b_k(n)f_{k-1}(n)\}$$

Joint process estimation
Linear part

$$e_{-1}^{(1)}(n) = e_b(n)$$

For $i = 0, 1, \ldots, N-1$

$$e_i^{(1)}(n) = e_{i-1}^{(1)}(n) - g_i^{(1)}(n)b_i(n)$$

$$\sigma_{1,i}^2(n) = \lambda\sigma_{1,i}^2(n-1) + (1-\lambda)b_i^2(n)$$

$$g_i^{(1)}(n+1) = g_i^{(1)}(n) + \frac{\mu}{\sigma_{1,i}^2(n)}e_i^{(1)}(n)b_i(n)$$

Nonlinear part

For $i = 0, 1, 2, \ldots, N-1$

$$e_{i,-1}^{(2)}(n) = \begin{cases} e_{N-1}^{(1)}(n); & i = 0 \\ e_{i-1,N-1}^2(n); & i \neq 0 \end{cases} \quad \text{For } j = i, i+1, \ldots, N-1$$

$$b_{i,j}^{(2)}(n) = b_i(n)b_j(n) - \sigma_{1,i}^2(n)\delta(i-j)$$

$$e_{i,j}^{(2)}(n) = e_{i,j-1}(n) - g_{i,j}^{(2)}(n)b_{i,j}^{(2)}(n)$$

$$\sigma_{2,i,j}^2(n) = \lambda\sigma_{2,i,j}^2(n-1) + (1-\lambda)(b_{i,j}^{(2)}(n))^2$$

$$g_{i,j}^{(2)}(n+1) = g_{i,j}^{(2)}(n) + \frac{\mu}{\sigma_{i,j}^2(n)}e_{i,j}^{(2)}(n)b_{i,j}^{(2)}(n)$$

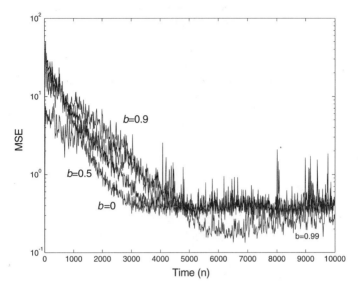

Figure 7.12   Mean-square estimation error of the adaptive lattice Volterra filter in Example 7.5 for four different input signals. (Copyright © 1996 IEEE.)

the system model is realized using the parallel-cascade structure may retain both these advantages. In this section, we develop algorithms for adapting the coefficients of a truncated Volterra filter realized using the parallel-cascade structure. We assume throughout our discussion that the system model is a homogeneous pth-order Volterra filter. Generalizations to inhomogeneous Volterra system models are straightforward extensions of the ideas developed here.

7.7.1   LMS Adaptation for the Parallel-Cascade Structure

Recall from the discussion of parallel-cascade structures in Section 3.4.1 that the output of a homogeneous pth-order parallel-cascade adaptive Volterra filter with N-sample memory implemented using lth- and $(p - l)$th-order branches may be written as

$$\hat{d}(n) = \sum_{i=1}^{r} \sigma_i(n) y_{l,i}(n) y_{p-l,i}(n)$$

$$= \sum_{i=1}^{r} \sigma_i(n) [\mathbf{X}_{N,l}^T(n) \mathbf{U}_i(n)][\mathbf{V}_i^T(n) \mathbf{X}_{N,p-l}(n)], \qquad (7.142)$$

where $y_{l,i}(n)$ and $y_{p-l,i}(n)$ represent the output of the lth- and $(p - l)$th-order Volterra filters, respectively, in the ith branch of the realization, $\mathbf{U}_i(n)$ is the coefficient vector corresponding to the lth-order Volterra filter in the ith branch, $\mathbf{V}_i(n)$ is the coefficient vector for the $(p - l)$th-order Volterra filter in the ith branch, and $\sigma_i(n)$ is the scaling factor for the product of $y_{l,i}(n)$ and $y_{p-l,i}(n)$. We have employed notation that is identical to that used in Chapter 3, with the exception that the time variations in the coefficient vectors and the scaling factors have been explicitly shown here. Let

$$e(n) = d(n) - \hat{d}(n) \tag{7.143}$$

denote the estimation error at time n. The coefficient update strategy for the adaptive LMS parallel-cascade Volterra filter can be derived as follows [149]:

$$\sigma_i(n+1) = \sigma_i(n) - \frac{\mu_1}{2} \frac{\partial}{\partial \sigma_i(n)} e^2(n)$$
$$= \sigma_i(n) + \mu_1 e(n) y_{l,i}(n) y_{p-l,i}(n), \tag{7.144}$$

$$\mathbf{U}_i(n+1) = \mathbf{U}_i(n) - \frac{\mu_2}{2} \frac{\partial}{\partial \mathbf{U}_i(n)} e^2(n)$$
$$= \mathbf{U}_i(n) + \mu_2 e(n) \sigma_i(n) y_{p-l,i}(n) \mathbf{X}_{N,l}(n), \tag{7.145}$$

and

$$\mathbf{V}_i(n+1) = \mathbf{V}_i(n) - \frac{\mu_3}{2} \frac{\partial}{\partial \mathbf{V}_i(n)} e^2(n)$$
$$= \mathbf{V}_i(n) + \mu_3 e(n) \sigma_i(n) y_{l,i}(n) \mathbf{X}_{N,p-l}(n), \tag{7.146}$$

where μ_1, μ_2, and μ_3 are small positive constants that control the speed of convergence and the steady-state characteristics of the adaptive filter.

7.7.1.1 *Parallel-Cascade Adaptation Using LU Decomposition*   Recall from Chapter 3 that the coefficients of the equivalent direct form structure have the form $\sum_{i=1}^{r} u_{i,j}(n) v_{i,l}(n) \sigma_i(n)$. Consequently, it is possible to scale $u_{i,j}(n)$, $v_{i,l}(n)$, and $\sigma_i(n)$ appropriately without affecting the values of the direct-form coefficients. As a result of the nonuniqueness of the coefficients observed above, we may choose to fix some of the coefficients in each path to some time-invariant value. One example is to choose $\sigma_i(n)$ to be one for all branches, select

$$u_{i,j}(n) = v_{i,j}(n) = 0; \qquad j = 1, 2, \dots, i-1, \tag{7.147}$$

and set $u_{i,i}(n)$ to be one at all values of n. This choice assumes that the coefficients of $x^2(n)$ in the direct-form representation is not zero. If this is not an appropriate assumption, we may choose some other constraint to obtain a unique set of coefficients. The representation in (7.147) corresponds to an LU decomposition of the coefficient matrix. When the adaptive filter employs the constraints defined above, the coefficient update equations may be modified according to

$$\mathbf{U}_i(n+1) = \mathbf{U}_i(n) + \mu_2 e(n) y_{p-l,i}(n) \mathbf{X}_{N,l}(n) \tag{7.148}$$

and

$$\mathbf{V}_i(n+1) = \mathbf{V}_i(n) + \mu_3 e(n) y_{l,i}(n) \mathbf{X}_{N,p-l}(n), \tag{7.149}$$

where μ_2 and μ_3 are positive step-size values for the update equations. Even though (7.148) and (7.149) employ matrix notation for simplicity, it should be understood that the first i coefficients of $\mathbf{U}_i(n)$ and the first $(i-1)$ coefficients of $\mathbf{V}_i(n)$, which are fixed to 0 or 1, respectively, are not updated during the adaptation process.

7.7.1.2 Parallel-Cascade Adaptation for LDL^T Decomposition

When the coefficient matrix is symmetric and we employ a decomposition in which $l = p - l = p/2$, it is possible to decompose the coefficient matrix so that $\mathbf{U}_i(n) = \mathbf{V}_i(n)$ in all branches. In this situation, the input–output relationship of (7.142) can be re-written as

$$\hat{d}(n) = \sum_{i=1}^{r} \sigma_i(n) y_{p/2,i}^2(n)$$

$$= \sum_{i=1}^{r} \sigma_i(n) (\mathbf{X}_{N,l}^T(n) \mathbf{U}_i(n))^2. \tag{7.150}$$

Furthermore, we can constrain the first $i-1$ elements of the coefficient vector $\mathbf{U}_i(n)$ to be zero, and set the ith element of $\mathbf{U}_i(n)$ to be one to obtain a unique representation of the system. The representation of the system given in (7.150) corresponds to an LDL^T decomposition of the coefficient matrix. We can derive the coefficient update strategy for such a realization as follows:

$$\sigma_i(n+1) = \sigma_i(n) + \mu_1 e(n) y_{p/2,i}^2(n) \tag{7.151}$$

and

$$\mathbf{U}_i(n+1) = \mathbf{U}_i(n) + \mu_2 e(n) \sigma_i(n) y_{p/2,i}(n) \mathbf{X}_{N,l}(n). \tag{7.152}$$

Once again, μ_1 and μ_2 are the convergence parameters of the adaptive filter and we do not update the first i entries of $\mathbf{U}_i(n)$. We leave the derivation of the two special cases described above as exercises for the reader.

7.7.1.3 Normalized LMS Adaptive Parallel-Cascade Volterra Filters

A normalized version of the coefficient update equations for the parallel-cascade realization has been derived in [216]. Experimental results demonstrating significant performance advantage of the normalized LMS parallel-cascade Volterra filter over its unnormalized counterpart were also presented in that work. The coefficient update equations for this system are

$$\sigma_i(n+1) = \sigma_i(n) + \frac{3\mu}{\xi(n)} e(n) y_{l,i}(n) y_{p-l,i}(n), \tag{7.153}$$

$$\mathbf{U}_i(n+1) = \mathbf{U}_i(n) + \frac{3\mu}{\xi(n)} e(n) \sigma_i(n) y_{p-l,i}(n) \mathbf{X}_{N,l}(n), \tag{7.154}$$

and

$$\mathbf{V}_i(n+1) = \mathbf{V}_i(n) \frac{3\mu}{\xi(n)} e(n) \sigma_i(n) y_{l,i}(n) \mathbf{X}_{N,p-l}(n), \tag{7.155}$$

where the normalization factor $\xi(n)$ is obtained as a smoothed estimate given by

$$\xi(n) = \alpha \xi(n-1) + (1-\alpha)[P_\sigma(n) + P_u(n) + P_v(n)]. \tag{7.156}$$

In these equations, the smoothing parameter α is a positive constant between zero and one, and the quantities $P_\sigma(n)$, $P_u(n)$ and $P_v(n)$ are defined as

$$P_\sigma(n) = \sum_{i=1}^{r} y_{l,i}^2(n) y_{p-l,i}^2(n), \tag{7.157}$$

$$P_u(n) = \|\mathbf{X}_{N,l}(n)\|^2 \sum_{i=1}^{r} \sigma_i^2(n) y_{p-l,i}^2(n), \tag{7.158}$$

and

$$P_v(n) = \|\mathbf{X}_{N,p-l}(n)\|^2 \sum_{i=1}^{r} \sigma_i^2(n) y_{l,i}^2(n), \tag{7.159}$$

respectively.

7.7.1.4 *Convergence and Stability Issues*   One issue that has not been solved fully is that of choosing the step size so that the adaptive filter operates in a stable manner. For quadratic systems, an approximate analysis is given in Lou et al. [149]. For higher-order systems, even approximate results are not available and the step-size selection is typically performed experimentally. For the normalized LMS adaptive parallel-cascade filters, Panicker et al. [216] derived a heuristic result that states that the algorithm is stable if the step size μ is chosen to be in the range (0,2).

Example 7.6   This example compares the performances of the NLMS adaptive parallel-cascade Volterra filter and the unnormalized LMS adaptive parallel-cascade Volterra filter employing the LDL^T decomposition, and the unnormalized LMS adaptive filter employing the direct-form realization of the system model in a stationary system identification problem. The normalized LMS adaptive Volterra filter implemented using the direct-form realization resulted in much poorer performance than did the other three structures, and therefore is not included in the comparisons presented here. The coefficients of the direct form LMS adaptive filter were initialized to zero. The scalar multipliers $\sigma_i(n)$ and the ith element of the coefficient vector $\mathbf{L}_i(n)$ were initialized to one in the parallel-cascade filter. The rest of the elements of $\mathbf{L}_i(n)$ were initialized to zero. This particular initialization ensures nonzero initial values for $P_\sigma(n)$ and $P_l(n)$. Furthermore, an all-zero coefficient set will not allow the parallel-cascade system to adapt. The initial values of the mean-square estimation error are slightly different because of the differences in the initialization process. The parameter α in (7.156) was chosen as 0.9 in the experiments involving normalized parallel-cascade structure. Since a relatively large value for $(1 - \alpha)$ was employed in the experiments, $\xi(0)$ was initialized to be zero. For values of α very close to one, it is advisable to chose $\xi(0)$ to be a small positive constant in order to avoid numerical problems during normalization in the early stages of adaptation.

The unknown system was a homogeneous fourth-order truncated Volterra system with five-sample memory whose coefficients were given by

$$h_4(m_1, m_2, m_3, m_4)$$
$$= \begin{cases} \dfrac{100}{2\pi[1.5^4 + a_1^4 + a_2^4 + a_3^4 + a_4^4]^{3/4}} + u(m_1, m_2, m_3, m_4); \\ \qquad\qquad\qquad\qquad\quad 0 \le m_1 \le m_2 \le m_3 \le m_4 \le 4 \\ 0; \qquad\qquad\qquad\qquad\qquad\text{otherwise,} \end{cases}$$

where $a_i = (m_i - 1)$, and $u(m_1, m_2, m_3, m_4)$ is a random variable that is uniformly distributed between -0.1 and $+0.1$. The number of branches in the parallel-cascade realization of the filter defined above is $\begin{pmatrix} 5 + 2 - 1 \\ 2 \end{pmatrix} = 15$. The input signal was generated as the output of a linear system with input–output relationship given by

$$x(n) = 0.6x(n - 1) + 0.8\epsilon(n),$$

where $\epsilon(n)$ was a white Gaussian signal with unit variance and zero mean value. The desired response signal was generated by processing this signal with the unknown system and then corrupting the output with an uncorrelated white Gaussian noise sequence with zero mean value and variance $\sigma_n^2 = 0.01$. Figure 7.13 displays a plot of the mean-square estimation error signals obtained using the direct-form LMS, parallel-cascade LMS, and parallel-cascade NLMS adaptive filters. The adaptive filters employed system models that were exactly matched to that of the unknown system. The step sizes for the three methods were selected so as to obtain approximately the same steady-state excess mean-square estimation error. The step sizes and the measured values of the corresponding excess mean-square errors are displayed in Table 7.6. It can be seen from Figure 7.14 that the NLMS

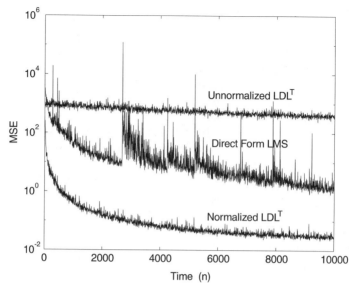

Figure 7.13 Mean-square error of the adaptive filters in Example 7.6. (Courtesy of T. M. Panicker. Copyright © 1998 IEEE.)

parallel-cascade adaptive filter converges significantly more rapidly than the direct-form LMS and the unnormalized LMS using the LDL^T decomposition. All the spikes in the figure are due to the direct-form LMS adaptive filter, indicating that this filter is operating near the stability bound for the step size.

Example 7.7   This example evaluates the effects of using a reduced number of branches in the parallel-cascade adaptive Volterra filters to reduce their complexity. Since this procedure is most useful in approximating nonlinear systems with long memory spans, we consider the identification of a homogeneous fourth-order Volterra system with infinite memory created as a cascade of a fourth-order Butterworth lowpass filter with a memoryless fourth-power operator followed by a fourth-order Chebyschev lowpass filter as shown in Figure 7.14. This system is similar to models of satellite communication systems in which the linear filters model the dispersive transmission paths to and from the satellite and the memoryless nonlinearity models amplifiers operating near the saturation region in the satellite [17]. The transfer functions of the lowpass Butterworth and Chebyschev filters in

TABLE 7.6   Parameters and Excess Mean-Square Errors in Example 7.6

Type	μ	Excess MSE
Direct-form LMS	8.3×10^{-7}	0.003694
Normalized LDL^T	0.074	0.003666
Unnormalized LDL^T	6.3×10^{-9}	0.003757

Source: Courtesy of T. M. Panicker. Copyright © 1998 IEEE.

Figure 7.14 System model in Example 7.7.

Figure 7.14 are given by

$$H_B(z) = \frac{(0.2851 + 0.5704z^{-1} + 0.2851z^{-2})(0.2851 + 0.5701z^{-1} + 0.2851z^{-2})}{(1.0 - 0.1024z^{-1} + 0.4475z^{-2})(1.0 - 0.0736z^{-1} + 0.0408z^{-2})}$$

and

$$H_C(z) = \frac{(0.2025 + 0.288z^{-1} + 0.0.2025z^{-2})(0.2025 + 0.0034z^{-1} + 0.2025z^{-2})}{(1.0 - 1.01z^{-1} + 0.5861z^{-2})(1.0 - 0.6591z^{-1} + 0.1498z^{-2})},$$

respectively. In the experiment, we attempt to identify this system using a memory span of 15 samples. The maximum number of branches required to implement the parallel-cascade system using a 15-sample memory is 120. We employed the NLMS adaptive filters with 120, 20, and 10-branches to identify the unknown system. The step size μ was chosen as 0.1 in all the experiments. Figure 7.15 shows the transient behavior of the mean-square estimation error for the three cases. It may appear from Figure 7.15 that the systems with fewer branches converge more rapidly. This discrepancy is due to the differences in the initial errors of the three estimates. Since one of the coefficients in each branch is fixed to 1, the initializations of the three filters are different, causing a relatively large difference in the initial errors. The three systems show almost the same convergence behavior after the initial phase.

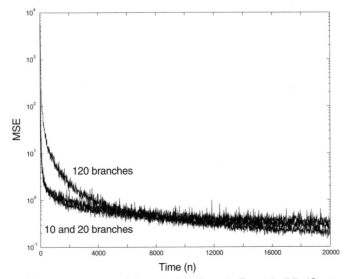

Figure 7.15 Mean-square error of the adaptive filters in Example 7.7. (Courtesy of T. M. Panicker. Copyright © 1998 IEEE.)

However, they converge to slightly different steady-state values where the system with the least number of branches converges to the highest steady-state value. The steady-state performance of the truncated systems was evaluated by running a single experiment over a long interval of time until the time average of the squared error over successive blocks of 100,000 iterations showed negligible variation. The excess mean-square error evaluated as a time average over 100,000 samples after the system has reached the steady state is shown in Table 7.7 for the three cases. It is clear from the results that an adaptive filter with as few as 20 branches may be adequate in this case.

7.8 BLOCK ADAPTIVE LMS VOLTERRA FILTERS

In many applications of adaptive filtering it is not necessary to update the coefficients at each time instant. Block adaptive filters update the coefficients once per block of L samples. In updating the coefficients only once per block, we assume that the operating environment does not vary significantly within each block. The greatest advantage of such methods is that they can be implemented efficiently using fast convolution techniques that employ fast Fourier transform (FFT) [208] or other computationally efficient algorithms. For simplicity of presentation, we consider block adaptive quadratic filters in this section.

7.8.1 Time-Domain Block Adaptation

Let $d(n)$ and $x(n)$ represent the desired response and input signals, respectively, of the adaptive filter. The objective of the block adaptive LMS quadratic filter is to change the coefficients during each iteration in a direction that attempts to reduce the sum of the squared estimation errors in each block, defined for the lth block as

$$
\begin{aligned}
J(l) = & \sum_{n=0}^{L-1} e^2(lR + n) \\
= & \sum_{n=0}^{L-1} \Bigg(d(lR + n) - \sum_{m_1=0}^{N-1} h(m_1; l)x(lR + n - m_1) \\
& - \sum_{m_1=0}^{N-1}\sum_{m_2=0}^{N-1} h(m_1, m_2; l)x(lR + n - m_1)x(lR + n - m_2) \Bigg)^2,
\end{aligned}
\tag{7.160}
$$

TABLE 7.7 Excess Mean-Square Errors in Example 7.7

No. of branches	Excess MSE
120	0.084246
20	0.087432
10	0.092697

Source: Courtesy of T. M. Panicker. Copyright © 1998 IEEE.

where the parameter R represents the number of samples between the starting points of successive blocks. It is easy to show that the stochastic gradient update equations for the above cost function is

$$h(m_1; l+1) = h(m_1; l) + \mu \sum_{n=0}^{L-1} e(lR+n)x(lR+n-m_1) \qquad (7.161)$$

and

$$h(m_1, m_2; l+1) = h(m_1, m_2; l) + \mu \sum_{n=0}^{L-1} e(lR+n)x(lR+n-m_1)x(lR+n-m_2),$$

$$(7.162)$$

where μ is the step-size parameter that controls the speed of convergence, and tracking and steady-state characteristics of the adaptive filter. If necessary, one can choose different values of μ to update different coefficients of the adaptive filter.

7.8.1.1 *Convergence Conditions*
It is a relatively straightforward task to show, using an analysis procedure similar to that employed in Section 7.1.2, that the mean values of the adaptive filter coefficients will converge to the optimal MMSE solution in stationary operating environments if

$$0 < \mu < \frac{2}{L\lambda_{\max}}, \qquad (7.163)$$

where $\lambda_{\max}$ is the maximum eigenvalue of the autocorrelation matrix of the input vector. Conditions that are easier to check than the one above may also be obtained in a manner similar to the derivations in Section 7.1.2.

7.8.1.2 *Computational Complexity of the Block Adaptive LMS Algorithm*
The computational complexity of the block LMS adaptive filter as described above is almost identical to that of the conventional LMS adaptive filter when there is no overlap between adjacent segments. A significant amount of complexity reduction is possible for low-order Volterra filters when the system is implemented using fast Fourier transform techniques. This topic is discussed next.

7.8.2 Frequency-Domain Adaptive LMS Volterra Filter

Frequency-domain adaptive Volterra filters were first developed by Mansour and Gray [152]. Our description of the system follows that by Im and Powers [95,96]. The derivation is a direct extension of frequency-domain adaptive linear filters [57,282] to the case of Volterra filters. The algorithm described below employs the overlap-save method of implementing a truncated Volterra filter. The overlap-save method was described in Section 3.2.5. We assume a symmetric implementation of the adaptive filter coefficients. Thus, the update equations for the coefficients also must be symmetric in its coefficients. Furthermore, as was the case for the overlap-save realization of time-invariant Volterra filters discussed in Section 3.2.5, we will see that successive blocks of the input signal must be overlapped. Let M denote the number of samples in each block of data. Then, $M - R$ denotes the number of samples that are overlapped in adjacent blocks. For ease of presentation, we assume throughout this subsection that $M = 2N$ and that $R = L = N$.

We start the derivations by writing the expressions for the gradient of $J(l)$ with respect to each coefficient as

$$\frac{\partial J(l)}{\partial h_1(m_1; l)} = -\sum_{n=0}^{N-1} e(n + lN)x(n + lN - m_1) \tag{7.164}$$

and

$$\frac{\partial J(l)}{\partial h_2(m_1, m_2; l)} = -\sum_{n=0}^{N-1} e(n + lN)x(n + lN - m_1)x(n + lN - m_2). \tag{7.165}$$

The gradient of the error function with respect to the linear coefficients in (7.164) is an estimate of the negative of the cross-correlation of the error signal and the input signal in the lth segment. Similarly, the gradient of the cost function with respect to the quadratic coefficients can also be regarded as the negative of an estimate of a third-order correlation function of the error signal with a product of the input signal with a delayed version of itself.

7.8.2.1 Gradient Calculation Using the Discrete Fourier Transform

Let us define two new sequences $\tilde{e}_l(n)$ and $\tilde{x}_l(n)$ as

$$\tilde{e}_l(n) = \begin{cases} 0; & 0 \leq n \leq N - 1 \\ e(lN + n); & N \leq n \leq 2N - 1 \end{cases} \tag{7.166}$$

and

$$\tilde{x}_l(n) = x((l - 1)N + n); \quad 0 \leq n \leq 2N - 1, \tag{7.167}$$

respectively. The initial segment of the input signal, denoted by the index $l = 0$, is defined as

$$\tilde{x}_0(n) = \begin{cases} 0; & 0 \leq n \leq N - 1 \\ x(n - N); & N \leq n \leq 2N - 1. \end{cases} \tag{7.168}$$

Let $\tilde{E}_l(k_1)$ and $\tilde{X}_l(k_1)$ represent the $2N$-point discrete Fourier transforms of $\tilde{e}_l(n)$ and $\tilde{x}_l(n)$, respectively. We leave it as an exercise for the reader to show, by direct evaluation, that

$$\frac{\partial J(l)}{\partial h_1(m_1; l)} = -\text{IDFT}\{\tilde{E}_l(k_1)\tilde{X}_l^*(k_1)\}$$

$$= -\frac{1}{2N}\sum_{k_1=0}^{2N-1}\tilde{E}_l(k_1)\tilde{X}_l^*(k_1)e^{j(2\pi/2N)k_1m_1}; \quad 0 \le m_1 \le N-1, \quad (7.169)$$

and that

$$\frac{\partial J(l)}{\partial h_2(m_1, m_2; l)} = -\text{IDFT}\{\tilde{E}_l(k_1 + k_2)\tilde{X}_l^*(k_1)\tilde{X}_l^*(k_2)\}$$

$$= -\frac{1}{(2N)^2}\sum_{k_1=0}^{2N-1}\sum_{k_2=0}^{2N-1}\tilde{E}_l(k_1 + k_2)\tilde{X}_l^*(k_1)\tilde{X}_l^*(k_2)$$

$$\times e^{j(2\pi/2N)(k_1m_1+k_2m_2)}; \quad 0 \le m_1, m_2 \le N-1. \quad (7.170)$$

The discrete Fourier transforms of the two sets of gradient functions can now be evaluated in the following manner. After computing the $2N$-point DFTs of $\tilde{e}_l(n)$ and $\tilde{x}_l(n)$, evaluate the products $\tilde{E}(k_1)\tilde{X}_l^*(k_1)$ and $\tilde{E}_l(k_1 + k_2)\tilde{X}^*(k_1)\tilde{X}^*(k_2)$. Recall that $\tilde{E}(k_1)$ is periodic with period $2N$ samples and therefore, the addition of the indices k_1 and k_2 can be performed using modulo-$2N$ arithmetic. Let $g_1(m_1; l)$ and $g_2(m_1, m_2; l)$ represent the one- and two-dimensional IDFTs of $\tilde{E}(k_1)\tilde{X}_l^*(k_1)$ and $\tilde{E}_l(k_1 + k_2)\tilde{X}^*(k_1)\tilde{X}^*(k_2)$, respectively. Then

$$\frac{\partial J(l)}{\partial h_1(m_1; l)} = \begin{cases} -g_1(m_1; l); & 0 \le m_1 \le N-1 \\ 0; & \text{otherwise} \end{cases} \quad (7.171)$$

and

$$\frac{\partial J(l)}{\partial h_2(m_1, m_2; l)} = \begin{cases} -g_2(m_1, m_2; l); & 0 \le m_1, m_2 \le N-1 \\ 0; & \text{otherwise.} \end{cases} \quad (7.172)$$

The frequency-domain representation of the gradients can now be found by computing the one- and two-dimensional DFTs of (7.171) and (7.172). Let $\hat{V}_1(k_1; l)$ and $\hat{V}_2(k_1, k_2; l)$ be given by

$$\hat{V}_1(k_1; l) = \text{DFT}\left\{\frac{\partial J(l)}{\partial h_1(m_1; l)}\right\}$$

$$= -\sum_{m_1=0}^{N-1}g_1(m_1; l)e^{-j(2\pi/2N)m_1k_1} \quad (7.173)$$

and

$$\hat{V}_2(k_1, k_2; l) = \mathrm{DFT}\left\{\frac{\partial J(l)}{\partial h_2(m_1, m_2; l)}\right\}$$

$$= -\sum_{m_1=0}^{N-1}\sum_{m_2=0}^{N-1} g_2(m_1, m_2; l)e^{-j(2\pi/2N)(m_1k_1+m_2k_2)}, \qquad (7.174)$$

respectively. Now the coefficients of the adaptive quadratic filter can be updated in the frequency domain as

$$H_1(k_1; l+1) = H_1(k_1; l) - \mu\hat{V}_1(k_1; l) \qquad (7.175)$$

and

$$H_2(k_1, k_2; l+1) = H_2(k_1, k_2; l) - \mu\hat{V}_2(k_1, k_2; l), \qquad (7.176)$$

where $H_1(k_1; l)$ represents the $2N$-point DFT of $h_1(m_1; l)$ and $H_2(k_1, k_2; l)$ is the $(2N \times 2N)$-point, two-dimensional DFT of $h_2(m_1, m_2; l)$.

Combining these results with the frequency-domain realization of truncated Volterra filters discussed in Section 3.2.5 results in the frequency-domain implementation of the block adaptive Volterra LMS filter. The steps involved in the implementation of the complete algorithm are depicted in Figure 7.16. The equations necessary to implement the block adaptive LMS quadratic filter are tabulated in Table 7.8.

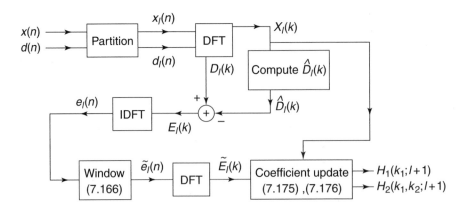

Figure 7.16 The frequency-domain adaptive LMS quadratic Volterra filter.

TABLE 7.8   The Frequency-Domain Adaptive LMS Quadratic Volterra Filter

Initialization

$H_1(k_1; 0)$ and $H_2(k_1, k_2; 0)$ may be chosen arbitrarily such that their inverse DFTs $h_1(m_1; 0)$ and $h_2(m_1, m_2; 0)$ are zero outside the range $0 \le m_1, m_2 \le N - 1$

Main iteration

$$\tilde{x}_l(n) = x((l - 1)N + n); \quad 0 \le n \le 2N - 1$$

$$d_l(n) = \begin{cases} 0; & 0 \le n \le N - 1 \\ d((l-1)N + n); & N \le n \le 2N - 1 \end{cases}$$

$$\tilde{X}_l(k) = \sum_{n=0}^{2N-1} \tilde{x}_l(n)e^{-j(2\pi/2N)kn}$$

$$D_l(k) = \sum_{n=0}^{2N-1} d_l(n)e^{-j(2\pi/2N)kn}$$

$$\hat{D}_l(k) = H_1(k; l)\tilde{X}_l(k) + \sum_{k_1+k_2=k} H_2(k_1, k_2; l)\tilde{X}_l(k_1)\tilde{X}_l(k_2)$$
$$\quad + \sum_{k_1+k_2=k+2N} H_2(k_1, k_2; l)\tilde{X}_l(k_1)\tilde{X}_l(k_2)$$

$$\tilde{e}_l(n) = \begin{cases} 0; & 0 \le n \le N - 1 \\ \dfrac{1}{2N} \displaystyle\sum_{k=0}^{2N-1}(D_l(k) - \hat{D}_l(k))e^{j(2\pi/2N)kn}; & N \le n \le 2N - 1 \end{cases}$$

$$\tilde{E}_l(k) = \sum_{n=0}^{2N-1} \tilde{e}_l(n)e^{-j(2\pi/2N)nk}$$

$$\frac{\partial J(l)}{\partial h_1(m_1; l)} = \begin{cases} -\dfrac{1}{2N} \displaystyle\sum_{k_1=0}^{2N-1} \tilde{E}_l(k_1)\tilde{X}_l^*(k_1)e^{j(2\pi/2N)m_1 k_1}; & 0 \le m_1 \le N - 1 \\ 0; & \text{otherwise} \end{cases}$$

$$\frac{\partial J(l)}{\partial h_2(m_1, m_2; l)} = \begin{cases} -\dfrac{1}{(2N)^2} \displaystyle\sum_{k_1=0}^{2N-1}\sum_{k_2=0}^{2N-1} \tilde{E}_l(k_1 + k_2)\tilde{X}_l^*(k_1)\tilde{X}_l^*(k_2)e^{j(2\pi/2N)(m_1 k_1 + m_2 m_2)}; \\ \qquad\qquad\qquad\qquad 0 \le m_1, m_2 \le N - 1 \\ 0; \qquad\qquad\qquad \text{otherwise} \end{cases}$$

$$\hat{V}_1(k_1; l) = \sum_{m_1=0}^{N-1} \frac{\partial J(l)}{\partial h_1(m_1; l)} e^{-j(2\pi/2N)m_1 k_1}$$

$$\hat{V}_2(k_1, k_2; l) = \sum_{m_1=0}^{N-1}\sum_{m_2=0}^{N-1} \frac{\partial J(l)}{\partial h_2(m_1, m_2; l)} e^{-j(2\pi/2N)(m_1 k_1 + m_2 k_2)}$$

$$H_1(k_1; l + 1) = H_1(k_1; l) - \mu\hat{V}_1(k_1; l)$$

$$H_2(k_1, k_2; l + 1) = H_2(k_1, k_2; l) - \mu\hat{V}_2(k_1, k_2; l)$$

7.8.3   Computational Complexity

Among the different steps in the adaptive filter, the calculations required to evaluate $\hat{D}_l(k)$ and $\partial J(l)/\partial h_2(m_1, m_2; l)$ in Table 7.8 are the most computation-intensive. Implementing both these steps require $O(N^2)$ complex arithmetical operations per block of N samples. Consequently, we can see that the computational complexity of the frequency-domain realization of the adaptive filter corresponds to $O(N)$ complex arithmetical operations per sample, which is an order of magnitude lower than the corresponding time-domain realizations. The computational advantage of the frequency-domain realization is significant for relatively large values of N.

Example 7.8   We consider a system identification problem in which the unknown system was a homogeneous quadratic filter whose coefficients were given by

$$h_2(m_1, m_2) = \frac{1.5}{2\pi[1.5^2 + (m_1 - \frac{63}{2})^2 + (m_2 - \frac{63}{2})^2]^{1.5}}; \qquad 0 \le m_1, m_2 \le 63.$$

The frequency-domain adaptive Volterra filter was implemented using a homogeneous quadratic system model with 64-sample memory, and the data block size M was chosen to be 128, twice the system memory length. The coefficients were initialized to some arbitrarily selected nonzero values. The input sequence was obtained as the output of a one-pole filter with transfer function $A(z) = 0.8/(1 - 0.6z^{-1})$ when its input was an IID Gaussian signal with zero mean value and variance equal to 0.1. The desired response signal $d(n)$ was obtained by adding to the output of the unknown system a zero-mean, white, Gaussian noise that was uncorrelated with the input signal and with variance equal to 0.001. The step size was selected to be 1.408×10^{-5}. Figure 7.17 shows the mean-square error (MSE) between the adaptive filter output and the desired response signal for the adaptive filter over the first 1000 blocks. The MSE curves were obtained by averaging the squared error associated with 50 independent estimates. The excess MSE was also measured after the system reached the steady state, and was found to be

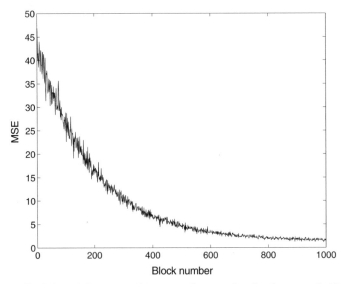

Figure 7.17 Evolution of the ensemble-averaged squared estimation error in Example 7.8. (Result courtesy of Linshan Li.)

7.881×10^{-7}. We can see from the results that the adaptive filter converges reasonably rapidly in spite of its reduced computational complexity, and that it eventually achieves coefficient values that are very close to the values of the coefficients of the system being identified.

7.9 EXERCISES

7.1 Develop a stochastic gradient adaptive filter that attempts to iteratively minimize the cost function

$$J(n) = E\left\{ \left| d(n) - \sum_{m_1=0}^{N-1} \sum_{m_2=m_1}^{N-1} h_2(m_1, m_2; n)x(n - m_1)x(n - m_2) \right| \right\}.$$

The resulting technique is commonly known as the *sign algorithm* for adaptive quadratic filtering.

7.2 A common method for improving the convergence characteristics of stochastic gradient adaptive filters is to use an adaptive step size sequence in place of the constant step size μ. One approach for adapting the step size is to employ an algorithm similar to the LMS adaptive filter for updating the step size also as [166,168]

$$\mu(n) = \mu(n - 1) - \frac{\rho}{2} \frac{\partial e^2(n)}{\partial \mu(n - 1)},$$

where ρ is a small positive constant that controls the adaptation properties of the step size.

(a) Show that the step size update equation for linear system models and the LMS adaptation is given by

$$\mu(n) = \mu(n - 1) - \rho e(n)e(n - 1)\mathbf{X}_1^T(n)\mathbf{X}_1(n - 1).$$

(b) Develop a stochastic gradient adaptive filter with gradient adaptive step size that attempts to reduce the absolute value of the estimation error during each iteration.

(c) Now assume that we have multiple step sizes and the adaptive filter is of the form

$$e(n) = d(n) - \vec{\mathbf{H}}_1^T(n)\mathbf{X}_1(n) - \vec{\mathbf{H}}_2^T(n)\mathbf{X}_2(n)$$

with the coefficient updates given by

$$\vec{\mathbf{H}}_1(n + 1) = \vec{\mathbf{H}}_1(n) + \mu_1(n)\mathbf{X}_1(n)e(n)$$

and

$$\vec{H}_2(n+1) = \vec{H}_2(n) + \mu_2(n)\mathbf{X}_2(n)e(n).$$

Derive the stochastic gradient update equations for $\mu_1(n)$ and $\mu_2(n)$ using the least-mean-square-error criterion.

7.3 *Computing Assignment:* In this problem, we investigate the effect of input signal correlation on the speed of convergence of an adaptive second-order Volterra filter. For this purpose, we consider a system identification problem in which the unknown system has an input–output relationship given by

$$y(n) = x(n) + x(n-1) + x(n-2) + 0.5x^2(n)$$

$$- 0.3x(n)x(n-1) + 0.2x(n)x(n-2) + 0.5x^2(n-1)$$

$$- x(n-1)x(n-2) + 0.4x^2(n-2).$$

Our objective is to use an LMS adaptive second-order Volterra filter employing an exactly matched system model and input signals with varying degrees of correlation to study the rate at which mean-square error converges to the steady-state values. Choose the input signal as the output of the linear system with input–output relationship

$$x(n) = bx(n-1) + \sqrt{1-b^2}\,\xi(n),$$

where ξ is a white, Gaussian random process with zero mean value and unit variance, $x(n)$ is the output of the system, and b is a parameter in the range $0 \le b < 1$. Choosing $b = 0$ results in an IID signal as the input to the adaptive filter, while choosing a positive value for b results in a correlated signal as its input. The closer b is to 1, the more correlated the signal $x(n)$ is. Create the desired response signal to the adaptive filter $d(n)$ by processing $x(n)$ with the unknown system and corrupting its output with additive white noise uncorrelated with $x(n)$ and having zero mean value and variance equal to 0.001. For choices of $b = 0$, 0.5, and 0.9, implement an LMS adaptive filter to identify the system using $x(n)$ and $d(n)$. Choose a step size $\mu = 0.01$, initialize each coefficient to zero value, and use 3000 samples in your experiments. Repeat the experiments 50 times using independent data sets, and plot the averages of the squared error over the 50 experiments as a function of time. Also, plot the averages of one of the coefficients of the linear part of the system model and one of the coefficients of the nonlinear part. Comment on the dependence of the convergence speed of the adaptive filter on the level of correlation in its input signal.

7.4 Consider an adaptive Volterra filter of the form

$$e(n) = d(n) - \vec{\mathbf{H}}^T(n)\mathbf{X}(n)$$
$$\vec{\mathbf{H}}(n+1) = \beta\vec{\mathbf{H}}(n) + \mu\mathbf{X}(n)e(n),$$

where β is a constant that is less than or equal to one, and is usually close to one. This algorithm, known as the *leaky LMS adaptive filter*, is useful in situations where the autocorrelation matrix of the input vector $\mathbf{X}(n)$ has an extremely large eigenvalue spread. To demonstrate the usefulness of this algorithm, perform a convergence analysis of the mean values of the adaptive filter coefficients using the same set of approximations and assumptions employed to analyze the LMS adaptive filter. Compare the rates at which the different modes of the leaky LMS adaptive filter converge to the rate of convergence of the different modes of the LMS adaptive filter for the same input signal. Is it possible to claim, on the basis of your analysis, that the leaky LMS adaptive filter is able to perform better in situations when the eigenvalue spread of the input signal is very large? Does the mean values of the coefficients converge to the optimal minimum MSE solution?

7.5 We wish to develop an adaptive Volterra filter employing the Hammerstein model as shown in Figure 7.18. Derive a least-mean-square adaptation algorithm for the coefficients of the linear filter and the static nonlinearity in the system model.

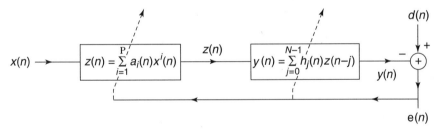

Figure 7.18 Adaptive Hammerstein filter.

7.6 We wish to develop an adaptive Volterra filter employing the Wiener model as shown in Figure 7.19. Derive a least-mean-square adaptation algorithm for the coefficients of the linear filter and the static nonlinearity in the system model.

7.7 An LNL model is a cascade connection of a linear system followed by a static nonlinearity followed by another linear system as shown in Figure 7.20, and is commonly employed to model nonlinear systems efficiently. Derive a least-mean-square adaptation algorithm for the coefficients of the linear filters and

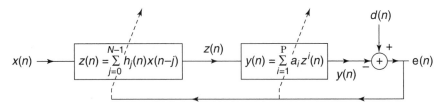

Figure 7.19 Adaptive nonlinear filter using the Wiener model.

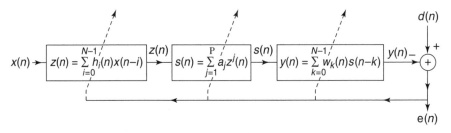

Figure 7.20 Adaptive LNL filter.

the static nonlinearity in the system model. Assume that the linear components have finite impulse response as shown in the figure.

7.8 Consider the block diagram of the system in Figure 7.21, which depicts the structure of an N-input, one-output artificial neuron. When several of these structures are cascaded together, they form a *feedforward artificial neural network*. The output of this system is

$$y(n) = f\left(\sum_{i=1}^{N} w_i(n)x_i(n)\right), \tag{7.177}$$

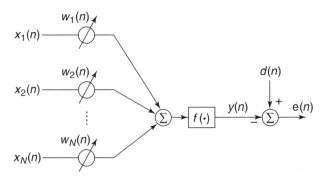

Figure 7.21 A single artificial neuron.

where $x_i(n)$ is the ith input signal and $w_i(n)$ is the ith neuron coefficient. A common choice for the function $f(u)$ is

$$f(u) = \frac{e^{\alpha u} - e^{-\alpha u}}{e^{\alpha u} + e^{-\alpha u}}$$

$$= \tanh(\alpha u),$$

which is also known as the *sigmoid function* in the neural network field. Derive a stochastic gradient algorithm for adjusting the ith coefficient of the artificial neuron to approximately minimize the mean-squared error $J(n) = E\{e^2(n)\}$, where $e(n) = d(n) - y(n)$. Express your answer in vector form.

7.9 Prove the following form of the matrix inversion lemma. If **A** is a non-singular matrix and **U** and **V** are column vectors of appropriate dimensions,

$$(\mathbf{A} + \mathbf{U}\mathbf{V}^T)^{-1} = \mathbf{A}^{-1} - \frac{\mathbf{A}^{-1}\mathbf{U}\mathbf{V}^T\mathbf{A}^{-1}}{1 + \mathbf{V}^T\mathbf{A}^{-1}\mathbf{U}},$$

provided $(\mathbf{A} + \mathbf{U}\mathbf{V}^T)$ is nonsingular. (*Hint:* Multiply the inverse of the left-hand-side of the expression with its right-hand side and show that the result is an identity matrix.)

7.10 Show that the least-squares predictor coefficients are updated as

$$\mathbf{A}(n) = \mathbf{A}(n-1) + \mathbf{k}(n-1)\mathbf{f}_{n-1}^T(n)$$

and

$$\mathbf{B}(n) = \mathbf{B}(n-1) + \mathbf{k}(n)\mathbf{b}_{n-1}^T(n)$$

7.11 Show that

$$\mathbf{P}_{\mathbf{vX}}(n)\mathbf{k}(n-1) = \mathbf{A}^T(n)\mathbf{X}(n-1),$$

where $\mathbf{P}_{\mathbf{vX}}(n)$ is as defined in (7.103).

7.12 Derive the relationships

$$\mathbf{f}_n(n) = \gamma(n-1)\mathbf{f}_{n-1}(n)$$

and

$$\mathbf{b}_n(n) = \gamma(n)\mathbf{b}_{n-1}(n)$$

7.13   Derive the recursive update equation for $\alpha^{-1}(n)$ given by

$$\alpha^{-1}(n) = \lambda^{-1}\left[\alpha^{-1}(n-1) - \frac{\alpha^{-1}(n-1)\mathbf{f}_{n-1}(n)\mathbf{f}_{n-1}^T(n)\alpha^{-1}(n-1)}{\lambda/\gamma(n-1) + \mathbf{f}_{n-1}^T(n)\alpha^{-1}(n-1)\mathbf{f}_{n-1}(n)}\right].$$

7.14   *Computing Assignment:* Repeat Exercise 7.3 using the conventional RLS adaptive second-order Volterra filter. Choose $\lambda = 0.99$ and $\delta = 0.01$ in your experiments.

7.15   Consider the matrix $\mathbf{Q}_2(n)$ employed for the bit-level decomposition of the input vector in the distributed arithmetic realization of the adaptive homogeneous quadratic filter. Assuming that the constituent bits of the signal samples are uncorrelated with each other and the input signal has zero mean value, show that the entries of the matrix $E\{\mathbf{Q}_2^T(n)\mathbf{Q}_2(n)\}$ take their values from the set $\{0, N, N^2\}$.

7.16   Derive the sign algorithm for an adaptive Volterra filter of order p implemented using the distributed arithmetic. Using an appropriate set of experiments, compare the convergence time and the residual error of your algorithm with those of the updating algorithm of (7.122).

7.17   Derive the updating equations for an adaptive Volterra filter implemented using the distributed arithmetic when a split-address technique is exploited employing two submemories.

7.18   Consider the system model for a homogeneous third-order Volterra system realized as shown in Figure 7.22. Derive an LMS adaptive filter to update the coefficients of the linear filter in the model.

7.19   *Computing Assignment:* In this exercise, we wish to compare the characteristics of the normalized LMS adaptive parallel-cascade Volterra filter and the LMS adaptive direct-form Volterra filter in a system identification problem. Repeat Exercise 7.3 using a parallel-cascade structure. In this case, it may be easiest to just compare the MSE curves, since a direct comparison of the coefficients requires converting the coefficients of the parallel-cascade system to those of an equivalent direct form system. While implementing the adaptive parallel-cascade filter, it is important to remember that initializing all the coefficients to zero will not allow the system to adapt.

7.20   Show that the mean values of the coefficients of the block adaptive LMS quadratic filter defined by (7.161) and (7.162) will converge if the step size is chosen such that

$$0 < \mu < \frac{2}{L\lambda_{max}}.$$

Use similar approximations and assumptions as in Section 7.1.2.

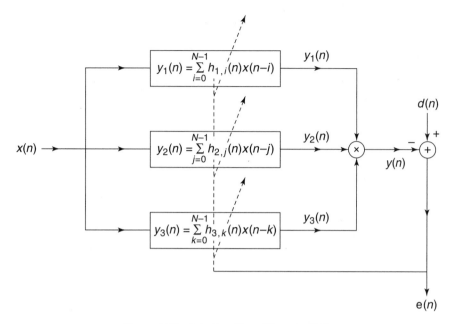

Figure 7.22 Adaptive filter of Exercise 7.18.

7.21 Show, by direct evaluation, that the gradients of the cost function can be calculated for the block adaptive LMS quadratic filter as in (7.169) and (7.170), i.e.,

$$\frac{\partial J(l)}{\partial h_1(m_1; l)} = -\text{IDFT}\{\tilde{E}_l(k_1)\tilde{X}_l^*(k_1)\}$$

$$= -\frac{1}{2N} \sum_{k=0}^{2N-1} \tilde{E}_l(k_1)\tilde{X}_l^*(k_1)e^{j(2\pi/2N)k_1 m_1}; \qquad 0 \le m_1 \le N-1$$

and

$$\frac{\partial J(l)}{\partial h_2(m_1, m_2; l)} = -\text{IDFT}\{\tilde{E}_l(k_1 + k_2)\tilde{X}_l^*(k_1)\tilde{X}_l^*(k_2)\}$$

$$= -\frac{1}{(2N)^2} \sum_{k_1=0}^{2N-1} \sum_{k_2=0}^{2N-1} \tilde{E}_l(k_1 + k_2)\tilde{X}_l^*(k_1)\tilde{X}_l^*(k_2)$$

$$\times e^{j(2\pi/2N)(k_1 m_1 + k_2 m_2)}; \qquad 0 \le m_1, m_2 \le N-1.$$

7.22 Extend the frequency-domain realization of the adaptive Volterra filter to a homogeneous system model of order $p > 2$. Compare the computational complexity of your method with the corresponding time-domain realizations.

8

RECURSIVE
POLYNOMIAL SYSTEMS

A significant problem associated with truncated Volterra series modeling of nonlinear systems is that many such systems require a very large number of parameters for accurate characterization. Consequently, it is important to seek alternate representations of nonlinear systems that are more parsimonious in the use of the coefficients. This chapter discusses one alternative that employs polynomial models with feedback. In general, such models may be represented using the input–output relationship given by

$$y(n) = \sum_{i=0}^{P} f_i(y(n-1), y(n-2), \ldots, y(n-M), x(n), x(n-1), \ldots, x(n-N)),$$

$$(8.1)$$

where $f_i(\cdot, \cdot, \ldots, \cdot)$ is an ith-order polynomial in the variables within the parentheses. Just as linear IIR filters can represent many linear systems with far fewer coefficients than their FIR counterparts, recursive polynomial models can accurately represent many nonlinear systems with greater efficiency than truncated Volterra series representations.

A special case of the recursive system model of (8.1) is the bilinear model [34,186]. Its input–output relationship is given by

$$y(n) = \sum_{i=0}^{N_1} a_i x(n-i) + \sum_{i=1}^{N_2} b_i y(n-i)$$

$$+ \sum_{i=0}^{N_3} \sum_{j=1}^{N_4} c_{i,j} x(n-i) y(n-j).$$

$$(8.2)$$

Figure 8.1 shows the signal flow diagram for a bilinear system for $N_1 = N_2 = N_3 = N_4 = 2$. This model may be considered as a straightforward extension of linear models. If $c_{i,j} = 0$ for all i and j, the resulting system is linear. Additionally, like truncated Volterra system models, the bilinear system model also is linear in the parameters; that is, the nonlinearity is completely due to the product terms involving the input and output samples.

Although bilinear system models are relatively simple, they possess some universal modeling properties [32]. Furthermore, they are representative of the larger class of recursive polynomial systems. Consequently, this chapter will emphasize bilinear system models over the more general models. However, many of the ideas discussed here can be easily extended to more general models.

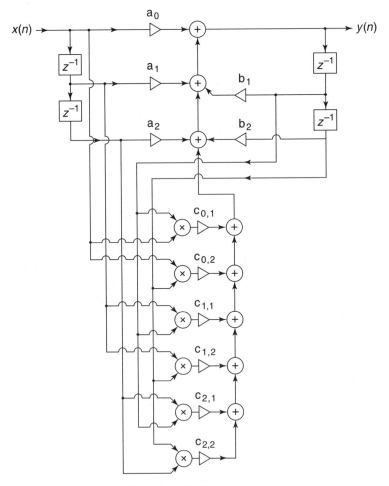

Figure 8.1 A bilinear system.

We begin our discussion by developing a Volterra series expansion for bilinear systems in Section 8.1. Convergence of the Volterra series expansion of its input–output relationship plays a crucial role in the stability analysis of bilinear systems. Section 8.2 discusses a set of sufficient conditions for the stability of bilinear systems. Adaptive filters equipped with bilinear system models are described in Section 8.3. Section 8.4 contains a comparison of the modeling capabilities of linear, quadratic, and bilinear systems in a decision feedback equalization problem.

8.1 VOLTERRA SERIES EXPANSION FOR BILINEAR SYSTEMS

To simplify the presentation, we adopt the following operator notation for our analysis. Let

$$A(q) = \sum_{i=0}^{N_1} a_i q^{-i}, \tag{8.3}$$

$$B(q) = \sum_{i=1}^{N_2} b_i q^{-i}, \tag{8.4}$$

and

$$C(q) = \sum_{i=0}^{N_3} \sum_{j=1}^{N_4} c_{i,j} [q^{-i}, q^{-j}], \tag{8.5}$$

where $q^{-i} x(n) = x(n-i)$ and $[q^{-i}, q^{-j}][x(n), y(n)] = x(n-i)y(n-j)$. Using these operators[1], the bilinear model in (8.2) can be represented as

$$(1 - B(q))y(n) = A(q)x(n) + C(q)[x(n), y(n)]. \tag{8.6}$$

Let us now define a sequence of systems as follows:

$$(1 - B(q))y_1(n) = A(q)x(n) \tag{8.7}$$

and

$$(1 - B(q))y_k(n) = C(q)[x(n), y_{k-1}(n)]; \qquad k = 2, 3, \ldots, n+1. \tag{8.8}$$

Figure 8.2 illustrates how to generate the outputs of the different systems in the sequence. It is left as an exercise for the reader to show that $y_k(n)$ can be written as a linear combination of only kth-order products of the input signal $x(n)$.

[1] Since we employ the operator notation here, we use q^{-1} rather than z^{-1} to denote the delay operator in this chapter.

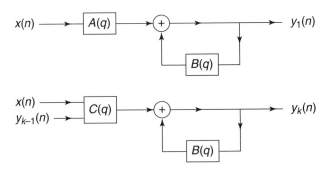

Figure 8.2   Generation of the Volterra kernels of a bilinear system.

Assume that the bilinear system as well as the sequence of systems presented above is initially at rest. Therefore

$$y(n) = y_1(n) = y_2(n) = \cdots = 0 \quad \text{for} \quad n \leq -1. \tag{8.9}$$

We now show that

$$y(n) = \sum_{k=1}^{n+1} y_k(n) \tag{8.10}$$

is the unique solution to the bilinear system defined in (8.6). An immediate consequence of this fact is that $y_k(n)$ is the kth-order term in the Volterra series expansion of the bilinear system.

We proceed to prove the preceding statements in the following manner. We first show that $y_k(l) = 0$ for $l \leq k - 2$. Using this result, we show that (8.10) is a solution to (8.6). Finally, we prove that this solution is unique. We discuss each of the above tasks next.

$y_k(l) = 0$ *for $l \leq k - 2$*: We prove this assertion using mathematical induction. Because $y_1(l) = 0$ for $l \leq -1$, this statement is true for $k = 1$. Assume that it holds for k greater than 1 and less than $(n + 1)$. By using (8.10) and the fact that $y_n(l - j) = 0$ for $l - j \leq n - 2$, it follows that $y_{n+1}(l) = 0$ for $l - 1 \leq n - 2$, i.e., for $l \leq n - 1$.

Equation (8.10) is a solution of (8.6): By definition

$$C(q)[x(n), y_{n+1}(n)] = \sum_{i=0}^{N_3} \sum_{j=1}^{N_4} c_{i,j} x(n-i) y_{n+1}(n-j). \tag{8.11}$$

Using (8.7) and (8.8), we can write

$$(1 - B(q)) \sum_{k=1}^{n+1} y_k(n) = A(q)x(n) + C(q)\left[x(n), \sum_{k=1}^{n} y_k(n)\right]. \tag{8.12}$$

Since $y_{n+1}(n-j) = 0$ for $j \geq 1$, we can add $C(q)[x(n), y_{n+1}(n)]$ to the right-hand-side of this equation to get

$$(1 - B(q)) \sum_{k=1}^{n+1} y_k(n) = A(q)x(n) + C(q)\left[x(n), \sum_{k=1}^{n+1} y_k(n)\right]. \tag{8.13}$$

Therefore, $y(n) = \sum_{k=1}^{n+1} y_k(n)$ is a solution to (8.6).

The solution given above is unique: To show the uniqueness, let us assume that $\dot{y}(n)$ and $\ddot{y}(n)$ are two different solutions of (8.6). Then

$$(1 - B(q))(\dot{y}(n) - \ddot{y}(n)) = C(q)[x(n), \dot{y}(n) - \ddot{y}(n)]. \tag{8.14}$$

Since by assumption, $\dot{y}(n) - \ddot{y}(n) = 0$ for $n < 0$, it follows that $\dot{y}(n) - \ddot{y}(n) = 0$ for $n \geq 0$ also. Because $y(n) = \sum_{k=1}^{n+1} y_k(n)$ is a solution to (8.6), it also has to be the unique solution.

Remark 8.1   The derivation presented above provides a means for recursively finding the Volterra series expansion of bilinear systems. There may be an infinite number of terms in such an expansion. The system not only has infinite memory, but also may accommodate very large orders of nonlinearity. This intuitively explains the ability of the bilinear systems to efficiently model many nonlinear systems.

8.2   STABILITY OF BILINEAR SYSTEMS

As is the case for all recursive systems, bilinear systems also have the potential to exhibit unstable behavior. The stability issues associated with recursive polynomial systems are more complicated than those associated with linear systems. To see this, let us set $N_1 = N_2 = N_3 = N_4 = N$ and rewrite (8.2) as

$$y(n) = \sum_{j=1}^{N}\left(b_j + \sum_{i=0}^{N} c_{i,j}x(n-i)\right)y(n-j) + \sum_{i=0}^{N} a_i x(n-i). \tag{8.15}$$

It is easy to infer from this equation that we can find bounded-input signals that can drive the system output unbounded for almost all nontrivial bilinear systems by forcing the instantaneous poles of the "time-varying linear" system to locations outside the unit circle. This indicates that most bilinear systems are unstable in the bounded-input bounded-output sense. Similar statements apply for a large class of recursive nonlinear systems.

Even though most recursive nonlinear systems are unstable from a purist's point of view, we can often define classes of input signals for which such systems will provide useful outputs and/or model signals and other real-world systems with good accuracy. In this section, we explore the notion of stability of bilinear systems from this perspective. In particular, we derive conditions on the parameters of a time-invariant bilinear system so that the system output is bounded whenever the input signal is bounded by a finite constant M_x. The key result is contained in the following theorem.

Theorem 8.1   For the bilinear system model represented by (8.6), let $p_1, p_2, \ldots, p_{N_2}$ denote the zeros of the polynomial $q^{N_2}(1 - B(q))$. Given a real, positive number M_x

$$
\begin{cases}
|p_i| < 1, \text{ for } i = 1, 2, \ldots, N_2, \text{ and} \\[2mm]
M_x \sum_{i=0}^{N_3} \sum_{j=1}^{N_4} |c_{i,j}| < \prod_{i=1}^{N_2}(1 - |p_i|)
\end{cases}
\tag{8.16}
$$

constitute a set of sufficient conditions for every $x(n)$ bounded by M_x to produce a bounded output $y(n)$. Furthermore, $y(n)$ is bounded by

$$
|y(n)| \leq \frac{\beta M_x \sum_{i=0}^{N_1} |a_i|}{1 - \left\{\beta M_x \sum_{i=0}^{N_3} \sum_{j=1}^{N_4} |c_{i,j}|\right\}},
\tag{8.17}
$$

where

$$
\beta = \frac{1}{\prod_{i=1}^{N_2}(1 - |p_i|)}.
\tag{8.18}
$$

Remark 8.2   When there is no nonlinearity in the system, (8.2) reduces to a recursive linear system and condition (8.16) reduces to the well-known necessary and sufficient condition for the resulting linear system to be stable in the bounded-input bounded-output sense.

PROOF OF THEOREM 8.1   Our proof depends heavily on the Volterra series expansion of bilinear systems. Furthermore, the conditions of the theorem form a sufficient set of conditions for the convergence of the Volterra series expansion given in the previous section.

Consider the signals $y_1(n), y_2(n), \ldots$ as defined in (8.7) and (8.8). It is straightforward to show that

$$
y_1(n) = \left\{\prod_{m=1}^{N_2}\left\{\sum_{l=0}^{n}(p_m)^l q^{-l}\right\}\right\} A(q)x(n)
\tag{8.19}
$$

and

$$y_k(n) = \left\{ \prod_{m=1}^{N_2} \left\{ \sum_{l=0}^{n} (p_m)^l q^{-l} \right\} \right\} C(q)[x(n), y_{k-1}(n)] \tag{8.20}$$

for $k = 2, 3, \ldots, n + 1$.

Let $x(n)$ be bounded by M_x. Also, assume that all the zeros of $q^{N_2}(1 - B(q))$ are inside the unit circle and that $\beta M_x \sum_{i=0}^{N_3} \sum_{j=1}^{N_4} |c_{i,j}| < 1$. It follows from (8.19) that

$$|y_1(n)| \leq \beta M_x \sum_{i=0}^{N_1} |a_i|. \tag{8.21}$$

We can immediately show that

$$|C(q)[x(n), y_1(n)]| \leq \left(M_x \sum_{i=0}^{N_3} \sum_{j=1}^{N_4} |c_{i,j}| \right) \left(\beta M_x \sum_{i=0}^{N_1} |a_i| \right). \tag{8.22}$$

In a similar manner, we can show that $y_2(n)$ is bounded by

$$|y_2(n)| \leq \left(\beta M_x \sum_{i=0}^{N_3} \sum_{j=1}^{N_4} |c_{i,j}| \right) \left(\beta M_x \sum_{i=0}^{N_1} |a_i| \right). \tag{8.23}$$

In general, $y_k(n)$ is bounded by

$$|y_k(n)| \leq \left(\beta M_x \sum_{i=0}^{N_3} \sum_{j=1}^{N_4} |c_{i,j}| \right)^{k-1} \left(\beta M_x \sum_{i=0}^{N_1} |a_i| \right) \tag{8.24}$$

for $k = 3, 4, \ldots, n + 1$. Since $y(n) = \sum_{k=1}^{n+1} y_k(n)$, we can bound $y(n)$ using

$$|y(n)| \leq \sum_{k=1}^{n+1} |y_k(n)|$$

$$\leq \left(\beta M_x \sum_{i=0}^{N_1} |a_i| \right) \sum_{k=1}^{n+1} \left(\beta M_x \sum_{i=0}^{N_3} \sum_{j=1}^{N_4} |c_{i,j}| \right)^{k-1}$$

$$= \left(\beta M_x \sum_{i=0}^{N_1} |a_i| \right) \frac{1 - \left\{ \beta M_x \sum_{i=0}^{N_3} \sum_{j=1}^{N_4} |c_{i,j}| \right\}^{n+1}}{1 - \left\{ \beta M_x \sum_{i=0}^{N_3} \sum_{j=1}^{N_4} |c_{i,j}| \right\}}$$

$$\leq \frac{\beta M_x \sum_{i=0}^{N_1} |a_i|}{1 - \left\{ \beta M_x \sum_{i=0}^{N_3} \sum_{j=1}^{N_4} |c_{i,j}| \right\}}. \tag{8.25}$$

The last inequality makes use of the condition that $\beta M_x \sum_{i=0}^{N_3} \sum_{j=1}^{N_4} |c_{i,j}| < 1$. This completes the proof.

8.2.1 Comments on the Stability theorem

Theorem 8.1 and the proof is based on the derivations in Lee and Mathews [138]. Similar results are also available [25,124,293]. The work in [293] discusses the time-varying case, and is considered in Exercise 8.3. Additional work on the stability of a different class of polynomial system models may be found in Mumolo and Carini [190].

The conditions provided by the theorem are sufficient, but not necessary, to guarantee the boundedness of the output signals. However, these appear to be the tightest set of conditions available at this time. One implication of the conditions derived above is that if the input signals have large magnitudes, then the nonlinearity should be mild. This is not a very surprising statement since the Volterra series terms for large input values grow with the order of nonlinearity, and therefore, the Volterra series expansion may not converge for such input signals.

8.3 ADAPTIVE BILINEAR FILTERS

This section presents algorithms for adapting the coefficients of a bilinear system model and the properties of such algorithms. Extensions of most of the results in this section to more general recursive polynomial system models are relatively straight-forward.

We begin our discussion with an extremely simple class of methods known as the *equation error methods.* We will find that these methods lead to biased steady-state solutions. We next discuss the *output error* adaptive bilinear filters. Output error algorithms have the potential to provide unbiased estimates. The stability of the time-varying systems estimated by the adaptive filters is of particular importance. We also discuss the stability of stochastic gradient as well as recursive least-squares adaptive bilinear filters in this section.

8.3.1 Equation Error Adaptive Bilinear Filters

There are two fundamentally different classes of methods for adaptive recursive filtering [104,257,281]. These classes of algorithms are known as equation error methods and output error methods. Figure 8.3 describes the differences between these two approaches. For simplicity of presentation, we assume that $N_1 = N_2 = N_3 = N_4 = N$ in our discussion. Given a desired response signal $d(n)$ and an input signal $x(n)$, the equation error bilinear adaptive filters find the estimate $\hat{d}(n)$ of the desired response signal as

$$\hat{d}(n) = \sum_{i=1}^{N} b_i(n)d(n-i) + \sum_{i=0}^{N} a_i(n)x(n-i) + \sum_{i=0}^{N}\sum_{j=1}^{N} c_{i,j}x(n-i)d(n-j). \quad (8.26)$$

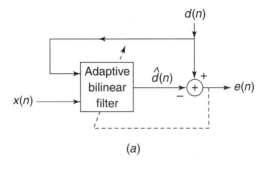

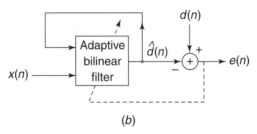

Figure 8.3 Equation error (a) and output error (b) adaptive filters.

The output error methods, on the other hand, find the estimate of the desired response signal as

$$\hat{d}(n) = \sum_{i=1}^{N} b_i(n)\hat{d}(n-i) + \sum_{i=0}^{N} a_i(n)x(n-i) + \sum_{i=0}^{N}\sum_{j=1}^{N} c_{i,j}x(n-i)\hat{d}(n-j). \quad (8.27)$$

The basic difference between the two approaches is that output error methods estimate the desired response signal using a truly recursive system model by feeding back the estimate $\hat{d}(n)$ to generate the adaptive estimates of the desired response signal. The equation error methods form the estimates using samples of the input and desired response signals. Consequently, they are not truly recursive estimators in the sense that the output of the estimator is not fed back to form the estimates.

To formulate the equation error bilinear adaptive filters, we define the input vector $\mathbf{Z}(n)$ as

$$\mathbf{Z}(n) = [x(n), \ldots, x(n-N), \ d(n-1), \ldots, d(n-N), \ x(n)d(n-1), \ldots,$$

$$x(n)d(n-N), \ldots, x(n-N)d(n-N)]^T \quad (8.28)$$

and the coefficient vector $\vec{\mathbf{H}}(n)$ as

$$\vec{\mathbf{H}}(n) = [a_0(n), \ldots, a_N(n), \ b_1(n), \ldots b_N(n), \ c_{0,1}(n), \ldots, c_{0,N}(n), \ldots, c_{N,N}(n)]^T. \quad (8.29)$$

The estimate $\hat{d}(n)$ of the desired response signal can be expressed using this notation as

$$\hat{d}(n) = \vec{\mathbf{H}}^T(n)\mathbf{Z}(n). \tag{8.30}$$

8.3.1.1 LMS Adaptive Equation Error Bilinear Filters

The equation error LMS adaptive bilinear filter updates the coefficient vector using the stochastic gradient adaptation algorithm that attempts to reduce the mean-square error cost function given by

$$J(n) = E\{(d(n) - \hat{d}(n))^2\} = E\{e^2(n)\} \tag{8.31}$$

at each time. Because $\mathbf{Z}(n)$ does not depend on $\vec{\mathbf{H}}(n)$, it is straightforward to derive an update equation for $\vec{\mathbf{H}}(n+1)$ as

$$\vec{\mathbf{H}}(n+1) = \vec{\mathbf{H}}(n) + \mu e(n)\mathbf{Z}(n), \tag{8.32}$$

where μ is a small positive constant that controls the rate at which the adaptive filter converges. The derivation of this algorithm is a straightforward extension of the derivation of LMS adaptive truncated Volterra filters.

8.3.1.2 Extended Least-Squares Adaptive Equation Error Bilinear Filters

There are several variations of least-squares adaptive filters. We consider only the exponentially weighted least-squares adaptive filter here. The objective of the extended least-squares (ELS) adaptive bilinear filter is to minimize the cost function $J(n)$ defined as

$$J(n) = \sum_{k=0}^{n} \lambda^{n-k}(d(k) - \hat{d}_n(k))^2 \tag{8.33}$$

at each time n, where λ is the forgetting factor $(0 < \lambda < 1)$ that controls the rate at which the adaptive filter tracks time-varying environments. In (8.33), $\hat{d}_n(k)$ is the output of the adaptive filter at time k based on the coefficients at time n, and is defined as

$$\hat{d}_n(k) = \sum_{i=0}^{N} a_i(n)x(k-i) + \sum_{i=1}^{N} b_i(n)d(k-i)$$

$$+ \sum_{i=0}^{N}\sum_{j=1}^{N} c_{i,j}(n)x(k-i)d(k-j) \tag{8.34}$$

and is the estimate of $d(k)$ obtained using the coefficients of the adaptive filter at time n. For notational simplicity, we define $\hat{d}(k)$ as

$$\hat{d}(k) = \hat{d}_k(k). \tag{8.35}$$

Let the input vector $\mathbf{Z}(n)$ and the coefficient vector $\vec{\mathbf{H}}(n)$ at time n be defined as

$$\mathbf{Z}(n) = [x(n), \ldots, x(n-N), \ d(n-1), \ldots, d(n-N), \ x(n)d(n-1),$$
$$\ldots, x(n)d(n-N), \ldots, x(n-N)d(n-N)]^T \tag{8.36}$$

and

$$\vec{\mathbf{H}}(n) = [a_0(n), \ldots, a_N(n), \ b_1(n), \ldots, b_N(n),$$
$$c_{0,1}(n), \ldots, c_{0,N}(n), \ldots, c_{N,N}(n)]^T, \tag{8.37}$$

respectively. The optimal solution $\vec{\mathbf{H}}(n)$, which minimizes $J(n)$ defined in (8.33), is given by

$$\vec{\mathbf{H}}(n) = \mathbf{C}^{-1}(n)\mathbf{P}(n), \tag{8.38}$$

where

$$\mathbf{C}(n) = \sum_{k=0}^{n} \lambda^{n-k} \mathbf{Z}(k)\mathbf{Z}^T(k) \tag{8.39}$$

and

$$\mathbf{P}(n) = \sum_{k=0}^{n} \lambda^{n-k} \mathbf{Z}(k)d(k). \tag{8.40}$$

Here, $\mathbf{C}(n)$ is the least-squares autocorrelation matrix of the input vector $\mathbf{Z}(n)$, and $\mathbf{P}(n)$ is the least-squares cross-correlation vector of the input vector $\mathbf{Z}(n)$ and the desired response signal $d(n)$.

Recursive Solution. A recursive solution for $\vec{\mathbf{H}}(n)$ can be obtained as in Chapter 7 using the matrix inversion lemma. The resulting algorithm is identical in form to the one in Table 7.1. The extended least-squares adaptive bilinear filter is presented in Table 8.1. This algorithm has a computational complexity that is proportional to N^4 arithmetical operations per iteration. Faster algorithms similar to the fast RLS Volterra filters can be derived as in Chapter 7. However, we do not discuss fast algorithms in this chapter. Such algorithms are described in Lee [135]. Extended least-squares lattice bilinear filters have also been derived, and may be found in Baik and Mathews [8].

8.3.1.3 A Critique of Equation Error Methods

The greatest advantage of the equation error adaptive filters is their inherent simplicity. In this approach, the recursive estimation problem is converted into a two-channel nonrecursive estimation problem. The error surface is quadratic in the coefficients, and it has a unique minimum unless the autocorrelation matrix of the input vector is singular. The

TABLE 8.1   The Extended Least-Squares Equation Error Adaptive Bilinear Filter

Initialization

$$\mathbf{C}^{-1}(0) = \frac{1}{\delta}\mathbf{I}; \; \delta > 0$$

$$\mathbf{P}(0) = \mathbf{0}$$

$$\vec{\mathbf{H}}(0) = \mathbf{0}$$

Main iteration

Do for $n = 1, 2, 3, \ldots$

$$\mathbf{Z}(n) = [x(n), \ldots, x(n-N), d(n-1), \ldots, d(n-N), x(n)d(n-1)$$

$$\ldots, x(n)d(n-N), \ldots, x(n-N)d(n-N)]^T$$

$$\epsilon(n) = d(n) - \mathbf{Z}^T(n)\vec{\mathbf{H}}(n-1)$$

$$\mathbf{k}(n) = \frac{\mathbf{C}^{-1}(n-1)\mathbf{Z}(n)}{\lambda + \mathbf{Z}^T(n)\mathbf{C}^{-1}(n-1)\mathbf{Z}(n)}$$

$$\mathbf{C}^{-1}(n) = \frac{1}{\lambda}\mathbf{C}^{-1}(n-1) - \frac{1}{\lambda}\mathbf{k}(n)\mathbf{Z}^T(n)\mathbf{C}^{-1}(n-1)$$

$$\vec{\mathbf{H}}(n) = \vec{\mathbf{H}}(n-1) + \mathbf{k}(n)\epsilon(n)$$

$$\hat{d}(n) = \mathbf{Z}^T(n)\vec{\mathbf{H}}(n)$$

$$e(n) = d(n) - \hat{d}(n)$$

---

coefficients of a properly designed adaptive filter should converge to this unique minimum. By casting the problem as that of a two-channel non-recursive estimator, we avoid the stability problems that are inherent in truly recursive estimators. The equation error LMS adaptive bilinear filter will operate in a stable manner as long as the step size μ is chosen to satisfy the stability requirements of the adaptation procedure. However, this statement does not imply that if the coefficients of the adaptive filter are copied into a bilinear filter implementation, the resulting recursive system is stable. The stability of such realizations must be determined separately.

Perhaps the greatest disadvantage of the equation error methods is that they result in biased estimates when the desired response signal in system identification problems contains noise. This happens because the statistics of $d(n)$ are different from those of the output of the unknown system. Consequently, the use of equation error methods are recommended only when the input signals are noise-free or when they contain only small amounts of noise.

Example 8.1   Since we have already seen the performance comparisons of the LMS and RLS adaptive filters in Chapter 7, we use the RLS adaptive filter to demonstrate the properties of the equation error estimation approach in this example. We consider a system identification problem in which the unknown system has input–output relationship given by

$$y(n) = \sum_{i=0}^{2} a_i x(n-i) + \sum_{i=1}^{2} b_i y(n-i)$$

$$+ \sum_{i=0}^{2} \sum_{j=1}^{2} c_{i,j} x(n-i) y(n-j),$$

where $a_0 = a_1 = a_2 = 1$, $b_1 = -b_2 = 0.5$, $c_{0,1} = 0.3$, $c_{1,1} = c_{1,2} = -0.2$, $c_{0,2} = 0.1$, $c_{2,1} = 0.1$, and $c_{2,2} = 0.3$. The input signal $x(n)$ to the adaptive filter was obtained as the output of a lowpass filter with transfer function

$$H(z) = \frac{1}{1 - 1.6z^{-1} + 0.95z^{-2}}$$

when its input was a white, Gaussian signal with zero mean value. The variance of the input signal to the lowpass filter was adjusted such that the variance of $x(n)$ was one. The desired response signal was obtained by corrupting the output of an unknown system with additive white noise that is uncorrelated with the input signal. The adaptive filter was run with the same number of coefficients as the unknown system. The equation error ELS adaptive filter used an exponential weighting factor corresponding to $\lambda = 0.995$. Figures 8.4a–8.4c show the average behavior of the coefficients of the adaptive filter for three different noise variances and three different coefficients. These results were obtained by averaging the coefficient estimates obtained over 50 independent estimates. We can see from these figures that equation error adaptive filters result in biased estimates when the desired response signal is corrupted with noise. The larger the noise level is, the greater the bias is.

8.3.2   Output Error Bilinear Filters

The output error adaptive filters attempt to alleviate the drawbacks of the equation error methods discussed in the previous subsection. They provide a truly recursive estimate with the hope that the resulting estimates will be unbiased. In a large number of situations, the output error algorithms are able to obtain unbiased estimates in the steady state. However, these algorithms also come with several disadvantages. In general, they are more complicated to derive and implement than the equation error algorithms. The error surfaces for the output error filters are nonlinear functions of the coefficient values. Consequently, they may contain local minima, and the adaptive filters may not necessarily converge to the global minimum of their error surfaces. Finally, the recursive nature of the system model may drive the adaptive filter to instability unless the algorithms are carefully designed to prevent them from becoming unstable. In spite of these difficulties, obtaining unbiased estimates is a desirable goal for any adaptive filter, and output error algorithms are the methods of choice in a large number of applications.

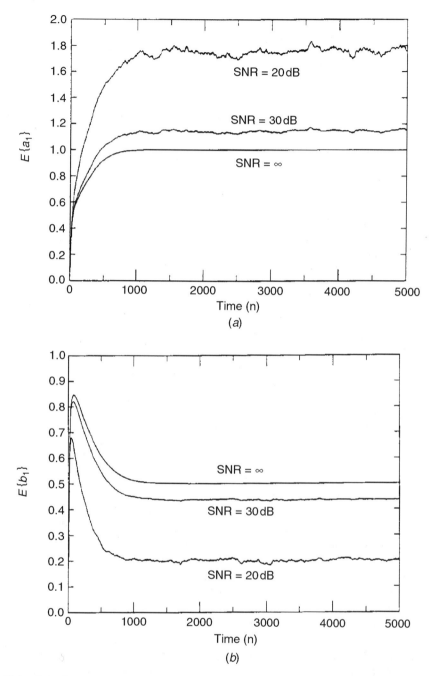

Figure 8.4   Evolution of the mean values of the coefficients of the equation error adaptive filter in Example 8.1: (a) evolution of $a_1(n)$; (b) evolution of $b_1(n)$. (Courtesy of H. K. Baik. Copyright © 1993 IEEE.)

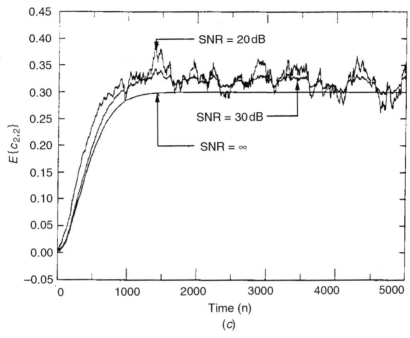

Figure 8.4 (*continued*) Evolution of the mean values of the coefficients of the equation error adaptive filter in Example 8.1: (*c*) evolution of $c_{2,2}(n)$. (Courtesy of H. K. Baik. Copyright © 1993 IEEE.)

8.3.2.1 *Extended Least-Squares Output Error Adaptive Bilinear Filters*

The derivation of the ELS output error adaptive bilinear filter is a straightforward extension of the derivation of the equation error algorithm. Since the difference between the two algorithms is in the way the estimate of the desired response signal is made, we simply define the input vector $\mathbf{Z}(n)$ in such a way that the estimated signal $\hat{d}(n)$ replaces $d(n)$ in the definition of (8.36):

$$\mathbf{Z}(n) = [x(n), \ldots, x(n-N), \ \hat{d}(n-1), \ldots, \hat{d}(n-N), \ x(n)\hat{d}(n-1),$$
$$\ldots, x(n)\hat{d}(n-N), \ldots, x(n-N)\hat{d}(n-N)]^T. \quad (8.41)$$

Substitution of this definition of $\mathbf{Z}(n)$ in Table 8.1 results in the extended least-squares output error adaptive bilinear filter.

Remark 8.3 The input vector $\mathbf{Z}(n)$ is defined using samples of the adaptive filter output for the output error algorithm. Consequently, it is impossible for us to compute the coefficient vector $\vec{\mathbf{H}}(n)$ prior to computing $\vec{\mathbf{H}}(n-1)$ in output error methods. In other words, the output error estimation can be performed only in a recursive manner.

Example 8.2 In this example, we consider the system identification problem in Example 8.1, with the variation that the adaptive filter employed is the extended least-squares output error algorithm. Figures 8.5a–8.5c display the evolution of the three coefficients corresponding to the same cases considered in Example 8.1. The parameter λ was chosen to be 0.995 for the output error filters also. Comparing the results presented in Figures 8.4 and 8.5, we can see that the output error algorithm is capable of estimating the parameters without bias even when a significant amount of noise is present in the desired response signal.

8.3.2.2 Stability of Extended Least-Squares Bilinear Filters
In this subsection, we consider the issue of stable adaptive filter operation under infinite precision implementation. Given that the input signals to the adaptive filter are "well-behaved" in some sense, what can we say about the output of the adaptive filter? We now show that both the equation error and output error extended least-squares adaptive bilinear filters are stable in the sense that the long-term average of the squared estimation error is bounded under fairly mild conditions. The result was derived by Lee and Mathews [139] and is an extension of a similar result for RLS adaptive recursive linear filters [295]. The following theorem describes the results more explicitly.

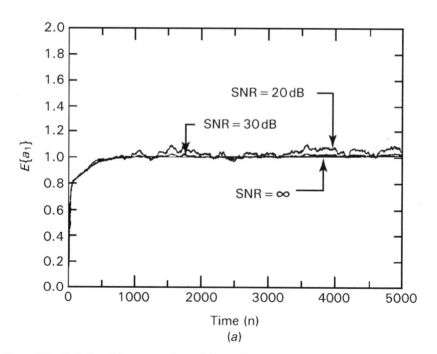

Figure 8.5 Evolution of the mean values of the coefficients of the output error adaptive filter in Example 8.2: (a) evolution of $a_1(n)$; (b) evolution of $b_1(n)$. (Courtesy of H. K. Baik. Copyright © 1993 IEEE.)

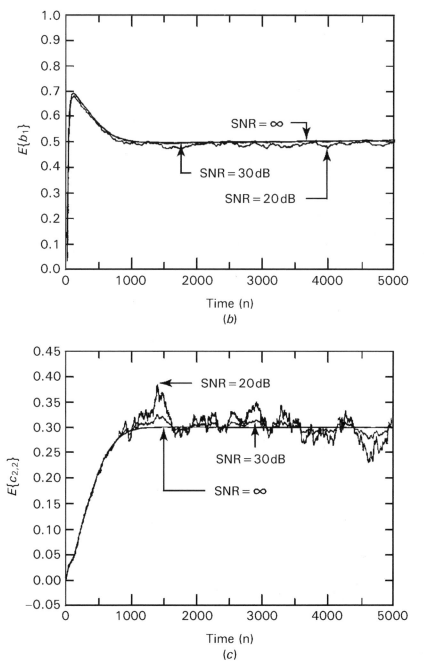

Figure 8.5 (*continued*) Evolution of the mean values of the coefficients of the output error adaptive filter in Example 8.2: (*c*) evolution of $c_{2,2}(n)$. (Courtesy of H. K. Baik. Copyright © 1993 IEEE.)

Theorem 8.2   Let $(1/n) \sum_{k=1}^{n} d^2(k)$ be bounded for all positive values of n. Then, the extended least-squares adaptive bilinear filters provide stable outputs in the sense that $(1/n) \sum_{k=1}^{n} e^2(k)$ is bounded for all positive values of n if $0 < \lambda < 1$.

PROOF   We prove the theorem by showing that

$$\frac{1}{n} \sum_{k=0}^{n} e^2(k) \leq \frac{1}{n} \sum_{k=0}^{n} d^2(k) + \frac{\lambda}{n} \vec{\mathbf{H}}^T(0)\mathbf{C}(0)\vec{\mathbf{H}}(0). \tag{8.42}$$

The proof is equally applicable to both equation error and output error methods. We simply use the appropriate definition of the input vector $\mathbf{Z}(n)$ for the case of interest. We assume that $\mathbf{C}^{-1}(n)$, the inverse of the autocorrelation matrix of the input vector, exists at each time. We will discuss the implications of this assumption later.

We start the proof by restating a few results we derived earlier. We leave it as an exercise for the reader to show that the a priori estimation error $\epsilon(n)$ and the a posteriori estimation error $e(n)$ are related to each other as

$$\epsilon(n) = \frac{e(n)}{1 - \mathbf{k}^T(n)\mathbf{Z}(n)}$$

$$= \frac{e(n)}{1 - \mathbf{Z}^T(n)\mathbf{C}^{-1}(n)\mathbf{Z}(n)}. \tag{8.43}$$

We also need to use the following update equations:

$$\mathbf{C}(n) = \lambda\mathbf{C}(n-1) + \mathbf{Z}(n)\mathbf{Z}^T(n), \tag{8.44}$$

$$\mathbf{P}(n) = \lambda\mathbf{P}(n-1) + d(n)\mathbf{Z}(n), \tag{8.45}$$

and

$$\vec{\mathbf{H}}(n) = \vec{\mathbf{H}}(n-1) + \mathbf{C}^{-1}(n)\mathbf{Z}(n)\epsilon(n). \tag{8.46}$$

Let us evaluate the quantity $\vec{\mathbf{H}}^T(n)\mathbf{C}(n)\vec{\mathbf{H}}(n)$ using the relationships presented above. Noting that $\mathbf{C}(n)\vec{\mathbf{H}}(n) = \mathbf{P}(n)$, and substituting (8.43)–(8.46) in the preceding expression, we get

$$\vec{\mathbf{H}}^T(n)\mathbf{C}(n)\vec{\mathbf{H}}(n) = \vec{\mathbf{H}}^T(n)\mathbf{P}(n)$$

$$= [\vec{\mathbf{H}}(n-1) + \mathbf{C}^{-1}(n)\mathbf{Z}(n)\epsilon(n)]^T [\lambda\mathbf{P}(n-1) + d(n)\mathbf{Z}(n)]$$

$$= \lambda\vec{\mathbf{H}}^T(n-1)\mathbf{C}(n-1)\vec{\mathbf{H}}(n-1) + \lambda\epsilon(n)\mathbf{Z}^T(n)\mathbf{C}^{-1}(n)\mathbf{P}(n-1)$$

$$+ \vec{\mathbf{H}}^T(n-1)\mathbf{Z}(n)d(n) + \epsilon(n)\mathbf{Z}^T(n)\mathbf{C}^{-1}(n)\mathbf{Z}(n)d(n). \tag{8.47}$$

The second and fourth terms of (8.47) can be evaluated to be

$$
\lambda\epsilon(n)\mathbf{Z}^T(n)\mathbf{C}^{-1}(n)\mathbf{P}(n-1) + \epsilon(n)\mathbf{Z}^T(n)\mathbf{C}^{-1}(n)\mathbf{Z}(n)d(n)
$$
$$
= \epsilon(n)\mathbf{Z}^T(n)\mathbf{C}^{-1}(n)[\lambda\mathbf{P}(n-1) + \mathbf{Z}(n)d(n)]
$$
$$
= \epsilon(n)\mathbf{Z}^T(n)\vec{\mathbf{H}}(n)
$$
$$
= \epsilon(n)(d(n) - e(n)). \tag{8.48}
$$

Similarly, we can show that

$$
\vec{\mathbf{H}}^T(n-1)\mathbf{Z}(n)d(n) = (d(n) - \epsilon(n))d(n). \tag{8.49}
$$

Substituting (8.48), (8.49), and (8.43) in (8.47) gives

$$
\vec{\mathbf{H}}^T(n)\mathbf{C}(n)\vec{\mathbf{H}}(n) = \lambda\vec{\mathbf{H}}^T(n-1)\mathbf{C}(n-1)\vec{\mathbf{H}}(n-1) + d^2(n)
$$
$$
- \frac{e^2(n)}{1 - \mathbf{Z}^T(n)\mathbf{C}^{-1}(n)\mathbf{Z}(n)}. \tag{8.50}
$$

We can manipulate this equation to express (8.50) for an arbitrary value of time $k \geq 1$ as

$$
\frac{e^2(k)}{1 - \mathbf{Z}^T(k)\mathbf{C}^{-1}(k)\mathbf{Z}(k)} = d^2(k) - \vec{\mathbf{H}}^T(k)\mathbf{C}(k)\vec{\mathbf{H}}(k)
$$
$$
+ \lambda\vec{\mathbf{H}}^T(k-1)\mathbf{C}(k-1)\vec{\mathbf{H}}(k-1). \tag{8.51}
$$

Summing both sides of (8.51) from $k = 1$ to $k = n$, and then dividing the result by n gives

$$
\frac{1}{n}\sum_{k=1}^{n}\left\{\frac{e^2(k)}{1 - \mathbf{Z}^T(k)\mathbf{C}^{-1}(k)\mathbf{Z}(k)}\right\} = \frac{1}{n}\sum_{k=1}^{n}d^2(k) + \frac{\lambda}{n}\vec{\mathbf{H}}^T(0)\mathbf{C}(0)\vec{\mathbf{H}}(0)
$$
$$
- \frac{(1-\lambda)}{n}\sum_{k=1}^{n-1}\vec{\mathbf{H}}^T(k)\mathbf{C}(k)\vec{\mathbf{H}}(k)
$$
$$
- \frac{1}{n}\vec{\mathbf{H}}^T(n)\mathbf{C}(n)\vec{\mathbf{H}}(n). \tag{8.52}
$$

It can be shown that

$$
0 \leq 1 - \mathbf{Z}^T(n)\mathbf{C}^{-1}(n)\mathbf{Z}(n) \leq 1 \tag{8.53}
$$

whenever $\mathbf{C}^{-1}(n)$ exists. Because $\mathbf{C}(n)$ is nonnegative definite and λ is no greater than unity, the last two terms on the right-handside of (8.52) are nonnegative. It is then easy to see that

$$\frac{1}{n}\sum_{k=1}^{n}e^2(k) \leq \frac{1}{n}\sum_{k=1}^{n}\left\{\frac{e^2(k)}{1 - \mathbf{Z}^T(n)\mathbf{C}^{-1}(n)\mathbf{Z}(n)}\right\}$$

$$\leq \frac{1}{n}\sum_{k=1}^{n}d^2(k) + \frac{\lambda}{n}\vec{\mathbf{H}}^T(0)\mathbf{C}(0)\vec{\mathbf{H}}(0). \tag{8.54}$$

This completes the proof.

8.3.2.3   A Discussion of the Stability Theorem   Theorem 8.2 implies that the extended least-squares adaptive bilinear filters can operate in a stable manner when they are implemented in a numerically stable manner. However, the bound itself is not a strong one. The result simply states that the estimation error power is bounded in some sense by a quantity slightly larger than the power in the desired response signal. It is possible for the coefficients to fluctuate a great deal even though the average squared error is bounded. Even such a result is a valuable one for recursive nonlinear systems because stronger results are not available at present.

Stable operation of the adaptive filter also requires that the least-squares autocorrelation matrix is not singular at any time. While this condition is satisfied for a large class of stochastic input signals and bilinear systems, it is possible that this requirement is violated for certain signals and certain bilinear systems. Signals for which the least-squares autocorrelation matrix of $\mathbf{Z}(n)$ is nonsingular for all times are said to *persistently excite* the system. Deriving conditions for persistence of excitation of bilinear systems is beyond the scope of this book. Dasgupta et al. derived some conditions for persistence of excitation of bilinear systems in [62].

An important issue that should be kept in mind during the study of the stability of recursive polynomial systems is that most such studies deal only with input signal–dependent stability of the systems. It is possible that a signal different from the input signal to a stable adaptive bilinear system may drive a time-varying bilinear system with the same set of coefficients as the adaptive filter to instability.

The theorem applies only to the exact realizations of the extended least-squares bilinear filter. Variations such as the recursive prediction error method (RPEM) that are not discussed in this chapter requires additional steps for their stabilization [69].

The stability result described in this subsection can easily be extended to problems involving adaptive bilinear prediction [137]. With some simple additional assumptions, we can also show that extended RLS bilinear predictors employed in predictive coding applications will operate in a stable manner [169].

8.3.3 LMS Adaptive Output Error Bilinear Filters

The objective of the output error LMS adaptive bilinear filter [134,140] is to update the coefficient vector $\vec{\mathbf{H}}(n)$ during each iteration in an attempt to reduce the mean-squared error defined as

$$J(n) = E\{(d(n) - \hat{d}(n))^2\}$$

$$= E\{(d(n) - \vec{\mathbf{H}}^T(n)\mathbf{Z}(n))^2\} \tag{8.55}$$

at each time. In the definition in (8.55), the coefficient vector $\vec{\mathbf{H}}(n)$ is defined as in (8.29). However, the input vector $\mathbf{Z}(n)$ is defined using the output signal $\hat{d}(n)$ rather than the desired response signal $d(n)$ as

$$\mathbf{Z}(n) = [x(n), \dots, x(n-N), \hat{d}(n-1), \dots, \hat{d}(n-N), x(n)\hat{d}(n-1), \dots,$$

$$x(n)\hat{d}(n-N), \dots, x(n-N)\hat{d}(n-N)]^T. \tag{8.56}$$

The stochastic gradient update equation for this problem is given by

$$\vec{\mathbf{H}}(n+1) = \vec{\mathbf{H}}(n) - \frac{\mu}{2} \frac{\partial (d(n) - \vec{\mathbf{H}}^T(n)\mathbf{Z}(n))^2}{\partial \vec{\mathbf{H}}(n)}$$

$$= \vec{\mathbf{H}}(n) - \mu e(n) \frac{\partial (d(n) - \vec{\mathbf{H}}^T(n)\mathbf{Z}(n))}{\partial \vec{\mathbf{H}}(n)}. \tag{8.57}$$

The estimator output $\hat{d}(n)$ as well as the input vector $\mathbf{Z}(n)$, which contains delayed versions of $\hat{d}(n)$, are functions of the coefficients of the adaptive filter. The gradient vector in (8.57) can be directly evaluated as

$$-\frac{\partial (d(n) - \vec{\mathbf{H}}^T(n)\mathbf{Z}(n))}{\partial \vec{\mathbf{H}}(n)} = \mathbf{Z}(n) + \sum_{i=1}^{N} b_i(n) \frac{\partial \hat{d}(n-i)}{\partial \vec{\mathbf{H}}(n)}$$

$$+ \sum_{i=0}^{N} \sum_{j=1}^{N} c_{i,j}(n) x(n-i) \frac{\partial \hat{d}(n-j)}{\partial \vec{\mathbf{H}}(n)}. \tag{8.58}$$

Equation (8.58) indicates the necessity for reevaluating the derivatives of past values of $\hat{d}(n)$ with respect to $\vec{\mathbf{H}}(n)$ during each iteration. An assumption commonly employed in the adaptive linear recursive filtering literature [281] is that the step size μ is sufficiently small such that $\vec{\mathbf{H}}(n) \approx \vec{\mathbf{H}}(n-1) \approx \cdots \approx \vec{\mathbf{H}}(n-N)$. Using this approximation and the fact that $d(n)$ is not a function of $\vec{\mathbf{H}}(n)$, we may write (8.58) as

$$\frac{\partial \hat{d}(n)}{\partial \vec{\mathbf{H}}(n)} = \mathbf{Z}(n) + \sum_{i=1}^{N} b_i(n) \frac{\partial \hat{d}(n-i)}{\partial \vec{\mathbf{H}}(n-i)} + \sum_{i=0}^{N} \sum_{j=1}^{N} c_{i,j}(n) x(n-i) \frac{\partial \hat{d}(n-j)}{\partial \vec{\mathbf{H}}(n-j)}. \tag{8.59}$$

TABLE 8.2   The Output Error LMS Adaptive Bilinear Filter

Initialization

$\vec{H}(0)$ may be selected as an arbitrary but stable bilinear system

$\Psi(0) = [0, 0, 0, \ldots, 0]^T$

Main iteration

$\mathbf{Z}(n) = [x(n), \ldots, x(n - N), \hat{d}(n - 1), \ldots, \hat{d}(n - N), x(n)\hat{d}(n - 1),$
$\ldots, x(n)\hat{d}(n - N), \ldots, x(n - N)\hat{d}(n - N)]^T$

$e(n) = d(n) - \vec{H}^T(n)\mathbf{Z}(n)$

$\Psi(n) = \mathbf{Z}(n) + \sum_{i=1}^{N} b_i(n)\Psi(n - i) + \sum_{i=0}^{N}\sum_{j=1}^{N} c_{i,j}(n)x(n - i)\Psi(n - j)$

$\vec{H}(n + 1) = \vec{H}(n) + \mu e(n)\Psi(n)$

Let us define the gradient vector $\Psi(n)$ to be

$$\Psi(n) = \frac{\partial \hat{d}(n)}{\partial \vec{H}(n)}.$$ (8.60)

Substituting this definition in (8.59) provides a more compact description of the calculation of the gradient vector as

$$\Psi(n) = \mathbf{Z}(n) + \sum_{i=1}^{N} b_i(n)\Psi(n - i) + \sum_{i=0}^{N}\sum_{j=1}^{N} c_{i,j}(n)x(n - i)\Psi(n - j).$$ (8.61)

This equation shows that the gradient vector can be computed during each iteration by processing the input vector $\mathbf{Z}(n)$ with a bilinear filter defined by the coefficients $b_i(n)$ and $c_{i,j}(n)$ for $0 \leq i \leq N$ and $1 \leq j \leq N$.

We now have all the information necessary to complete the iterations of the output error LMS adaptive bilinear filter at each time. They are summarized in Table 8.2.

8.3.3.1   A Simpler Algorithm   The algorithm of Table 8.2 has a computational complexity that corresponds to $O(N^4)$ multiplications per iteration. This high computational burden arises from the calculation of the gradient vector because each of the $N^2 + 3N + 1$ elements of the gradient vector is computed with a bilinear filter with $N(N + 2)$ coefficients at each time instant. We can simplify the algorithm by an order of magnitude by making another approximation in the calculation of the gradient vector. We note from the definition of the input vector in (8.56) that the elements of $\mathbf{Z}(n)$ are delayed versions of the $2(N + 1)$ elements of the vector $\mathbf{z}(n)$

defined as

$$\mathbf{z}(n) = [x(n), \hat{d}(n-1), x(n)\hat{d}(n-1), x(n-1)\hat{d}(n-1), \ldots, x(n-N)\hat{d}(n-1).$$

$$x(n)\hat{d}(n-2), x(n)\hat{d}(n-3), \ldots, x(n)\hat{d}(n-N)]^T. \tag{8.62}$$

This motivates the approximation of computing $2N + 2$ elements of the gradient vector $\Psi(n)$ as

$$\Psi_s(n) = \mathbf{z}(n) + \sum_{i=1}^{N} b_i(n)\Psi_s(n-i) + \sum_{i=1}^{N}\sum_{j=0}^{N} c_{i,j}(n)x(n-j)\Psi_s(n-i), \tag{8.63}$$

and approximating the remaining elements of $\Psi(n)$ with appropriately delayed versions of the elements of $\Psi_s(n)$. The delayed versions are obtained using the same set of operations employed to construct $\mathbf{Z}(n)$ from $\mathbf{z}(n)$. The output error LMS bilinear filter implemented in this manner requires storage of N past values of the $(2N + 2)$-element vector $\Psi_s(n)$. The computational complexity of this method can be shown to correspond to $O(N^3)$ multiplications per iteration, and thus, the approximation provides significant savings in the computations required to implement the filter. We will see shortly that the simplification does not cause significant performance degradation.

8.3.3.2 A Highly Simplified Output Error Adaptive Filter Feintuch proposed an adaptive algorithm that mimics the form of the non-recursive LMS adaptive filter for recursive linear filtering [68]. When adapted to the bilinear filtering case, the corresponding algorithm is described by the following equations:

$$e(n) = d(n) - \vec{\mathbf{H}}^T(n)\mathbf{Z}(n) \tag{8.64}$$

and

$$\vec{\mathbf{H}}(n+1) = \vec{\mathbf{H}}(n) + \mu e(n)\mathbf{Z}(n). \tag{8.65}$$

Because the dependence of $\mathbf{Z}(n)$ on the coefficient vector is ignored in the gradient calculation, this algorithm satisfies no stochastic gradient cost function. Consequently, we refer to this method as the *pseudo-LMS adaptive bilinear filter.* The computational complexity of this adaptive filter is only $O(N^2)$ multiplications per iteration. Experiments have shown that the algorithm works well in most situations. However, it has been shown theoretically and experimentally for the linear case that the method diverges for all choices of the step size for certain input signals. Because of its potential to diverge for all positive step sizes, extreme care must be taken if this method is employed in practice.

Example 8.3 In this example, we compare the performance of the three different variations of the output error adaptive bilinear filters described in this section. The

TABLE 8.3   Mean and Variances of the Filter Coefficients in Example 8.3

		LMS		Simplified LMS		Peudo-LMS	
σ_η^2		0.001	0.1	0.001	0.1	0.001	0.1
a_1	Mean	0.9999	1.0057	1.0001	1.0089	0.9994	1.0077
	Variance	4.78×10^{-6}	3.80×10^{-4}	5.12×10^{-6}	4.64×10^{-4}	4.85×10^{-6}	4.77×10^{-4}
b_1	Mean	0.5004	0.4943	0.5003	0.4914	0.5001	0.4957
	Variance	4.79×10^{-6}	5.93×10^{-4}	5.02×10^{-6}	5.97×10^{-4}	3.23×10^{-6}	3.98×10^{-4}
$c_{2,2}$	Mean	0.3001	0.2941	0.2996	0.2927	0.2999	0.2949
	Variance	5.15×10^{-6}	5.77×10^{-4}	5.74×10^{-6}	7.63×10^{-4}	6.57×10^{-6}	4.80×10^{-4}

Source: Results courtesy of Junghsi Lee.

adaptive filters were employed in the system identification mode in the experiments. The unknown system, the input signal and the noise sequences were identical to those employed in Example 8.1. All the results presented are averages over 50 independent experiments. Table 8.3 tabulates the steady-state values and the steady-state estimation variances associated with the coefficients $a_1(n)$, $b_1(n)$, and $c_{2,2}(n)$ of the LMS, simplified LMS, and pseudo-LMS output error adaptive bilinear filters for two different measurement noise variances. For all three methods, the step size μ was set equal to 0.009. The steady-state values were evaluated as time averages of the coefficient sequences in the range [9000, 10000] and over the 50 independent experiments. The input signal in these experiments was a zero-mean and white Gaussian sequence with variance 0.18, which resulted in an output variance of approximately one. The measurement noise was also a white Gaussian sequence with variance σ_η^2 equal to either 0.001 or 0.1. We observe from the results tabulated in the table that all three algorithms perform in approximately the same manner, and that they provide unbiased estimates in the presence of measurement noise. Even though the pseudo-LMS filter performs very well in this example, we do not advocate its use in practice because of documented stability problems in certain situations. The simplified LMS meets the test of good performance as well as computational simplicity, and therefore, we recommend this as the method of choice among the three algorithms considered in this example.

8.3.3.3   Stability Monitoring of LMS Adaptive Bilinear Filters   A key
drawback of the adaptive filters described above is that they are not guaranteed to operate in a stable manner. Consequently, the algorithms must be modified to monitor the adaptive filter for stability. If the coefficients are found to belong to the class that causes the bilinear system to be unstable, they must also be projected to a space that defines stable bilinear filters. We now briefly discuss a method that accomplishes this.

Even though we are concerned with the stability of a time-varying bilinear system model, our approach employs the sufficient stability condition derived in Section 8.2. The method consists of the following steps:

1. Let $\vec{\mathbf{H}}_0' = \vec{\mathbf{H}}(n)$ and $\mu_1' = \mu$. Set an iteration index $k = 1$.

2. In the kth iteration at time n, find the new coefficient vector using the update equation

$$\vec{\mathbf{H}}_k' = \vec{\mathbf{H}}_{k-1}' + \mu_k' \Psi(n)e(n). \tag{8.66}$$

Now, test the updated coefficients using the result of Theorem 8.1. The maximum value of the input is taken as the maximum of all input samples over the past K samples. A rule of thumb to select K is to choose it in the neighborhood of $10N$. Increment the iteration index k by 1.

3. If the test of stability is successful, set $\vec{\mathbf{H}}(n+1) = \vec{\mathbf{H}}_k'$ and wait for the next data to come in and begin the iterations for the next time. If the test fails and k is smaller than some preselected threshold, reduce the step size by a preselected factor and update the coefficients again using the new step size. It is typical to choose the step size to be in the range $[0.5, 0.9]$ of the previous step size. Let μ_k' be the new step size. If k is larger than or equal to the threshold, set

$$\vec{\mathbf{H}}(n+1) = \vec{\mathbf{H}}(n-1) \tag{8.67}$$

and wait for the next data to come in and begin the iterations for the next time.

4. Go to step (2).

8.3.3.4 *A Discussion of the Stabilization Approach*

As we will soon see in Example 8.4, the stabilization approach works reasonably well in many applications. However, it is important to keep in mind that the stability condition employed in the system was derived for time-invariant bilinear systems, and may not be adequate for time-varying systems. Consequently, the algorithm is not guaranteed to work in a stable manner with the projection mechanism described above, even though it appears to give satisfactory results in general.

One problem associated with the scheme presented above is that the roots of a polynomial must be evaluated after each coefficient update. Several root tracking methods are available in the literature [298]. However, they tend to be very computation intensive. In fact, the most efficient root-tracking method available at this time requires $0(N^3)$ arithmetical operations at each time instant.

Example 8.4 In this example, we consider the same system identification problem of Examples 8.3. The input signal in the previous examples was a zero-mean signal with variance $\sigma_x^2 = 0.05$. We saw in Example 8.3 that all the output error methods performed in a stable manner without any stability monitoring mechanism in that example. We now repeat the experiment with an input signal with variance equal to 0.20 obtained by multiplying the input signal of the previous examples by two. Figure 8.6 displays the evolution of the coefficients obtained using the simplified output error LMS bilinear filter during one run when the step size was selected as

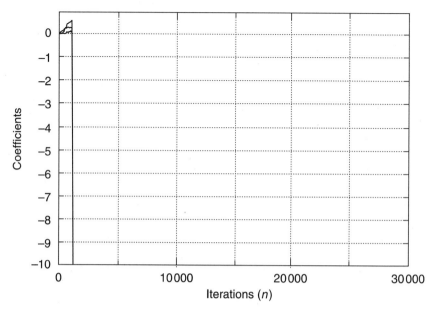

Figure 8.6 Evolution of the mean values of the coefficients in Example 8.4 for the simplified output error LMS adaptive bilinear filter without stability monitoring. (Courtesy of J. Lee. Copyright © 1995 IEEE.)

$\mu = 0.001$. We can see that the coefficients exhibit highly erratic behavior for this situation. The coefficient evolution of the adaptive filter when the stability monitoring mechanism described in this section is shown in Figure 8.7 for the same set of algorithm parameters as before. We can see that the evolution of the coefficients is much better behaved with the stability monitoring and projection system, indicating that it is a useful system in practice even though the stability tests employed were accurate only for time-invariant bilinear systems. As stated in the description of the tests, we should not expect the stability monitoring tests to work well for large step sizes, and therefore must be careful to use conservatively small values of the step size in adaptive LMS bilinear filters.

8.3.4 Additional Work on Adaptive Recursive Polynomial Filters

Much additional work has been done in the area of adaptive recursive polynomial filtering beyond that presented in this chapter. Interested readers may refer to the papers cited below for additional information on these topics.

Billings and his colleagues have done considerable amount of work on parameter estimation using recursive polynomial models [22]. The tools they have employed are primarily the extended least-squares algorithm and its variations. The stability theorem proved in the previous section is applicable only to extended least-squares

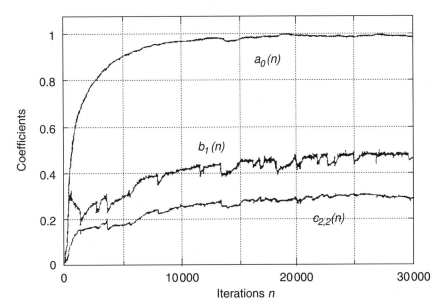

Figure 8.7 Evolution of the mean values of the coefficients in Example 8.4 for the simplified output error LMS adaptive bilinear filter with stability monitoring. (Courtesy of J. Lee. Copyright © 1995 IEEE.)

methods. Many of its variations may not operate in a stable manner as described in Theorem 8.2. In such cases, additional steps must be taken to ensure that those algorithms operate in a stable manner. Fnaiech and Ljung [69] have presented a method that employs a Kalman filter to stabilize such algorithms. They have proved, using a theorem in Jazwinsky [103] that the resulting filters are stable. Unfortunately, the Kalman filter employed to stabilize the system has a computational complexity that is an order of magnitude larger than that of efficient realizations of the basic adaptive filter itself.

Bose and Chen [26] have presented a conjugate gradient adaptive filter for bilinear systems. They used a preconditioning technique to improve the convergence characteristics of the adaptive filter. They employed a stability test and a projection technique similar to the one presented in Section 8.3.3.3 for their system [25]. Bose has also developed a two-dimensional, adaptive bilinear filter [105]. A stochastic gradient adaptive filter with a different class of recursive polynomial system models and a stability test appropriate for those system models was presented by Mumolo and Carini [190]. An adaptive least-squares lattice bilinear filter has been described in [8]. This system belongs to the class of exact recursive least-squares algorithms and therefore is stable in the sense of the result in Theorem 8.2.

Adaptive recursive polynomial filters have found use in several practical applications recently. Examples include analysis of sonar signals [180] and analysis of harmonics generated by a vibrating motor [263].

8.4   A PERFORMANCE COMPARISON OF THREE SYSTEM MODELS IN AN EQUALIZATION PROBLEM

A finite-memory communication channel can be modeled in a general manner using the input–output relationship

$$y(n) = f(x(n + M), x(n + M - 1), \ldots, x(n), \ldots, x(n - N)) + \eta(n), \qquad (8.68)$$

where $x(n)$ is the input data sequence to the channel, $f(\cdot)$ is a function that models the behavior of the channel, $\eta(n)$ is a noise sequence that models the disturbances introduced into the information symbols as well as the errors in the channel model employed, and $y(n)$ is the channel output at time n. Because of the amplitude and delay distortions of transmission channels, we seldom can make efficient use of the available channel bandwidth unless an appropriate equalizer is employed. The task of a channel equalizer is to reduce the effects of channel distortion from the received signals. This task is accomplished by processing the received data as well as the estimated input symbol sequence with a possibly adaptive model of the channel.

When the transmission channel can be modeled as a linear filter, i.e., when $f(\cdot)$ is a linear function of $x(n)$, a conventional decision feedback equalizer (DFE) with a linear system model is typically employed for eliminating the resulting intersymbol interference (ISI) [232]. The behavior of such a DFE may be characterized by the input–output relationship

$$\hat{y}(n) = \sum_{i=-r}^{0} b_i(n - 1)y(n - i) + \sum_{i=1}^{s} a_i(n - 1)\bar{x}(n - i), \qquad (8.69)$$

where r and s are positive integers, $y(n)$ is the received signal and $\bar{x}(n)$ denotes either the true values of the transmitted signals $x(n)$ or the estimated channel output $\bar{x}(n)$, depending on whether the equalizer is operating in the training mode or the decision feedback mode, respectively. The estimated channel output $\tilde{x}(n)$ is produced by passing $\hat{y}(n)$ through a decision device (quantizer). Because of the bulk delay in the communication channel, the system in (8.69) represents a causal system. The problem of adaptive decision feedback equalization is illustrated in Figure 8.8.

When $f(\cdot)$ is a nonlinear function of $x(n)$, it is known that the performance of the conventional DFE with linear system model is severely limited. Özgunel has employed a multichannel adaptive lattice algorithm to remove the nonlinear ISI

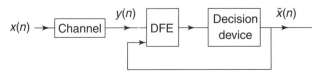

Figure 8.8   Decision feedback equalization.

caused by a Volterra channel [211]. Other types of adaptive nonlinear equalizers including Volterra DFEs have also been proposed [17]. In this section, we compare the performance of a DFE employing a linear system model with those of a DFE that uses a second-order Volterra system model of the form

$$\hat{y}(n) = \sum_{i=-r}^{0} b_{1,i}(n-1)y(n-i) + \sum_{i=1}^{s} a_{1,i}(n-1)\bar{x}(n-i)$$
$$+ \sum_{i=1}^{s}\sum_{j=i}^{s} a_{2,i,j}(n-1)\bar{x}(n-i)\bar{x}(n-j) \tag{8.70}$$

and a DFE that employs a bilinear system model of the form

$$\hat{y}(n) = \sum_{i=-r}^{0} b_i(n-1)y(n-i) + \sum_{i=1}^{s} a_i(n-1)\bar{x}(n-i)$$
$$+ \sum_{j=-r}^{0}\sum_{i=1}^{s} c_{i,j}(n-1)\bar{x}(n-i)y(n-j). \tag{8.71}$$

These results are reproduced from Ma et al. [151].

Example 8.5   We consider a satellite communication channel that is modeled as a linear system followed by a memoryless nonlinearity as shown in Figure 8.9. The input–output relationship of the linear system is given by

$$q(n) = 0.08x(n+2) - 0.12x(n+1) + x(n) + 0.18x(n-1)$$
$$- 0.1x(n-2) + 0.09x(n-3)$$
$$- 0.05x(n-4) + 0.04x(n-5) + 0.03x(n-6) + 0.01x(n-7).$$

The noisy output of the memoryless nonlinearity is given by

$$y(n) = q(n) + 0.036q^2(n) - 0.011q^3(n) + \eta(n).$$

The input signal to the channel, $x(n)$, is an independent, identically distributed, four-level ($\pm1, \pm3$) sequence. The additive noise $\eta(n)$ is a Gaussian sequence with zero

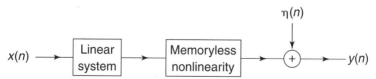

Figure 8.9   Model of the satellite communication channel of Example 8.5.

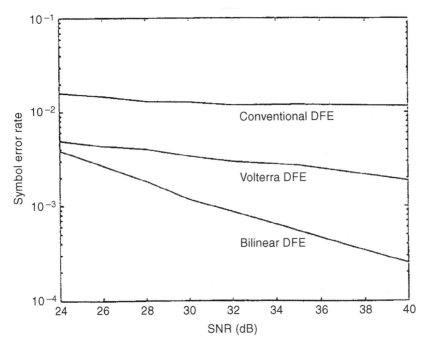

Figure 8.10   Bit error rates for the three DFEs in Example 8.5. (Courtesy of J. Lee. Copyright © 1994 IEEE.)

mean value and variance determined by the desired output signal-to-noise ratio (SNR) in each experiment.

The adaptive bilinear DFE was run with $r = 10$ and $s = 3$, i.e., the number of equalizer coefficients is 47. The conventional DFE used here was described by (8.69), and it was run with $r = 23$ and $s = 23$. The second-order Volterra DFE was implemented as defined in (8.70) with $r = 10$ and $s = 7$. Note that the complexity associated with each of the three models above is identical. The DFEs were adapted using the recursive (extended) least-squares algorithm with $\lambda = 0.9995$. A training sequence of length 5000 symbols was used in the systems. After the initial 5000 samples, the adaptive filters were run in the decision-directed mode.

Figure 8.10 shows the symbol error rate performance of the three DFEs for a range of output SNRs. It can be seen that the performance of the conventional DFE with the linear system model is very limited, just as one would expect for a nonlinear channel. The bilinear equalizer outperformed the Volterra DFE, especially when the SNR is relatively high. The results demonstrate the potential of recursive polynomial system models as effective tools for nonlinear channel equalization.

8.5 EXERCISES

8.1 Show that $y_k(n)$, defined in (8.7) and (8.8), can be written as a linear combination of only the kth-order products of current and past samples of the input signal.

8.2 Find the linear and second-order kernel of the bilinear system given by the input–output relationship

$$y(n) = 0.6y(n-1) + 0.2x(n) + 0.25x(n-1) + 0.05x(n)y(n-1).$$

Find a bound on $x(n)$ for which $y(n)$ will also be bounded.

8.3 Consider a time-varying bilinear system whose input–output relationship is given by

$$y(n) = \sum_{i=1}^{N} b_i(n)y(n-i) + \sum_{i=0}^{N} a_i(n)x(n-i)$$

$$+ \sum_{i=0}^{N}\sum_{j=1}^{N} c_{i,j}(n)x(n-i)y(n-j).$$

Let the input signal $x(n)$ be bounded by M_x, and also let $h_i(n)$ denote the unit impulse response signal of the linear system given by

$$y(n) = \sum_{i=1}^{N} b_i(n)y(n-i) + \sum_{i=0}^{N} a_i(n)x(n-i)$$

at time n. Let

$$\alpha(n) = \sum_{i=0}^{\infty} |h_i(n)| \le M_h < \infty.$$

That is, the impulse response signal is absolutely summable at all times. Derive a sufficient condition on the time-varying coefficients for $y(n)$ to be bounded.

8.4 We consider a recursive nonlinear system model with input–output relationship [190]

$$y(n) = x(n) + \sum_{m=1}^{N_1} b_m y(n-m)$$

$$+ \sum_{m_1=1}^{N_2}\sum_{m_2=1}^{N_2} c_{m_1,m_2} y(n-m_1)y(n-m_2),$$

where $x(n)$ is the input signal and $y(n)$ is the output signal. Let the poles of the system

$$w(n) = x(n) + \sum_{m=1}^{N_1} b_m w(n-m)$$

be $p_1, p_2, \ldots, p_{N_1}$, and let

$$\alpha = \prod_{m=1}^{N_1} (1 - p_m)$$

and

$$\gamma = \sum_{m_1=1}^{N_2} \sum_{m_2=1}^{N_2} |c_{m_1,m_2}|.$$

Show, using techniques similar to those employed for deriving the stability condition for bilinear systems, that $y(n)$ is bounded for all values of n if $|p_m| < 1$ for all m and $|x(n)| < (\alpha^2/4\gamma)$ for all n.

8.5 *Computing Assignment:* In this assignment, we consider the identification of a recursive nonlinear system from input–output measurements. The system to be identified satisfies the input–output relationship [263]

$$
\begin{aligned}
y(n) = {} & 0.8833x(n) + 0.0393x(n-1) + 0.8546x(n-2) \\
& + 0.8258x^2(n) + 0.7582x(n)x(n-1) + 0.175x(n)x(n-2) \\
& + 0.0864x^2(n-1) + 0.4916x(n-1)x(n-2) + 0.07111x^2(n-1) \\
& - 0.0375y(n-1) - 0.0598y(n-2) - 0.0370y(n-3) \\
& - 0.0468y(n-4) - 0.0476y^2(n-1) - 0.0781y(n-1)y(n-2) \\
& - 0.0189y(n-1)y(n-3) - 0.0626y(n-1)y(n-4) \\
& - 0.0221y^2(n-2) - 0.0617y(n-2)y(n-3) \\
& - 0.0378y(n-2)y(n-4) - 0.0041y^2(n-3) \\
& - 0.0543y(n-3)y(n-4) - 0.0603y^2(n-4).
\end{aligned}
$$

For your assignment, perform the following tasks:

(a) Generate 10,000 samples of an IID input signal that is uniformly distributed in the range $(-0.1, 0.1)$. Process this signal with the system described above and add IID Gaussian noise with zero mean value and variance such that the output SNR is 40 dB. Let the input signal be $x(n)$ and the measured output signal be $d(n)$. Our objective is to estimate the coefficients of the unknown system using $x(n)$ and $d(n)$. Since the measurement noise is relatively small in this case, we can use equation error estimators here.

(b) Estimate the coefficients of the unknown system using a least-squares optimization criterion over the 10,000 samples. You can assume perfect knowledge of the system model, that is, you may use the correct model orders in the estimation problem. Compare the average value of the squared estimation error for your solution with the mean-square value of the measurement noise. Compare the result of the estimator employing the recursive polynomial model with the average value of the squared estimation error when the system model employed by the estimator was a second-order Volterra system with ten sample memory.

(c) Derive a recursive least-square adaptive filter for the recursive system model. For an exponential window with parameter $\lambda = 0.99$, implement the filter with $x(n)$ and $d(n)$ as the input signal and the desired response signal, respectively. Compare the time-averaged values of each coefficient of the adaptive filter, obtained by computing the averages after the system has converged, with the coefficients of the unknown system. Also, compare the time-averaged value of the squared estimation error, after the system has converged, with the mean-square value of the measurement noise.

(d) Repeat part **c** using a least-mean-square adaptive filter. You will need to experiment with the step size of the algorithm to achieve satisfactory convergence results. You may also consider employing multiple step sizes for the adaptive filter.

8.6 Derive the algorithm shown in Table 8.1 for extended least-squares adaptive bilinear filters.

8.7 In this exercise, we analyze the bias in the estimate of the coefficients in the equation error methods using a simple linear system model. We consider only the minimum mean-square error estimates in this exercise. Consider the system identification problem shown in Figure 8.11. The input–output

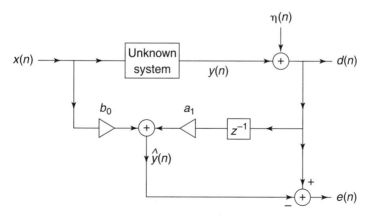

Figure 8.11 System identification problem of Exercise 8.7.

relationship of the unknown system is given by

$$y(n) = 0.6y(n-1) + 0.8x(n).$$

The desired response signal $d(n)$ is generated by corrupting $y(n)$ with $\eta(n)$, an IID signal with zero mean value and variance σ_η^2 such that $d(n) = y(n) + \eta(n)$. Assume that the input signal is an IID signal with zero mean value and unit variance, and that $x(n)$ and $\eta(n)$ are mutually independent. Derive an expression for the difference between the minimum mean-square error estimate of the coefficients and their true values.

8.8 Show that the a priori estimation error and the a posteriori estimation error are related to each other in the extended least-squares bilinear filters as

$$e(n) = \frac{\epsilon(n)}{1 - \mathbf{k}^T(n)\mathbf{Z}(n)} = \frac{\epsilon(n)}{1 - \mathbf{Z}^T(n)\mathbf{C}^{-1}(n)\mathbf{Z}(n)}.$$

[*Hint:* Substitute the coefficient update equation for $\vec{\mathbf{H}}(n)$ in the expression for directly evaluating $e(n)$ and simplify.]

9

INVERSION AND TIME
SERIES ANALYSIS

There are many applications in which we need to compensate for nonlinear distortions introduced into the signals of interest. Examples of such applications include equalization of digital communication channels in which the amplifiers operate at or near the saturation region to conserve energy, and in the process inflict severe nonlinear distortions into the signals transmitted through them [17], magnetic and optical data storage systems that introduce nonlinear distortions into the recorded bit sequences [1,21,91], and compensation of harmonic distortion in audio signals introduced by loudspeaker and microphone nonlinearities [70,76] etc. Many such applications are discussed in more detail in Chapter 10. In each of these applications, it is necessary to find a transformation of the signals that, in combination with the system that introduced the distortion, will result in an identity system that produces the signal of interest without distortion at the output of the overall system. Such a transformation is said to be the *inverse* of the nonlinear system that produced the distortions.

In applications such as forecasting or prediction, it is necessary to model the input signal using a possibly nonlinear function of its past samples. The process of characterizing signals using appropriate models is known as *time series analysis.* In this chapter, we describe some methods for inverting nonlinear systems and characterizing signals using polynomial system models. In the next section, we obtain results for the exact inverse of a class of nonlinear system models that include polynomial systems. Because the exact inverse may not always be realizable, or may not be stable, we also describe the theory of pth-order inverses of nonlinear systems in this section. Such systems are approximations of exact inverses, and are formally defined in the next section. The last section of this chapter deals with modeling signals using truncated Volterra system models and bilinear system models.

9.1 INVERSION OF NONLINEAR SYSTEMS

Unlike linear, time-invariant systems, a cascade of two nonlinear systems may result in distinctly different characteristics depending on the order in which the two systems are interconnected. Consequently, the definition of the inverse of a nonlinear system must also include a statement of whether the inverse system should be connected before or after the nonlinear system to cancel out the nonlinearities. This idea leads to the definition of the *preinverse* and the *postinverse* of a nonlinear system.

9.1.1 Exact Inverses

Let $\mathcal{H}$ and $\mathcal{G}$ represent two nonlinear systems. Then $\mathcal{G}$ is said to be the preinverse of $\mathcal{H}$ if for every input signal $x(n)$

$$\mathcal{H}\{\mathcal{G}\{x(n)\}\} = x(n). \tag{9.1}$$

In a similar manner, we say that $\mathcal{G}$ is the postinverse of $\mathcal{H}$ if

$$\mathcal{G}\{\mathcal{H}\{x(n)\}\} = x(n). \tag{9.2}$$

Figures 9.1 and 9.2 describe these definitions in a graphical manner.

The following theorem presents a relatively general result for the exact inversion of a class of nonlinear systems with input–output relationship [40,41]:

$$y(n) = g[x(n)]h[x(n-1), y(n-1)] + f[x(n-1), y(n-1)], \tag{9.3}$$

where $g[\cdot]$, $h[\cdot, \cdot]$ and $f[\cdot, \cdot]$ are causal, discrete-time, and nonlinear operators, and the inverse of the operator $g[\cdot]$ exists. We also assume throughout this section that the input signal $x(n) = 0$ for $n < 0$.

Theorem 9.1   The exact pre- and postinverse of the system in (9.3) is described by the input–output relationship

$$z(n) = g^{-1}\left[\frac{u(n) - f[z(n-1), u(n-1)]}{h[z(n-1), u(n-1)]}\right], \tag{9.4}$$

Figure 9.1   The system represented by $\mathcal{G}$ is the preinverse of the system $\mathcal{H}$.

$$x(n) \longrightarrow \boxed{\mathcal{H}} \longrightarrow \boxed{\mathcal{G}} \longrightarrow x(n)$$

Figure 9.2   The system represented by $\mathcal{G}$ is the postinverse of the system $\mathcal{H}$.

where $u(n)$ and $z(n)$ are the input and output signals, respectively, of the inverse system.

PROOF   We demonstrate first that the system in (9.4) is the postinverse of (9.3). We proceed by mathematical induction. Let $x(n)$ and $y(n)$ represent the input and output signals, respectively, of the system in (9.3). Since $x(n)$ is a causal signal and since $\mathcal{G}$ and $\mathcal{H}$ are causal systems, $x(n) = z(n)$ for $n < 0$. Let us assume that

$$z(n - i) = x(n - i) \qquad \forall i > 0. \tag{9.5}$$

We must now show using (9.5) that

$$z(n) = x(n) \tag{9.6}$$

when $u(k) = y(k)$ for $k \le n$. Now

$$z(n) = g^{-1}\left[\frac{y(n) - f[z(n-1), y(n-1)]}{h[z(n-1), y(n-1)]}\right]$$

$$= g^{-1}\left[\frac{\{g[x(n)]h[x(n-1), y(n-1)] + f[x(n-1),}{h[z(n-1), y(n-1)]}\frac{y(n-1)] - f[z(n-1), y(n-1)]\}}{}\right]. \tag{9.7}$$

By substituting $z(n - i) = x(n - i)$ from (9.5) into (9.7), it follows in a straightforward manner that $z(n) = x(n)$. We can prove in a similar manner that the system in (9.4) is also the preinverse of the system in (9.3). This completes the proof.

Example 9.1   We wish to find the inverse of the bilinear system

$$y(n) = x(n) + \sum_{i=1}^{N-1} a_i x(n-i) + \sum_{i=1}^{N-1} b_i y(n-i) + \sum_{i=0}^{N-1}\sum_{j=1}^{N-1} c_{ij} x(n-i) y(n-j).$$

Let us define f, h, and g to be

$$f[x(n-1), y(n-1)] = \sum_{i=1}^{N-1} a_i x(n-i) + \sum_{i=1}^{N-1} b_i y(n-i) + \sum_{i=1}^{N-1}\sum_{j=1}^{N-1} c_{ij} x(n-i) y(n-j),$$

$$h[x(n-1), y(n-1)] = 1 + \sum_{j=1}^{N-1} c_{0j} y(n-j)$$

and

$$g[x(n)] = x(n),$$

respectively. Then, we can utilize Theorem 9.1 to find the inverse of the bilinear system to be

$$z(n) = \frac{\left\{ u(n) - \sum_{i=1}^{N-1} b_i u(n-i) - \sum_{i=1}^{N-1} a_i z(n-i) - \sum_{i=1}^{N-1}\sum_{j=1}^{N-1} c_{ij} z(n-i) u(n-j) \right\}}{1 + \sum_{j=1}^{N-1} c_{0j} y(n-j)}.$$

9.1.2 *p*th-Order Inverses

Not all nonlinear systems possess an inverse and many nonlinear systems admit an inverse only for a certain subset of input signals. Furthermore, the inverse of the nonlinear system defined in (9.4) may not exist or may not be stable. For these reasons, we now develop the theory of *p*th-order inverses of nonlinear systems [272,275]. The *p*th-order pre- and postinverses of a nonlinear system $\mathcal{H}$ are defined as depicted in Figures 9.3 and 9.4. A nonlinear system $\mathcal{G}_p$ is said to be the *p*th-order preinverse of $\mathcal{H}$ if for every input signal $x(n)$

$$\mathcal{H}\{\mathcal{G}_p\{x(n)\}\} = x(n) + T_p\{x(n)\}, \tag{9.8}$$

where T_p represents a nonlinear system whose Volterra kernels of order zero through p are zero. Thus, the *p*th-order inverse represents an approximation to the exact inverse. Assuming that the sequence of *p*th-order inverses $\mathcal{G}_p$ converges, it will converge to the exact inverse as p goes to ∞. In a similar manner, we say that $\mathcal{G}_p$ is the *p*th-order postinverse of $\mathcal{H}$ if

$$\mathcal{G}_p\{\mathcal{H}\{x(n)\}\} = x(n) + S_p\{x(n)\}, \tag{9.9}$$

where S_p represents a nonlinear system whose Volterra kernels of order zero through p are zero.

9.1.2.1 *A Recursive Synthesis Procedure for Inverting Volterra Systems*   Schetzen was the first to define and derive a method for finding the *p*th-order inverse of a nonlinear system [275]. The original definition constrained the

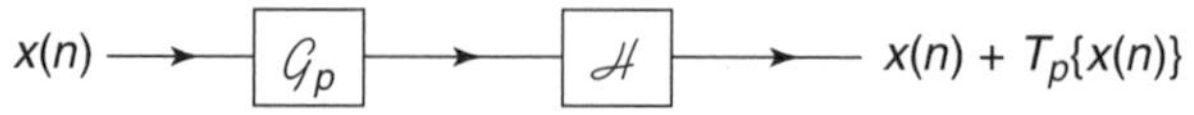

$$x(n) \longrightarrow \boxed{\mathcal{G}_p} \longrightarrow \boxed{\mathcal{H}} \longrightarrow x(n) + T_p\{x(n)\}$$

Figure 9.3 The system represented by $\mathcal{G}_p$ is the *p*th-order preinverse of the system $\mathcal{H}$.

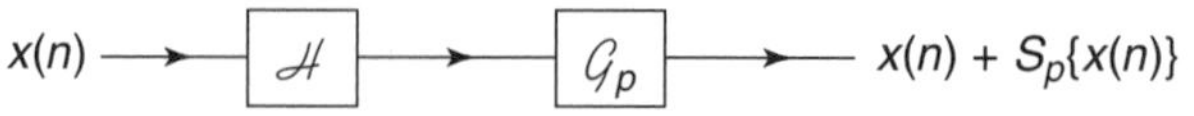

$$x(n) \longrightarrow \boxed{\mathcal{H}} \longrightarrow \boxed{\mathcal{G}_p} \longrightarrow x(n) + S_p\{x(n)\}$$

Figure 9.4 The system represented by $\mathcal{G}_p$ is the *p*th-order postinverse of the system $\mathcal{H}$.

pth-order inverse to be pth-order Volterra systems, i.e., the pth-order inverses do not have Volterra kernels of order higher than p. The definition above does not impose such a constraint, and therefore is somewhat more general. This relaxed definition is due to Sarti and Pupolin [272]. This definition also leads to the derivation of simpler and computationally more efficient expressions for the inverse system than the original definition. However, because of the presence of higher-order components, their definition of the pth-order inverse [272] does not result in a unique inverse system. The synthesis procedure described below follows the derivation in [272]. The pth-order inverses are formally treated using the contraction mapping theorem in [203].

9.1.2.2 Preinversion
Consider the cascade connection of the nonlinear systems $\mathcal{G}_p$ and $\mathcal{H}$ shown in Figure 9.3. Let the Volterra series expansions for the two systems be given by

$$\mathcal{G}_p\{x(n)\} = \sum_{k=1}^{p} \bar{g}_k[x(n)] + \sum_{k=p+1}^{\infty} \bar{r}_{p,k}[x(n)], \qquad (9.10)$$

and

$$\mathcal{H}\{y(n)\} = \sum_{k=1}^{\infty} \bar{h}_k[y(n)] \qquad (9.11)$$

respectively, where we have used the special notation $\bar{r}_{p,k}[x(n)]$ to denote the components of $\mathcal{G}_p[x(n)]$ of order k larger than p. Let $\mathcal{Q}$ represent the cascade connection of the two systems. Our objective is to choose Volterra kernels of the system $\mathcal{G}$ such that Volterra kernels $\bar{q}_k$ of $\mathcal{Q}$ are

$$\bar{q}_1[x(n)] = x(n) \qquad (9.12)$$

and

$$\bar{q}_k[x(n)] = 0; \ k = 2, 3, \ldots, p. \qquad (9.13)$$

The definition of the pth-order inverses does not constrain the kernels of order higher than p of the cascade connection. Therefore, we are free to choose the residual kernels of $\mathcal{G}$ in a manner that results in an efficient implementation of the pth-order preinverse of the system $\mathcal{H}$.

Remark 9.1 The definitions of the systems $\mathcal{G}_p$ and $\mathcal{H}$ do not include the zeroth-order kernels. However, this does not cause any loss of generality in our discussions since the effect of a nonzero bias term can be easily removed from the output of the systems.

Selection of the Linear Component. It is easy to show that the linear component of $\mathcal{Q}$ is given by

$$\bar{q}_1[x(n)] = \bar{h}_1[\bar{g}_1[x(n)]]. \qquad (9.14)$$

For the first-order Volterra kernel of the cascade connection to be an identity system, we must choose the kernel $\bar{g}_1$ such that it is the inverse of the linear system defined by $\bar{h}_1$, i.e.,

$$\bar{g}_1[x(n)] = \bar{h}_1^{-1}[x(n)], \tag{9.15}$$

where we have used the notation $\bar{h}_1^{-1}$ for the inverse system. For the overall system to be stable, it is important that the first-order component of $\mathcal{G}_p$ be stable. Therefore, we will assume that the linear component of $\mathcal{H}$ is such that it is stable and has minimum phase property.

Synthesis of the pth-Order Inverse. The kth-order term in the Volterra series expansion for the cascade connection of $\mathcal{G}_p$ followed by $\mathcal{H}$ may be expressed as

$$\bar{q}_k[x(n)] = \left(\sum_{j=1}^{k} \bar{h}_j \left[\sum_{i=1}^{k-j+1} \bar{g}_i[x(n)] \right] \right)_k$$

$$= \bar{h}_1[\bar{g}_k[x(n)]] + \left(\sum_{j=2}^{k} \bar{h}_j \left[\sum_{i=1}^{k-j+1} \bar{g}_i[x(n)] \right] \right)_k. \tag{9.16}$$

Here, $(\cdot)_k$ indicates the homogeneous kth-order term of the Volterra series expansion of $(\cdot)$. The kernels of $\mathcal{G}_p$ and $\mathcal{H}$ not included in the above equation do not contribute to the kth-order component at the output of the cascade connection. Since $\bar{q}_k[x(n)] = 0$ for $2 \le k \le p$, we can set the right-hand side of (9.16) to zero and solve for the homogeneous kth-order component of $\mathcal{G}_p[x(n)]$ for $k \ge 2$ to get

$$\bar{g}_k[x(n)] = -\bar{h}_1^{-1} \left[\left(\sum_{j=2}^{k} \bar{h}_j \left[\sum_{i=1}^{k-j+1} \bar{g}_i[x(n)] \right] \right)_k \right]. \tag{9.17}$$

Since we have seen in (9.15) that $\bar{h}_1^{-1}[x(n)] = \bar{g}_1[x(n)]$, this expression can be rewritten as

$$\bar{g}_k[x(n)] = -\bar{g}_1 \left[\left(\sum_{j=2}^{k} \bar{h}_j \left[\sum_{i=1}^{k-j+1} \bar{g}_i[x(n)] \right] \right)_k \right]; \qquad 2 \le k \le p. \tag{9.18}$$

We can use this expression to write a general expression for $\mathcal{G}_p[x(n)]$ by adding together the components due to the kth-order Volterra kernel over all values of k. This procedure gives

$$\mathcal{G}_p[x(n)] = \bar{g}_1[x(n)] + \sum_{k=2}^{p} -\bar{g}_1 \left[\left(\sum_{j=2}^{k} \bar{h}_j \left[\sum_{i=1}^{k-j+1} \bar{g}_i[x(n)] \right] \right)_k \right] + \sum_{k=p+1}^{\infty} \bar{r}_{p,k}[x(n)]. \tag{9.19}$$

Recall that the notation $(\cdot)_k$ is employed to indicate the homogeneous kth-order component of $(\cdot)$. Therefore, adding components that are of higher order than k to terms within the parentheses of this function on the right-hand side of (9.19) does not change the overall system. Let us add $-\left[\sum\limits_{j=2}^{k}\overline{h}_j\left(\sum\limits_{i=k-j}^{p-1}\overline{g}_i[x(n)] + \sum\limits_{k=p}^{\infty}\overline{r}_{p-1,k}[x(n)]\right)\right]$

to the expression within the parentheses of this function. This process results in an expression that is functionally identical to but structurally different from (9.19). The structural difference will allow us to obtain a recursive expression for calculating the pth-order inverse of $\mathcal{H}$. The addition of the terms presented above gives

$$\mathcal{G}_p\{x(n)\} = \overline{g}_1[x(n)] + \sum_{k=2}^{p}$$

$$-\overline{g}_1\left[\left(\sum_{j=2}^{k}\overline{h}_j\left[\sum_{i=1}^{k-j+1}\overline{g}_i[x(n)] + \sum_{i=k-j+2}^{p-1}\overline{g}_i[x(n)] + \sum_{k=p}^{\infty}\overline{r}_{p-1,k}[x(n)]\right]\right)\right]_k$$

$$+ \sum_{k=p+1}^{\infty}\overline{r}_{p,k}[x(n)], \tag{9.20}$$

Since the pth-order inverse is necessarily a $(p-1)$th-order inverse also, we can express $\mathcal{G}_p$ using the Volterra series expansion

$$\mathcal{G}_{p-1}[x(n)] = \sum_{i=1}^{k-j+1}\overline{g}_i[x(n)] + \sum_{i=k-j+2}^{p-1}\overline{g}_i[x(n)] + \sum_{k=p}^{\infty}\overline{r}_{p-1,k}[x(n)]. \tag{9.21}$$

Substituting this result in (9.20) gives a recursive expression for $\mathcal{G}_p$ as

$$\mathcal{G}_p\{x(n)\} = \overline{g}_1[x(n)] + \sum_{k=2}^{p}\sum_{j=2}^{k}-\overline{g}_1[(\overline{h}_j[\mathcal{G}_{p-1}\{x(n)\}])_k] + \sum_{k=p+1}^{\infty}\overline{r}_{p,k}[x(n)]. \tag{9.22}$$

Since adding terms involving nonlinearities of order higher than k does not affect the functionality of the operator $(\cdot)_k$ in this expression, we can replace the above recursive system by

$$\mathcal{G}_p\{x(n)\} = \overline{g}_1[x(n)] + \sum_{k=2}^{p}\sum_{j=2}^{k}-\overline{g}_1[(\overline{h}_j[\mathcal{G}_{p-1}\{x(n)\}])_k] + \sum_{k=p+1}^{\infty}\overline{r}_{p,k}[x(n)]$$

$$= \overline{g}_1[x(n)] + \sum_{k=2}^{p}-\overline{g}_1[(\mathcal{H}'_p\{\mathcal{G}_{p-1}\{x(n)\}\})_k] + \sum_{k=p+1}^{\infty}\overline{r}_{p,k}[x(n)], \tag{9.23}$$

where $\mathcal{H}'_p$ denotes the pth-order truncation of the nonlinear system $\mathcal{H}$ from which its linear component $\overline{h}_1$ has been removed. In deriving (9.23), we changed the order in which the summation over the index j and the linear transformation $\overline{g}_1$ are implemented. To complete the synthesis of the pth-order preinverse, we need to

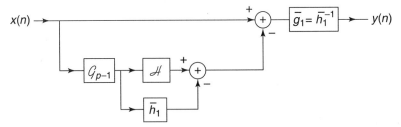

Figure 9.5 Recursive synthesis of the pth-order inverse of a Volterra system.

specify how to select the higher-order residual terms $\bar{r}_{p,k}$ for $k > p$. These components do not affect the first p kernels of the cascade connection of the $\mathcal{G}_p$ and $\mathcal{H}$, and therefore can be selected in a way that will provide other benefits such as ease of implementation. In our synthesis procedure, we will select the residual components as

$$\bar{r}_{p,k}[x(n)] = -\bar{g}_1[(\mathcal{H}'_p\{\mathcal{G}_{p-1}\{x(n)\}\})_k]; \qquad k > p. \tag{9.24}$$

Substituting this choice of $\bar{r}_{p,k}$ in (9.23) gives a relatively easy-to-implement recursive structure for the pth-order inverse as

$$\mathcal{G}_p[x(n)] = \bar{g}_1[x(n)] + \sum_{k=2}^{\infty} -\bar{g}_1[(\mathcal{H}'_p\{\mathcal{G}_{p-1}\{x(n)\}\})_k]$$

$$= \bar{g}_1[x(n)] - \bar{g}_1[\mathcal{H}'_p\{\mathcal{G}_{p-1}\{x(n)\}\}]. \tag{9.25}$$

Figure 9.5 shows a block diagram for synthesizing $\mathcal{G}_p$ recursively. Since $\mathcal{G}_1$ is the inverse of the linear component of $\mathcal{H}$, we can easily initialize this recursion, and therefore implement the pth-order inverse.

9.1.2.3 *Synthesis of the postinverse* We now show that the pth-order preinverse is also the pth-order postinverse. This result is stated and proved in the form of a theorem below.

Theorem 9.2 Let $\mathcal{G}_p$ represent a Volterra system such that

$$(\mathcal{H}\{\mathcal{G}_p\{x(n)\}\})_k = \begin{cases} x(n); & k = 1 \\ 0; & 2 \leq k \leq p. \end{cases} \tag{9.26}$$

Then

$$(\mathcal{G}_p\{\mathcal{H}[x(n)]\})_k = \begin{cases} x(n); & k = 1 \\ 0; & 2 \leq k \leq p. \end{cases} \tag{9.27}$$

PROOF We can prove this result using mathematical induction. Let us express $\mathcal{H}\{\mathcal{G}_p\{x(n)\}\}$ using a Volterra series expansion as

$$\mathcal{H}\{\mathcal{G}_p\{x(n)\}\} = x(n) + \sum_{j=p+1}^{\infty} \bar{s}_{p,j}[x(n)], \tag{9.28}$$

where $\bar{s}_{p,j}[x(n)]$ denotes the homogenous jth-order component of the output of the cascade system for orders $j > p$. Then

$$(\mathcal{G}_p\{\mathcal{H}\{\mathcal{G}_p\{x(n)\}\}\})_k = \left(\mathcal{G}_p\left\{x(n) + \sum_{j=p+1}^{\infty} \bar{s}_{p,j}[x(n)]\right\}\right)_k$$

$$= (\mathcal{G}_p\{x(n)\})_k = \bar{g}_k[x(n)]; \ \ k \le p. \tag{9.29}$$

The second equality is true because the input components of the form $\bar{s}_{p,j}[x(n)]$ to the system $\mathcal{G}_p$ contribute nonlinear terms of order larger than p when j is greater than p.

For $k = 1$, we have

$$(\mathcal{G}_p\{\mathcal{H}\{\mathcal{G}_p\{x(n)\}\}\})_1 = \bar{g}_1[\bar{h}_1[\bar{g}_1[x(n)]]]. \tag{9.30}$$

Since $\bar{g}_1$ and $\bar{h}_1$ are linear operators, it is straightforward to see that $\bar{g}_1$ is the first-order preinverse as well as the postinverse of $\mathcal{H}$. Let us denote the cascade of the system $\mathcal{H}$ followed by the system $\mathcal{G}_p$ with $\mathcal{Q}$, and let

$$\mathcal{Q}\{x(n)\} = \sum_{k=1}^{\infty} \bar{q}_k[x(n)]. \tag{9.31}$$

Recall that we have already shown that $\bar{q}_1[x(n)] = x(n)$. For $k = 2$, we have

$$(\mathcal{G}_p\{\mathcal{H}\{\mathcal{G}_p\{x(n)\}\}\})_2 = (\mathcal{Q}\{\mathcal{G}_p\{x(n)\}\})_2$$

$$= (\mathcal{G}_p\{x(n)\} + \bar{q}_2[\mathcal{G}_p\{x(n)\}])_2$$

$$= \bar{g}_2[x(n)] + \bar{q}_2[\bar{g}_1[x(n)]]$$

$$= \bar{g}_2[x(n)]. \tag{9.32}$$

The last equality came from (9.29). Since $\bar{g}_1[x(n)]$ is in general not zero and (9.32) is true for arbitrary choices of $x(n)$, the preceding result implies that $\bar{q}_2[x(n)] = 0$, indicating that $\mathcal{G}_p$ is a second-order postinverse of $\mathcal{H}$, as claimed in the theorem.

Now, let us assume that

$$\bar{q}_i[x(n)] = 0; \quad i = 2, 3, \ldots, k - 1. \tag{9.33}$$

Then

$$
\begin{aligned}
(\mathcal{G}_p\{\mathcal{H}\{\mathcal{G}_p\{x(n)\}\}\})_k &= (\mathcal{Q}\{\mathcal{G}_p\{x(n)\}\})_k \\
&= (\mathcal{G}_p\{x(n)\} + \bar{q}_k[\mathcal{G}_p\{x(n)\}])_k \\
&= \bar{g}_k[x(n)] + \bar{q}_k[\bar{g}_1[x(n)]] \\
&= \bar{g}_k[x(n)].
\end{aligned}
\tag{9.34}
$$

Again, the last equality follows from (9.29). It immediately follows that $\bar{q}_k[x(n)] = 0$, proving the statement of the theorem by induction that the pth-order pre- and postinverses have identical structures.

Example 9.2 In this example, we explicitly show the structure of the first-, second-, and third-order inverses of an arbitrary Volterra system with input–output relationship

$$
y(n) = \sum_{k=1}^{\infty} \bar{h}_k[x(n)],
$$

synthesized using the method described above. Figure 9.6 shows block diagrams of the inverse systems. The first-order inverse is the inverse of the linear system described by $\bar{h}_1$:

$$
\bar{g}_1[x(n)] = \bar{h}_1^{-1}[x(n)].
$$

The second-order inverse makes use of the structure of the first-order inverse as shown in the figure. Similarly, the implementation of the third-order inverse of $\mathcal{H}$ uses the structure of the second-order inverse as explicitly shown in the figure.

9.1.2.4 *Inversion of recursive polynomial systems*
We can combine the synthesis procedure described above with the method for deriving Theorems 9.1 to develop procedures for constructing the pth-order inverses of nonlinear systems implemented using recursive system models. The following theorem [41] presents an efficient method of computing a pth-order inverse of the system

$$
y(n) = g[x(n)] + f[x(n-1), y(n-1)].
\tag{9.35}
$$

This system is a special case of the system in (9.3) when $h[\cdot, \cdot] = 1$.

Theorem 9.3 Let $g[\cdot]$ and $f[\cdot, \cdot]$ be causal, discrete-time nonlinear operators with convergent Volterra series expansions with respect to all the arguments. Moreover, let the pth-order inverse $g_p^{-1}[\cdot]$ of the system $g[\cdot]$ exist. Then a pth-order inverse of the causal, discrete-time nonlinear system described in (9.35) is given by the following input–output relationship:

$$
z(n) = g_p^{-1}[u(n) - f[z(n-1), u(n-1)]].
\tag{9.36}
$$

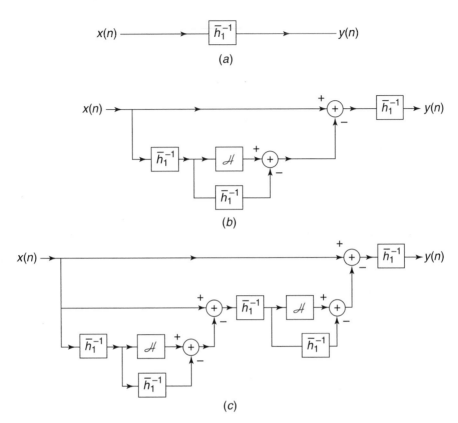

Figure 9.6 Block diagrams of pth-order inverses for $p = 1, 2, 3$: (a) first-order inverse; (b) second-order inverse; (c) third-order inverse.

PROOF   As was the case for the Theorem 9.1, we first show that the system in (9.36) is the pth-order postinverse of the system in (9.35). Using the same variables as in the derivation of Theorem 9.1, we express $z(n)$ as

$$z(n) = g_p^{-1}[y(n) - f[z(n-1), y(n-1)]]$$
$$= g_p^{-1}[g[x(n)] + f[x(n-1), y(n-1)]$$
$$-f[z(n-1), y(n-1)]]. \tag{9.37}$$

We proceed by mathematical induction. We assume that, for any k greater than zero, the output $z(n-k)$ differs from $x(n-k)$ only by $v_p(n-k)$, a term whose Volterra series expansion in $x(n)$ contains only kernels of order larger than p, i.e.,

$$z(n-k) = x(n-k) + v_p(n-k) \qquad \forall k > 0. \tag{9.38}$$

We have to prove that the Volterra series expansion of $z(n) - x(n)$ have zero kernels of order less than or equal to p. Since $f[\cdot, \cdot]$ admits a convergent Volterra series expansion, we have from (9.38) that the Volterra series expansion of the difference $f[x(n-1), y(n-1)] - f[z(n-1), y(n-1)]$ contains only kernels of order greater than p, i.e.,

$$f[x(n-1), y(n-1)] - f[z(n-1), y(n-1)] = 0 + v'_p(n), \qquad (9.39)$$

where the Volterra kernels of $v'_p(n)$ of order less than or equal to p are zero. Substituting (9.39) in (9.37), we get

$$z(n) = g_p^{-1}\big[g[x(n)] + v'_p(n)\big]. \qquad (9.40)$$

The pth-order inverse of the operator $g[\cdot]$ derived in [275] is given by a pth-order truncated Volterra series whose kernels depend only on the first p kernels of the Volterra series expansion of $g[\cdot]$. The pth-order inverse derived in Sarti and Pupolin [272] may have Volterra kernels of order greater than p. However, the inverse still has a Volterra series expansion with finite order of nonlinearity, and it depends only on the first p kernels of the Volterra series expansion of $g[\cdot]$. Consequently, it immediately follows from (9.40) that

$$z(n) = x(n) + v_p(n) \qquad (9.41)$$

and that the system in (9.37) is the pth-order postinverse of the system in (9.35). We can prove in a similar manner that it is also a preinverse of the system in (9.37).

Remark 9.2   Because of the rational structure of the system in (9.4), a similar expression for the pth-order inverse of the system in (9.3) does not exist.

Example 9.3   We wish to derive a pth-order inverse for the second-order Volterra filter given by the following expression:

$$y(n) = \sum_{i=0}^{N-1} a_i x(n-i) + \sum_{i=0}^{N-1}\sum_{j=i}^{N-1} b_{ij} x(n-i) x(n-j).$$

Let

$$g[x(n)] = a_0 x(n) + x(n) \sum_{j=0}^{N-1} b_{0j} x(n-j)$$

and

$$f[x(n-1)] = \sum_{i=1}^{N-1} a_i x(n-i) + \sum_{i=1}^{N-1}\sum_{j=i}^{N-1} b_{ij} x(n-i) x(n-j).$$

According to Theorem 9.3, a pth-order inverse for the second-order Volterra filter is

$$z(n) = g_p^{-1}\left[u(n) - \sum_{i=1}^{N-1} a_i z(n-i) - \sum_{i=1}^{N-1} \sum_{j=i}^{N-1} b_{ij} z(n-i)z(n-j) \right].$$

The pth-order inverse $g_p^{-1}[\cdot]$ can be computed iteratively as in the Sarti–Pupolin article [272] and is given by

$$g_p^{-1}[u(n)] = -g_1^{-1}[q_p[g_{p-1}^{-1}[u(n)]]] - u(n)],$$

where $g_1^{-1}[\cdot]$ is the inverse of the first Volterra operator of $g[\cdot]$ (i.e., a_0^{-1} in our case) and $q_p[\cdot]$ is the truncated Volterra series expansion of the system $g[\cdot]$ that contains only the second through pth-order Volterra kernels. While it is possible to compute the pth-order inverse of the whole system as in [272], using this technique for inverting a smaller subsystem and then applying Theorem 9.3 is a more efficient procedure in most situations.

9.2 POLYNOMIAL TIME SERIES ANALYSIS

A large class of parametric spectrum estimation methods are based on linear system models [162]. Spectrum estimation algorithms based on autoregressive (AR), moving-average (MA), and autoregressive moving-average (ARMA) models are quite popular and are relatively simple to implement. However, as we have seen throughout this book, many real-world signals are generated by physical systems containing inherent nonlinearities, and therefore, nonlinear signal models are appropriate in many applications. This section briefly describes some approaches for signal modeling or time-series analysis using polynomial system models.

The basic problem of nonlinear time series modeling is depicted in Figure 9.7. The input signal $\xi(n)$ that is processed by the system $\mathcal{H}$ is not available for measurement. Rather, the problem involves estimating the parameters of the system model $\mathcal{H}$ from measurements of a finite number of samples of the output $x(n)$ of the system, as well as knowledge of some statistical information regarding $\xi(n)$. It is usually assumed that $\xi(n)$ is an IID sequence. Many methods that use closed-form expressions to solve for the unknown system coefficients also assume that $\xi(n)$ is Gaussian-distributed.

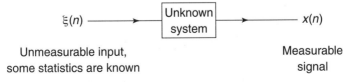

Figure 9.7 A typical time-series analysis problem.

One difficulty with nonlinear signal models is that there is often no direct method of calculating the parameters. Consequently, it is necessary to devise iterative schemes to estimate the parameters of such models. For ease of presentation, we describe the methods using a particular type of bilinear system model. Closed-form expressions for direct estimation of the parameters are available for this model. The iterative schemes described in the section for this model are applicable to more general polynomial system models.

9.2.1 A Specific Bilinear Model

In this section, we consider modeling a signal as being generated by a bilinear system with input–output relationship

$$x(n) = \sum_{i=1}^{p} a_i x(n-i) + \sum_{i=1}^{p} b_i x(n-i)\xi(n-1) + \xi(n). \tag{9.42}$$

We have employed the same parameter p in both the summations for the sake of simplicity. We assume that $\xi(n)$ is an unmeasurable input to the system to be identified, that it is an IID process with zero mean value and unknown but fixed variance σ_ξ^2. Our task is to identify the parameters of the system $a_1, a_2, \ldots, a_p, b_1, b_2, \ldots, b_p$, and the variance σ_ξ^2 of $\xi(n)$. We will first describe an iterative procedure that is applicable to most nonlinear models. Then we will also discuss a direct method for estimating these parameters through solving closed-form expressions that relate the higher-order statistics of the input signal to the parameters of the model. Throughout our discussions we will assume that the model order as well as the specific model we wish to use in the problem accurately matches the situation on hand.

9.2.1.1 An Iterative Method Closed-form expressions that relate certain measurable statistics of the signal to the parameters that must be estimated are not typically available for most nonlinear time series models. Consequently, the most common approach to estimating the parameters of the model is to resort to some form of numerical search algorithm that operates in an iterative manner. Iterative algorithms for estimating the parameters of bilinear time series models are described in Subba Rao [302]. In this section, we describe the principle of iterative estimation for bilinear time series analysis using one of the many such techniques available.

Let

$$\mathbf{A}_k = [a_1 \quad a_2 \quad \cdots \quad a_p]^T \tag{9.43}$$

and

$$\mathbf{B}_k = [b_1 \quad b_2 \quad \cdots \quad b_p]^T \tag{9.44}$$

denote the coefficient vectors containing the parameters of the bilinear time series model in (9.42), obtained after the ith iteration of the estimation procedure. Let K denote the number of samples available for the estimation procedure, and let the coefficients be initialized using the vectors $\mathbf{A}_0$ and $\mathbf{B}_0$. We may also initialize the estimation error $\hat{\xi}_k(0)$ to be zero during the kth iteration. We can now derive an iterative algorithm that processes the complete set of K samples during each iteration in the following manner.

Let

$$\mathbf{X}(n) = [x(n) \quad x(n-1) \quad \cdots \quad x(n-p+1)]^T \tag{9.45}$$

denote a vector of input samples at time n. During the kth iteration, we compute the estimate of $x(n)$ as

$$\hat{x}(n) = \mathbf{A}_{k-1}^T \mathbf{X}(n-1) + \mathbf{B}_{k-1}^T \mathbf{X}(n-1)\hat{\xi}_k(n-1); \qquad 1 \leq n \leq K, \tag{9.46}$$

and where the unmeasurable input process is estimated as

$$\hat{\xi}_k(n) = x(n) - \hat{x}(n). \tag{9.47}$$

We will use a Gauss–Newton recursion during each iteration to update the coefficients in an effort to reduce the sum of the squared estimation errors given by

$$J_k = \frac{1}{K} \sum_{n=1}^K \hat{\xi}_k^2(n). \tag{9.48}$$

In the Gauss–Newton recursion, the coefficients are updated as

$$\begin{bmatrix} \mathbf{A}_k \\ \mathbf{B}_k \end{bmatrix} = \begin{bmatrix} \mathbf{A}_{k-1} \\ \mathbf{B}_{k-1} \end{bmatrix} - \frac{\mu}{2} \mathbf{R}_k^{-1} \frac{\partial J_k}{\partial \begin{bmatrix} \mathbf{A}_{k-1} \\ \mathbf{B}_{k-1} \end{bmatrix}}, \tag{9.49}$$

where $\mathbf{R}_k$ is the least-squares autocorrelation matrix of the regression vector $[\mathbf{X}^T(n-1) \quad \mathbf{X}^T(n-1)\hat{\xi}_k(n-1)]^T$ and μ is a small positive constant that controls the speed of convergence of the algorithm. The least-squares autocorrelation matrix of the regression vector may be estimated as

$$\mathbf{R}_k = \frac{1}{K} \sum_{n=1}^K \begin{bmatrix} \mathbf{X}(n-1) \\ \mathbf{X}(n-1)\hat{\xi}_k(n-1) \end{bmatrix} [\mathbf{X}^T(n-1) \quad \mathbf{X}^T(n-1)\hat{\xi}_k(n-1)]. \tag{9.50}$$

It is relatively easy to show that the gradient of the cost function is given by

$$\frac{\partial J_k}{\partial \begin{bmatrix} \mathbf{A}_{k-1} \\ \mathbf{B}_{k-1} \end{bmatrix}} = \frac{1}{K} \sum_{n=1}^{K} -2\hat{\xi}_k(n) \begin{bmatrix} \mathbf{X}(n-1) \\ \mathbf{X}(n-1)\hat{\xi}_k(n-1) \end{bmatrix}. \tag{9.51}$$

Substituting (9.50) and (9.51) in (9.49), we get the coefficient update equation employed in the Gauss–Newton recursion as

$$\begin{bmatrix} \mathbf{A}_k \\ \mathbf{B}_k \end{bmatrix} = \begin{bmatrix} \mathbf{A}_{k-1} \\ \mathbf{B}_{k-1} \end{bmatrix} - \mu \mathbf{R}_k^{-1} \frac{1}{k} \sum_{n=1}^{k} \hat{\xi}_k(n) \begin{bmatrix} \mathbf{X}(n-1) \\ \mathbf{X}(n-1)\hat{\xi}_k(n-1) \end{bmatrix}. \tag{9.52}$$

We note here that the cost function J_k is the estimate of the mean-square value σ_ξ^2 of the unmeasurable input sequence. The iterations should be continued until J_k does not change significantly from one iteration to the next. A commonly employed stopping criterion is to stop the iterations if the ratio of the most recent change in the cost function to the current value of the cost function is smaller than some preselected, positive threshold ϵ, i.e., we will select $\mathbf{A}_k$ and $\mathbf{B}_k$ as the final estimates if

$$\frac{|J_{k-1} - J_k|}{J_k} < \epsilon. \tag{9.53}$$

Remark 9.3 The method described above is one of the many iterative methods possible for estimating the parameters of a nonlinear time series model. A common approach is to employ some form of the output error adaptive filtering algorithms described in Chapter 7 to predict the signal of interest using past samples of the input signal and the past samples of the prediction error signal. If the algorithm converges to the global minimum of the performance surface, the parameters of the adaptive filter will correctly estimate the model parameters.

9.2.1.2 A Direct Method
Using the basic assumption in the time series analysis problem that the unmeasurable input signal belongs to an IID process, we can show that the mean value, the autocorrelation function and the third-order correlation function satisfy the statement of the following theorem [167].

Theorem 9.4 Let $x(n)$ be a stationary random process satisfying the evolution equation as given by (9.42). Let the input process $\xi(n)$ be an IID process with zero mean value and variance σ_ξ^2. Let $c(n) = x(n) - E\{x(n)\}$. Let $r_{cc}(m) = E\{c(n)c(n-m)\}$ and $\Gamma_{ccc}(l, m) = E\{c(n)c(n-l)c(n-m)\}$ denote the autocorrelation function and third-order correlation function, respectively, of $c(n)$.

Then

$$\mu = E\{x(n)\} = \left(1 - \sum_{i=1}^{p} a_i\right)^{-1} b_1 \sigma_\xi^2. \tag{9.54}$$

$$r_{cc}(m) = \begin{cases} \sum_{i=1}^{p} a_i r_{cc}(m-i); & m > 1 \\[2ex] \sum_{i=1}^{p} a_i r_{cc}(m-i) + \sigma_\xi^2 \sum_{i=1}^{p} \sum_{j=1}^{p} b_i b_j r_{cc}(i-j) \\[1ex] \quad + \mu\left(\sum_{i=1}^{p} b_i\right)\sigma_\xi^2\left(a_1 + \mu \sum_{i=1}^{p} b_i\right) + b_1^2 \sigma_\xi^4 + \sigma_\xi^2; & m = 0 \\[2ex] \sum_{i=1}^{p} a_i r_{cc}(m-i) + \mu\left(\sum_{i=1}^{p} b_i r_{cc}(m-i)\right)\sigma_\xi^2; & m = 1, \end{cases} \tag{9.55}$$

and

$$\Gamma_{ccc}(1, m) = \sum_{i=1}^{p} a_i \Gamma_{ccc}(i-1, m-1) + \mu\left(\sum_{i=1}^{p} b_i r_{cc}(m-i)\right)\sigma_\xi^2; \qquad m \geq 2. \tag{9.56}$$

Before proving the results of the theorem, we will make several observations.

Comparison with Linear Models. The correlation function of $c(n)$ for lags greater than one can be shown to be identical to the correlation function of an autoregressive moving-average (ARMA) process with p autoregressive parameters and one moving-average parameter. This means that the bilinear time series cannot be distinguished from an ARMA time series by knowledge of the second-order statistics alone. We need knowledge of the higher-order statistics of the signal to identify the parameters of the bilinear model.

Estimation of the Parameters. We can see from the expression in (9.55) for the autocorrelation function for lags larger than one that the functional relationship is linear in the parameters $a_1, a_2, \ldots, a_p$. Similarly, the functional relationship of the third-order correlation function to the third-order correlation functions at other lag values and the autocorrelation function is linear in the parameters $\sigma_\xi^2 b_1, \sigma_\xi^2 b_2, \ldots, \sigma_\xi^2 b_p$. Consequently, we can estimate the parameters a_i and $\sigma_\xi^2 b_i$ using p relationships for $r_{cc}(m)$ for $2 \leq m \leq p+1$ and another p relationships for $\Gamma_{ccc}(1, m)$ for $2 \leq m \leq p+1$. The expression for $r_{cc}(0)$ in (9.55), after the estimates of a_i and $\sigma_\xi^2 b_i$ values are substituted in it, becomes a quadratic function in σ_ξ^2. Consequently, we can solve for σ_ξ^2 by finding the roots of the quadratic equation and retaining the positive root. Additional improvements in the performance of the estimator can be obtained by solving an overdetermined set of equations that results from the use of a larger number of relationships than the minimum number necessary [167].

Sketch of the Proof of Theorem 9.4. It is straightforward to show using the IID characteristics of $\xi(n)$ and the causality of the bilinear system under consideration that

$$E\{x(n)\xi(n)\} = \sigma_\xi^2. \tag{9.57}$$

It is left as an exercise for the reader to show, using direct calculations and this result, that

$$\mu = E\{x(n)\} = \left(1 - \sum_{i=1}^p a_i\right)^{-1} b_1 \sigma_\xi^2. \tag{9.58}$$

Subtracting μ from $x(n)$, and simplifying gives

$$c(n) = \sum_{i=1}^p a_i c(n-i) + \sum_{i=1}^p b_i c(n-i)\xi(n-1) + \xi(n)$$

$$+ \mu\left(\sum_{i=1}^p b_i\right)\xi(n-1) - b_1\sigma_\xi^2, \tag{9.59}$$

where we have used (9.58) and substituted for μ as $\mu = \mu\left(\sum_{i=1}^p a_i\right) + b_1\sigma_\xi^2$. To compute the autocorrelation function of $c(n)$, we multiply both sides of (9.59) with $c(n-m)$, and take the expectation of the resulting expression to get

$$r_{cc}(m) = \sum_{i=1}^p a_i E\{c(n-i)c(n-m)\} + \sum_{i=1}^p b_i E\{c(n-i)c(n-m)\xi(n-1)\}|$$

$$+ E\{c(n-m)\xi(n)\} + \mu\left(\sum_{i=1}^p b_i\right)E\{\xi(n-1)c(n-m)\}$$

$$- b_1\sigma_\xi^2 E\{c(n-m)\}. \tag{9.60}$$

We can show in a straightforward manner that if $m > 1$, all terms except the first one are zero, i.e.,

$$r_{cc}(m) = \sum_{i=1}^p a_i r_{cc}(m-i); \qquad m > 1. \tag{9.61}$$

The last term of (9.60) is always zero because $c(n-m)$ has zero mean value. By replacing $c(n)$ with the evolution equation in (9.59) and taking the statistical expectations, we can show that

$$E\{c(n)\xi(n)\} = \sigma_\xi^2. \tag{9.62}$$

It is also not difficult to show that

$$E\{\xi(n-1)c(n)\} = \sigma_\xi^2 \left(a_1 + \mu \left(\sum_{i=1}^p b_i \right) \right) \tag{9.63}$$

and that

$$E\{c(n-i)c(n-m)\xi(n-1)\} = 0; \qquad i, m \geq 1. \tag{9.64}$$

Using these results and some additional lengthy but relatively straightforward calculations, we can get the expressions for the autocorrelation function at lags zero and one. These derivations may be found in Mathews [167]. More general results for a larger class of bilinear models may be found in Sesay and Subba Rao [280].

Expressions for $\Gamma_{ccc}(1, m)$ for values of $m \geq 2$ can be found in a similar manner. Multiplying both sides of (9.59) with $c(n-1)c(n-m)$, we get

$$\begin{aligned}
E\{c(n)c(n-1)c(n-m)\} &= \sum_{i=1}^p a_i E\{c(n-i)c(n-1)c(n-m)\} \\
&\quad + \sum_{i=1}^p b_i E\{c(n-i)c(n-1)c(n-m)\xi(n-1)\} \\
&\quad + E\{c(n-1)c(n-m)\xi(n)\} \\
&\quad + \mu \left(\sum_{i=1}^p b_i \right) E\{c(n-1)c(n-m)\xi(n-1)\} \\
&\quad - b_1 \sigma_\xi^2 E\{c(n-1)c(n-m)\}. \tag{9.65}
\end{aligned}$$

The first and the last terms of the right-hand side (RHS) of these expressions can be left without additional simplification as the corresponding third-order correlation function and the autocorrelation function, respectively. By virtue of (9.64), the third and fourth terms of (9.65) are zero. To evaluate the second term, we can expand $c(n-1)$ as before and show that

$$E\{c(n-i)c(n-1)c(n-m)\xi(n-1)\} = \sigma_\xi^2 r_{cc}(m-i); \qquad i, m > 1. \tag{9.66}$$

For $i = 1$ and $m > 1$, we can expand $c^2(n-1)$ using (9.59), multiply each term in the expansion with $c(n-m)\xi(n-1)$, and take the expectation of the result. Since $\xi(n-1)$ is correlated with only signal samples that occur at time $n-1$ in this expansion and since $m > 1$ for our calculations, the result simplifies to

$$E\{c^2(n-1)c(n-m)\xi(n-1)\} = 2\sum_{l=1}^{p} a_l E\{c(n-l-1)c(n-m)\xi^2(n-1)\}$$

$$+ 2\sum_{l=1}^{p} b_l E\{c(n-l-1)c(n-m)\xi^2(n-1)\}$$

$$+ 2\mu\left(\sum_{l=1}^{p} b_l\right)E\{\xi(n-2)\xi^2(n-1)c(n-m)\}.$$

(9.67)

The second term can be shown to be zero for $m \geq 2$ using (9.64). The third term is zero for $m > 2$. The first expectation on the RHS of (9.67) simplifies to

$$2\sum_{l=1}^{p} a_l E\{c(n-l-1)c(n-m)\xi^2(n-1)\} = 2\sigma_\xi^2 \sum_{l=1}^{p} a_l r_{cc}(m-l-1); \qquad m \geq 2.$$

(9.68)

To evaluate the third term for the case when $m = 2$, we can expand $c(n-2)$ using (9.59), and show in the usual manner that

$$E\{\xi(n-2)\xi^2(n-1)c(n-m)\} = \sigma_\xi^4; \qquad m = 2.$$

(9.69)

Substituting the intermediate results derived in (9.65) results in (9.56), thus completing the proof of Theorem 9.4.

Example 9.4 In this example, we consider the identification of a bilinear time series generated as

$$x(n) = -0.2x(n-1) - 0.1x(n-2) - 0.5x(n-3)$$
$$+ (0.1x(n-1) + 0.2x(n-2) + 0.3x(n-3))\xi(n-1) + \xi(n),$$

where $\xi(n)$ is a white, Gaussian random process with zero mean value and variance $\sigma_\xi^2 = 0.1$. There are seven parameters to be estimated here: $a_1, a_2, a_3, b_1, b_2, b_3$, and σ_ξ^2. One thousand estimates were obtained from 2000 samples of pseudorandom signals satisfying the preceding equation. While generating these signals, the first 50,000 samples of each sequence were discarded in order to eliminate the influence of the initial conditions from the signals. The estimates were obtained using an iterative technique employing the Gauss–Newton search method and the direct approach employing the closed-form expressions relating the higher-order statistics of the input signal to the parameters of the system model. The higher-order statistics of the signals were estimated as time averages over the 2000 samples available for each estimate. The mean values of the parameters as well as the variance of the

TABLE 9.1 Statistics of the parameter estimates in Example 9.4

Parameter	True Value	Iterative Method		Direct Method	
		Mean	Variance	Mean	Variance
a_1	−0.2	−0.200	3.95×10^{-4}	−0.196	1.47×10^{-3}
a_2	−0.1	−0.101	4.02×10^{-4}	−0.101	4.46×10^{-4}
a_3	−0.5	−0.500	5.84×10^{-4}	−0.499	6.02×10^{-4}
b_1	0.1	0.099	2.14×10^{-3}	0.103	2.80×10^{-2}
b_2	0.2	0.203	3.73×10^{-3}	0.200	2.68×10^{-2}
b_3	0.3	0.302	3.70×10^{-3}	0.301	6.60×10^{-3}

Results courtesy of Shayne Wissler.

estimates are tabulated in Table 9.1. It can be seen that the mean values of the coefficients obtained from both sets of estimates are close to the actual parameter values. In this set of experiments, the variance of the parameter estimates were smaller for the iterative method than the direct approach. This improved performance of the iterative method comes at a significant increase in the computational complexity. Attempts to initialize the iterations using the estimates obtained from the closed-form expressions did not result in a significant reduction in the number of iterations required for the iterative method to converge to its solution [333].

9.3 EXERCISES

9.1 Derive expressions for the exact inverse for each of the following systems:

(a) $y(n) = x(n) - 0.5x(n-1) + 0.2x^2(n) + 0.1x(n)x(n-1)$

(b) $y(n) = x(n) - 0.5x(n-1) + 0.2x(n-1)y(n-1)$

(c) $y(n) = x(n) + 0.75x(n-1) + 0.1y(n-1) + 0.05y^2(n-1)$

9.2 Show that the system in (9.4) is a preinverse of the system in (9.3).

9.3 Show that the system in (9.36) is a pth-order preinverse of the nonlinear system in (9.35).

9.4 Find the first-, second- and third-order inverses of the truncated Volterra system whose input–output relationship is given by

$$y(n) = x(n) + 1.2x(n-1) + 0.32x(n-2) + 0.1x^2(n) + 0.15x(n)x(n-1)$$
$$+ 0.2x(n)x(n-2).$$

9.5 Find the first-, second- and third-order inverses of the systems in Exercise 9.1.

9.6 Consider a first-order bilinear system with input–output relationship

$$y(n) = x(n) - 0.2x(n-1) + 0.05x(n)y(n-1).$$

Find a bound on the input amplitude for which the exact inverse of this system is stable in the sense that the output is bounded whenever the input is smaller in magnitude than the bound you derive.

9.7 Derive a recursive procedure to estimate the parameters of the nonlinear time-series model

$$x(n) = a_1 x(n) + a_2 x(n-1) + \xi(n) + b_1 \xi^2(n),$$

where $\xi(n)$ is an unmeasurable input to the system. You may assume that $\xi(n)$ is an IID signal.

9.8 Develop an adaptive algorithm employing the output error LMS approach to estimate the parameters of the bilinear time-series model

$$x(n) = \sum_{i=1}^{p} a_i x(n-i) + \sum_{i=1}^{p} b_i x(n-i)\xi(n-1) + \xi(n)$$

on line.

9.9 Consider the bilinear time series described by (9.42). Show that its mean value is given by

$$E\{x(n)\} = \left(1 - \sum_{i=1}^{p} a_i\right)^{-1} b_1 \sigma_\xi^2.$$

9.10 Find numerical values of the mean, the autocorrelation function, and the third-order correlation function of the bilinear time series

$$x(n) = 0.8x(n-1) + 0.1x(n-1)\xi(n-1) + \xi(n),$$

where $\xi(n)$ is an IID signal with zero mean value and variance $\sigma_\xi^2 = 0.05$.

10

APPLICATIONS OF POLYNOMIAL FILTERS

The previous chapters developed the theory for analyzing nonlinear systems characterized using polynomial models, and also for the design and realization of polynomial systems. In this chapter, we describe several applications in which polynomial filters have found uses. Because truncated Volterra filters often suffer from drawbacks associated with the need for a large number of parameters, they are typically employed in applications where the nonlinearities are relatively mild, and can be modeled using a few lower-order kernels with finite planes of support. In the rest of this chapter, we describe applications in image, video, and speech processing; communication channel equalization; and compensation of loudspeaker nonlinearities. Several other applications in which polynomial filters have found use are described with appropriate references in the final section.

10.1 CHOICE OF SYSTEM MODELS

Most of the applications described in this chapter require the selection of a system model and the identification of its parameters. There is a large volume of literature dealing with identification of nonlinear systems using truncated Volterra and other related system models [56,67,92,194,195,262]. Many such methods are described in Chapters 5–8. In many instances, polynomial system models are employed because of the conceptual simplicity they provide. However, considerations of the physical systems naturally lead to low-order Volterra system models in several situations. For example, it has been shown that a quadratic transfer function is appropriate for modeling nonlinear three-wave interactions associated with turbulence in both fluids and plasmas [110,230,258,259]. Other such applications include electromagnetic scattering from nonlinear targets [229], modeling the nonlinear dynamic behavior of offshore structures subject to random sea wave excitation [111,116], and cancellation of intersymbol intereference for voiceband data transmission [19].

10.2 COMPENSATION OF NONLINEAR DISTORTIONS

There are a large number of applications in which the underlying systems introduce nonlinear distortions into the signals of interest. In such situations, it is necessary to compensate for such distortions using an appropriate linearization technique. Examples of applications where linearization is important include

1. MOS D/A converters [2] characterized by input–output relationships that deviate from linearity required for accurate reproduction of the input signals at the output.
2. Integrated-circuit realizations of continuous-time filters [317] where transistors are affected by nonlinearities at large-signal dynamics.
3. Optical communication systems [127], where nonlinearities of LEDs or lasers can be significant.
4. Digital communication channels in which the amplifiers operate at or near the saturation region to conserve energy. Saturation nonlinearities introduce distortions into the received signals in such situations [109,233].
5. Sound reproduction systems in which nonlinear distortions are introduced by the loudspeakers because of their nonideal characteristics [76,114].
6. Magnetic recording systems in which the nonlinear characteristics of the media introduce distortions that reduce the storage density of the media if no compensations are introduced to combat such nonlinearities [21,91].
7. High-density optical recording systems in which the readout process is affected by intersymbol interference and crosstalk among adjacent tracks because of the high linear density of the pits and the reduced track pitch [1].

10.2.1 Techniques for Linearization

Linearization of nonlinear systems is typically achieved in one of two ways. The first approach is useful when the output signal is available for further processing. In such situations, we can apply a nonlinear filter that compensates for distortions to the output signals. Examples where postprocessing techniques are appropriate include equalization of nonlinear distortions in digital communications systems [17]. In applications such as loudspeaker linearization, the output of the loudspeaker is not available for further processing. In such situations, it is appropriate to predistort the signals before they are presented to the systems that introduce the distortions. The characteristics of the prefilter in cascade with those of the nonlinear system should be such that the overall effect is ideally that of a linear system. Figure 10.1 shows the two approaches for linearization. In applications in which the nonlinearities are relatively mild, truncated Volterra systems can be profitably employed for the

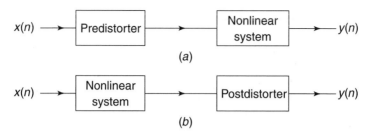

Figure 10.1 Two methods for compensating for nonlinear distortions: (a) predistortion method; (b) postdistortion method.

linearization procedure. Furthermore, adaptive nonlinear filters may be employed to identify and track the nonlinearities in the system. To illustrate the principles behind the compensation techniques, we describe an application involving the linearization of loudspeaker nonlinearities below.

10.2.2 Linearization of Loudspeaker Nonlinearities

Nonlinear distortions in a loudspeaker arise from nonlinearity of the suspensions and from nonuniform flux density. These distortions severely affect the quality of the sound reproduced by the loudspeaker. Nonlinear distortions arising from both of these sources can be controlled by a careful design that imposes expensive constraints or by limiting the output power. Another approach that is often less expensive and also does not require limiting the output power is to use digital linearization techniques. Such methods have been described in [70,76].

Figure 10.2 describes an approach [76] for reducing the distortions due to the nonlinearities in the loudspeaker characteristics. In this method, the inverse function of the input output relationship of the loudspeaker is estimated using an adaptive filter employing the output of the loudspeaker as its input and the input to the loudspeaker as its desired response signal. The coefficients of this adaptive filter are

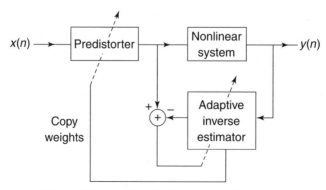

Figure 10.2 A method for loudspeaker linearization.

then copied to the predistortion filter. The basic idea is that the predistortion filter and the loudspeaker connected in cascade should have less nonlinear distortion than the loudspeaker itself. This approach assumes that the postinverse of the nonlinear loudspeaker characteristics identified by the adaptive filter is a good approximation to the preinverse of the system. We note that the predistorter in Figure 10.2 is implemented as a preinverse. We also recall that, unlike with linear, time-invariant systems, interchanging the order of nonlinear components in a cascade connection may change the properties of the interconnected system. The technique for linearization also assumes that the noise levels in the measured output signal is low so that there is not significant bias in the estimated inverse filter parameters.

We can also employ the method described in Chapter 9 for synthesizing pth-order inverses of nonlinear systems for linearization of the loudspeaker nonlinearities. We present an example below that utilizes this approach. The example also contains some approaches for simplifying the calculations involved by use of subband techniques [42].

Example 10.1 In this example, we consider the linearization of the nonlinearities associated with a synthetic loudspeaker using a pth-order linearizer. One cell of the pth-order linearizer is shown in Figure 10.3. Frank [70] has shown that loudspeaker nonlinearities can be efficiently modeled with good accuracy using low-order, truncated Volterra systems. The loudspeaker in our experiment was modeled using a quadratic filter with 100 linear coefficients and a 40-sample memory for the second-order component. The second-order harmonic distortion of the system is shown in Figure 10.5. Since the distortions were primarily in the range $[0, f_N/3]$ Hz, where f_N denotes the Nyquist frequency, the system of Figure 10.4 was employed to perform a pth-order linearization. The upper branch of the system contained a PCAS lowpass filter [132] with cutoff frequency $f_N/3$. The output of the lowpass filter was subsampled by a factor of three. The parameters of the loudspeaker were estimated

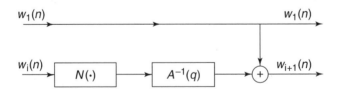

Figure 10.3 The ith cell in the pth-order linearizer.

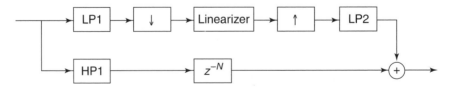

Figure 10.4 The subband linearizer.

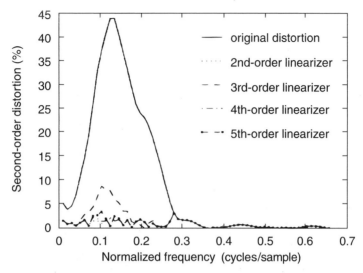

Figure 10.5 Second-order distortion measured in the experiments of Example 10.1. (Courtesy of A. Carini. Copyright © 1998 IEEE.)

using a quadratic filter with 51-sample memory length for the linear component and 40-sample memory for the second-order nonlinearity from subsampled versions of the input to the loudspeaker and its output in the presence of uncorrelated, 30-dB measurement noise. The estimated model was then used to prelinearize the system using second-, third-, fourth-, and fifth-order linearizers.

Figure 10.5 also shows the second-order harmonic distortion measured at the output of the linearized systems. We note that the second-order distortion is smallest in the case of the second-order linearizer. This linearizer is sufficient to correct for the second-order distortions, and it produces the most compact spectrum for the predistorted signal. The third-order linearizer exhibits a higher second-order distortion in this experiment. This is due to model mismatch, our approximations and the wider band of the predistorted signal whose intermodulation contributions alter the amplitude of the fundamental frequency components. The higher-order linearizers exhibit second-order distortions comparable to those of the second-order linearizer. The improvement due to the use of the linearizer is evident from all our experiments.

10.3 APPLICATIONS IN COMMUNICATION SYSTEMS

Polynomial system models have found a large number of applications in communications. Examples of such applications include modeling highly distorted reference channels [297], nonlinear transmission amplifiers [233], and nonlinear bandpass channels [17,20] in digital transmission systems. Other applications include nonlinear echo cancellation [2,58,59,283] and nonlinear discrete-time prediction [174]. We consider two of these applications in this section.

10.3.1 Nonlinear Echo Cancellation

Figure 10.6 shows a simplified block diagram of a satellite communication system. The communication between the base station and the satellite typically employs "four-wire" transmission, in which the transmission to the satellite occurs at a different carrier frequency than the transmission from the telephones. However, communication between the base station and individual telephones takes place over "two wires," where transmission to and from the individual telephones takes place over the same channel. The isolation of the traffic to and from the base station is accomplished using *hybrid circuits*. Unfortunately, the hybrid circuits are not ideal, and a portion of the signals received from the satellite gets transmitted back to the satellite and from the satellite to the *far-end* station. This component represents the echo in the received signals. In situations where the echo arrives more than 20 ms after the initial transmission, the intelligibility of the received signals is affected adversely. The amplifiers in the satellite introduce nonlinear distortions since they often operate at or near saturation to conserve energy. Consequently, it is necessary to use nonlinear echo cancellation to achieve satisfactory echo rejection in such systems. A reasonably detailed description of the techniques for combating amplifier

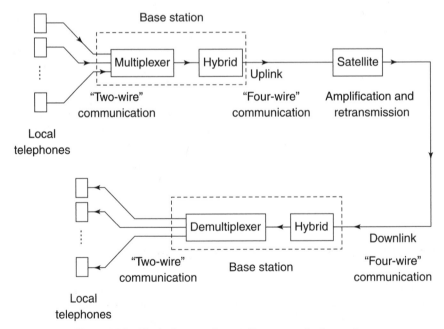

Figure 10.6 Block diagram of a satellite communication system.

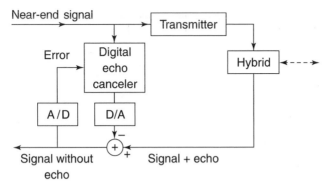

Figure 10.7 Echo canceler with digital filtering and analog cancellation.

nonlinearities in satellite communication channels using Volterra system models is available in [17].

Another situation requiring nonlinear echo cancellation arises in modern digital subscriber loop modems, where echo cancelers are employed to ensure full-duplex transmission with adequate channel separation. The purpose of the canceler is to remove the near-end crosstalk, or echo signal, that interferes with the signal coming from the distant transmitter. Since the latter signal may be highly attenuated (40–50-dB attenuation is typical) and the attenuation of the hybrid can be as high as 10 dB, a 50–60-dB rejection of the echo is required to achieve an acceptable signal-to-echo interference ratio. In order to guarantee this level of echo attenuation, it is often necessary to take into account the effects of nonlinearities. In practical systems, nonlinear distortions occur as a result of asymmetries in the transmitted pulses, saturation effects in transformers, and deviations from linearity in analog-to-digital (A/D) and digital-to-analog (D/A) converters.

A method for expanding an arbitrary nonlinear function of a number of bits in a series similar to the Volterra series expansion in the bits has been proposed in [2]. The echo canceler is then implemented as an adaptive filter whose output must compensate for the actual echo. A typical echo cancellation system is shown in Figure 10.7. The nonlinearity $d(\cdot)$ that exists in the echo cancellation path can be compensated for by incorporating the inverse function $d^{-1}(\cdot)$ or an approximation of this function in the echo canceler.

Example 10.2 Nonlinear distortions introduced by MOS (metal oxide semiconductor) D/A converters are often modeled as an even symmetric function with input–output relationship

$$d(x) = x + b|x|,$$

where the parameter b is given to be equal to -0.005 [2]. The inverse function d^{-1} can be approximated using a polynomial expansion. Because of the even symmetry of the nonlinear part, the coefficients of the odd powers in the expansion are zero.

Since the nonlinearity is relatively mild, we may truncate this expansion to have only second-order terms. We also assume that the input data are binary and take values from the set $\{1, -1\}$ with equal probability. For simplicity, we consider an echo path with impulse response function given by

$$h(n) = e^{-0.8n}; \qquad n = 0, 1, \ldots, 7.$$

Figure 10.8 displays a graph of the mean-square value of the residual echo cancellation error as a function of the number of nonlinear coefficients in the system. The system model includes an offset term, a linear term, and a quadratic term. For each system configuration we used the LMS adaptation algorithm with adaptation constants equal to 0.2, 0.025, and 0.0046875, respectively. In all the cases 4000 iterations were performed and the system converged during that time span. The mean-square error during the last 200 iterations is displayed in Figure 10.8. These results show that the nonlinear components consisting of an offset term as well as the second-order terms are able to reduce the echo by about 40 dB when the system was implemented with the maximum precision available on the computer. Similar simulations reported by Agazzi et al. [2] show that the importance of the taps of the second-order term depends on the precision of the arithmetic representation of the variables within the implementation. In the results given in that article [2], a

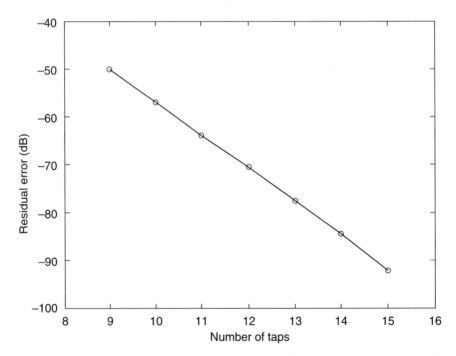

Figure 10.8   Mean-square value of the residual error in the echo canceler as a function of the number of nonlinear taps. (Copyright © 1992 IEEE.)

modest number of nonlinear coefficients provided an improvement of 20–30 dB with respect to purely linear cancellation when the system was implemented with 13 bits of precision. However, the authors of [2] found that there was relatively no advantage to using the nonlinear components in the echo canceler when it was implemented with only 10 bits of precision. Echo cancellation using quadratic system models has been considered by several other researchers also [58,59,283,285].

Remark 10.1 In many applications involving communication systems, including channel equalization, it is convenient to describe the system using complex-valued signals and filters. In such cases, it is necessary to model the nonlinear channels and nonlinear cancelers using a complex-valued Volterra series representation. In complex-valued representations of Volterra series expansions, the coefficients of the Volterra kernel are complex-valued, and they act on products of complex-valued data symbols. Such expansions are relatively straightforward extensions of the real-valued representations.

10.3.2 Equalization of Communication Channels

Many digital communication channels are subject to substantial amplitude and delay distortions. These distortions vary from one connection to another and with time. One approach for compensating for these distortions is to apply an appropriate adaptive filtering scheme to the received signal. The purpose of the adaptive filter is to construct an approximation of the inverse function of the channel characteristics. As described earlier, many communication channels introduce nonlinear distortions into the transmitted signals. The individual sources of channel nonlinearities can often be modeled as memoryless. Since these effects take place in a network where dispersive linear filtering operations also occur, the overall effect of the channel on the input signal is a nonlinear mapping with memory. Such nonlinearities can be effectively described using discrete Volterra series expansions. Falconer has described an example of such a transmission system for passband quadrature amplitude modulation (QAM) data transmission where the dispersive nonlinear channel is modeled as a cascade of linear bandpass filters and memoryless quadratic and cubic nonlinearities [66]. In this case, the transmitted signal is described mathematically as the real part of a complex signal and the Volterra series expansion is given in terms of complex data symbols and coefficients. The Volterra series technique has been applied for adaptive equalization of channel nonlinearities [66] and for nonlinear equalization of digital satellite channels [16].

One disadvantage of the inverse modeling technique is that while attempting to compensate for the characteristics of the channel, the adaptive filter can amplify the additive noise components in the received signal. Furthermore, in the case of digital communication channels, the adaptive filter does not in general exploit the fact that the transmitted sequence is drawn from a finite alphabet. An approach to equalization that takes into account the finite input symbol set is to consider the equalization as a classification problem. If the transmitted sequence is chosen from a finite set of symbols with equal probability and the channel is modeled as an FIR filter, the

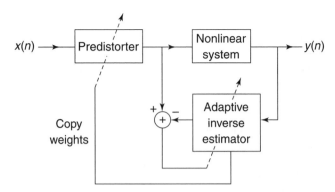

Figure 10.9 The learning structure of a truncated Volterra predistorter in [65].

channel output assumes a finite set of values. In the presence of additive noise, the received signals form clusters around their ideal values. The classification problem involves assigning regions within the space spanned by the noisy outputs to symbols in the input alphabet. The use of a Volterra series classifier has been shown in [188] to be more powerful than a linear classifier in selecting the decision boundaries.

10.3.3 Predistortion for Nonlinear Channels

An approach to nonlinear predistortion was proposed in [65]. To overcome the difficulties of the pth-order inverse method, the *indirect learning architecture* is used. While the pth-order inverse method first requires a Volterra series model of the system to be estimated prior to compensating for the distortions, the indirect method derives a truncated Volterra system model directly for the predistorter. Eun and Powers used the learning structure shown in Figure 10.9 [65]. This approach is essentially the same as the technique used in Section 10.2.2 for linearization of loudspeaker nonlinearities. In this approach, two identical truncated Volterra system models are used for training and predistortion. The inverse of the channel characteristics is directly adapted using the RLS algorithm, and at each step the training model is copied to the predistortion block. As the error $e(n)$ approaches zero, the overall output of the system $y(n)$ approaches the system input $x(n)$, that is, the desired output of the overall system since the inputs to the identical networks are equal. When convergence is reached, the training network is removed. Experimental results for this approach are provided in [65] for a problem involving compensation of the nonlinearities of a baseband satellite communication channel. Experiments using sixteen-level phase-shift keying (PSK) and 16-level QAM were conducted and the results included in [65] demonstrate the good performance of the predistortion scheme.

10.4 THE FREQUENCY AND AMPLITUDE ESTIMATION

Consider the application of Newton's law of motion to the motion of a mass m suspended by a spring of force constant c. The displacement $x(t)$ is related to the other parameters of the system through the second-order differential equation

$$\frac{d^2x(t)}{dt^2} + \frac{c}{m}x(t) = 0. \tag{10.1}$$

The solution to this equation is the simple harmonic motion

$$x(t) = A\cos(\Omega t + \phi), \tag{10.2}$$

where A is the amplitude of the oscillation, $\Omega = \sqrt{c/m}$ is the angular frequency of the oscillation, and ϕ is an arbitrary initial phase. The total instantaneous energy $E(t)$ of the system is the sum of the potential energy in the spring and the kinetic energy of the mass, and is given by

$$E(t) = \frac{1}{2}cx^2(t) + \frac{1}{2}m\left(\frac{dx(t)}{dt}\right)^2. \tag{10.3}$$

Substituting for $x(t) = A\cos(\Omega t + \phi)$ in this equation and solving, we get the following relationship for the instantaneous energy:

$$E(t) = \frac{1}{2}mA^2\Omega^2. \tag{10.4}$$

A nonlinear, continuous-time operator measuring such energy assumes the form [107]

$$\Psi[x(t)] = \left(\frac{dx(t)}{dt}\right)^2 - x(t)\frac{d^2x(t)}{dt^2} \tag{10.5}$$

This step can be easily derived substituting for $x(t) = A\cos(\Omega t + \phi)$ in the preceding equation. An approximate discrete version of this algorithm that uses only three consecutive samples of the input signal at a time is given by the input–output relationship

$$y_T(n) = \Psi_d[x(n)] = x^2(n) - x(n+1)x(n-1). \tag{10.6}$$

The most important characteristic of this system is its ability to track the instantaneous frequency of the signal in those cases where the input signal contains single sinusoidal components with time-varying frequency. Therefore, this algorithm is useful for rapid formant tracking in speech analysis. In fact, the algorithm was originally applied to the nonlinear analysis of speech by Teager [309, 310]. For this reason, this nonlinear system is usually termed *Teager's algorithm* or *model*. This operator has also been used in algorithms for the estimation of the amplitude envelope and the instantaneous frequency of discrete-time amplitude and frequency modulated signals [155–157].

To see how Teager's algorithm can be employed to estimate the instantaneous energy, consider a sinusoidal input signal of the form

$$x(n) = A \cos(\omega_0 n + \phi). \tag{10.7}$$

Substituting for $x(n)$ as above, and rearranging the terms in (10.6), we get

$$\Psi_d[x(n)] = A^2 \sin^2(\omega_0) \approx A^2 \omega_0^2 \tag{10.8}$$

for small values of ω_0. This result indicates that Teager's algorithm gives a measure, per half unit mass, of the energy necessary to generate a single sinusoid [107]. The approximation gives a relative error below 11% for $\omega < (\pi/4)$.

Maragos and colleagues have extended the above result to energy operators with longer memory spans and applied the new operators to cochannel demodulation and separation of AM–FM signal mixtures, and in the analysis and modeling of speech [158,159]. Another approach to estimating the amplitude of a signal as a polynomial function of its samples is given in [24].

10.5 SPEECH MODELING

Efficient strategies for speech modeling and coding are of fundamental importance in applications involving speech recognition, synthesis, analysis, and transmission. It is known that the acoustical mechanism involved in speech production is nonlinear. Signal modulations due to separate airflows in the vocal tract have been documented in [309]. A nonlinear model for speech was derived in [155] using the Teager energy operator. Other operators based on Teager's energy operator have been employed for the analysis of the speech resonances [156,157]. Sparse quadratic detectors for nonlinear preprocessing of speech aimed at pitch tracking and burst recognition were presented by Atlas and Fang [7]. Finally, it was shown in [189] that a nonlinear quadratic predictor gives better prediction gains than conventional short and long-term linear predictors. In what follows, we consider the last application involving nonlinear prediction of speech.

Traditional methods of speech prediction typically employ a short-term linear predictor to exploit the correlations between successive samples of the signal, and a long-term linear predictor to exploit the quasi-periodic nature of voiced speech. We consider a nonlinear predictor of the form

$$\hat{x}(n) = \sum_{m_1=1}^{N_1} h_1(m_1)x(n - m_1) + \sum_{m_1=1}^{N_2} \sum_{m_2=m_1}^{N_2} h_2(m_1, m_2)x(n - m_1)x(n - m_2). \tag{10.9}$$

By defining the instantaneous prediction error as

$$e(n) = x(n) - \hat{x}(n), \tag{10.10}$$

we can create a nonlinear analysis model given by

$$e(n) = x(n) - \sum_{m_1=1}^{N_1} h_1(m_1)x(n - m_1) - \sum_{m_1=1}^{N_2} \sum_{m_2=m_1}^{N_2} h_2(m_1, m_2)x(n - m_1)x(n - m_2) \tag{10.11}$$

and a corresponding nonlinear synthesis model given by

$$x(n) = e(n) + \sum_{m_1=1}^{N_1} h_1(m_1)x(n - m_1) + \sum_{m_1=1}^{N_2} \sum_{m_2=m_1}^{N_2} h_2(m_1, m_2)x(n - m_1)x(n - m_2). \tag{10.12}$$

We can employ any of the techniques described in Chapters 5 and 6 that are applicable to non-Gaussian signals or an appropriate adaptive technique given in Chapter 7 to estimate the parameters of the system model. In typical applications, the parameters are estimated over contiguous blocks of duration ranging from 5 to 40 ms using least-squares algorithms.

Example 10.3   The speech sequences analyzed in this example were obtained by sampling several utterances of four speakers at the rate of 10,000 samples per second. Figure 10.10 shows the least-squares prediction error values, normalized by the average value of the squared sample values in each block, and averaged over a number of frames, utterances, and speakers as a function of the orders of the linear and second-order terms of the system model. Based on these results, we conclude that choosing $N_1 = 10$ and $N_2 = 8$ provides a good balance between complexity and performance of the model [189]. Figure 10.11a shows a 50-ms segment of voiced speech. Figure 10.11c displays the prediction error signal obtained from the analysis of this segment using the nonlinear model. To provide a comparison with the performance of a linear predictor, we plot the prediction error signal obtained using a linear predictor with 46 coefficients in Figure 10.11b. We note that both the predictors employ the same number of coefficients. We can see from the figures that the height of the pitch pulses are lower and the ringing following each spike is reduced with the nonlinear predictor. We conclude from these results that a polynomial predictor is able to better model the speech segment in this example than the linear model considered. We attribute this improved performance to the nonlinear phenomena that produces speech. It was also shown in [189] that a Volterra predictor is able to predict the signal periodicity more efficiently than a conventional long-term linear predictor.

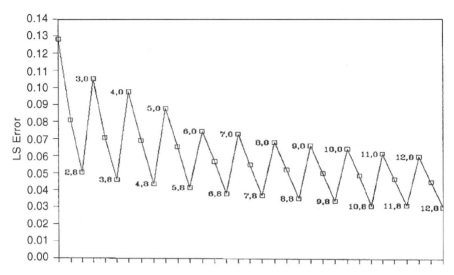

Figure 10.10 Least-square prediction errors as a function of the orders of the linear and quadratic terms. (Courtesy of E. Mumolo.)

Remark 10.2   One of the primary uses of speech prediction arises in speech compression. Simplified block diagrams of the encoder for compression and the decoder for reconstruction of the signals are shown in Figure 10.12. The predictor may be adapted using block least-squares techniques, in which the coefficient values are transmitted as side information. One can also use a *backward adaptive* prediction for continuously adapting the coefficients. In this case, the system employs the reconstructed signal $\tilde{x}(n)$ for prediction. Since $\tilde{x}(n)$ is available at both the encoder and the decoder, no side information need be sent. Jayant and Noll have provided additional information on predictive compression in [101].

One issue of importance in the reconstruction process is the fact that the synthesis filter of (10.12) is a recursive system. Consequently, the stability of the system plays a crucial role in the proper operation of the decoder. Sufficient conditions for the stability of the system model of (10.12) have been derived [39].

10.6 POLYNOMIAL FILTERS FOR IMAGE PROCESSING

In several situations in image processing it is necessary to resort to some form of nonlinear filtering to overcome the drawbacks of linear techniques. Most images we deal with are not Gaussian-distributed or stationary. Furthermore, they are often corrupted by signal-dependent and/or multiplicative noises or additive noise distortions with impulsive distributions. It is well-known that linear filters perform poorly in such situations. For example, linear filters are not able to remove impulsive noise without blurring the edges.

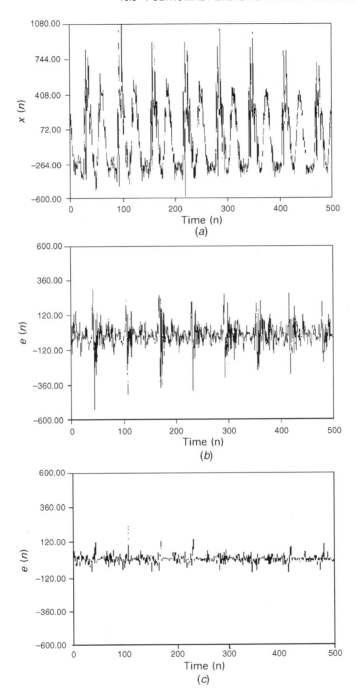

Figure 10.11 Analysis of a 50-ms segment of voiced speech: (*a*) original signal; (*b*) residual error of a 46-order linear predictor; (*c*) residual error of a quadratic predictor with $N_1 = 10$ and $N_2 = 8$. (Courtesy of E. Mumolo.)

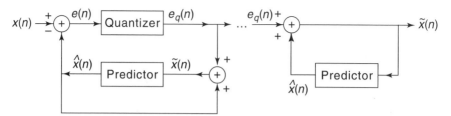

Figure 10.12   Predictive coding of speech.

Another rationale for employing nonlinear filters in many image processing applications is that human perception is a complex and highly nonlinear system. Therefore, it is desirable to apply some nonlinear processing to images so that the result of such processing is more tuned to the perceptual quality assessment performed by human observers. In this section, we describe the applications of polynomial filters in the problems of edge extraction, image enhancement, document image processing, and texture discrimination. We will see that it is possible to design nonlinear filters with relatively small computational complexity and obtain significant performance increases over traditional approaches for solving these problems.

Performance evaluation of many image processing techniques in which the final judges of quality of the processed images are human observers is a difficult problem. One possible approach for performance evaluation in such situations is to have a panel of human observers rate the quality of the processed images, and obtain a quality measure based on an average rating provided by the panel. However, this is a time-consuming and expensive procedure, and cannot be performed on a large scale. It is well-known that performance measures such as mean-square error (MSE) and signal-to-noise ratio (SNR) are not accurate measures of subjective quality. Consequently, we will often use adhoc measures that match the subjective quality better than MSE and SNR in the applications discussed below.

10.6.1 Edge Extraction

Edge detection is a basic tool in image processing, having applications in machine vision, automated inspection systems, image coding, and several other problems. A number of algorithms for edge detection have been proposed in the literature. One example of such a system is the Sobel operator. This operator has been widely used in practice because of its simple structure and good performance. Let

$$
\mathbf{X}(n_1, n_2) = \begin{bmatrix} x_1 & x_2 & x_3 \\ x_4 & x_5 & x_6 \\ x_7 & x_8 & x_9 \end{bmatrix}
$$

$$
= \begin{bmatrix} x(n_1 - 1, n_2 - 1) & x(n_1 - 1, n_2) & x(n_1 - 1, n_2 + 1) \\ x(n_1, n_2 - 1) & x(n_1, n_2) & x(n_1, n_2 + 1) \\ x(n_1 + 1, n_2 - 1) & x(n_1 + 1, n_2) & x(n_1 + 1, n_2 + 1) \end{bmatrix} \quad (10.13)
$$

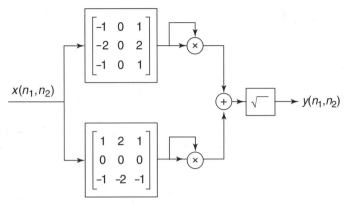

Figure 10.13 Realization of the Sobel operator.

be the matrix of nine pixels used during the filtering operation at location (n_1, n_2). The output $y(n_1, n_2)$ of the Sobel operator is

$$y(n_1, n_2) = [(x_3 - x_1 + 2(x_6 - x_4) + x_9 - x_7)^2$$
$$+ (x_1 - x_7 + 2(x_2 - x_8) + x_3 - x_9)^2]^{1/2}. \qquad (10.14)$$

It is not difficult to see that the presence of edges in the 3×3-element input block at any location causes the output to assume high values. The input-output relationship of the Sobel operator clearly shows that it is a quadratic operator followed by a square-root operation. Moreover, the quadratic filter can be efficiently implemented using the parallel-cascade structure with two parallel branches as shown in Figure 10.13.

In many applications of edge extraction, the input signals may be corrupted by additive noise. In such cases, the Sobel operator may amplify the noise at its input. The next example provides the description and experimental results associated with a quadratic filter that is capable of edge extraction and impulsive noise reduction simultaneously.

Example 10.4 We consider the problem of extracting the edges from a gray-scale input image corrupted by additive impulsive noise. Such distortions occur frequently in satellite images because of the harsh conditions through which the images are transmitted to the earth. The linear filters in the two parallel branches of the Sobel operator have band pass characteristics that tend to increase the noise present in the images. To avoid this annoying effect, the noisy input images are often preprocessed by a median filter before applying the Sobel operator. This approach requires two sequential filtering steps. The use of the two-step procedure can be avoided by designing a homogeneous quadratic filter that allows us to perform noise elimination and edge extraction simultaneously. Such a filter was designed in [234] by imposing a penalty function on isolated impulses present in the input image. The coefficient matrix $\mathbf{H}_2$ so obtained is given by

$$
\mathbf{H}_2 = \begin{bmatrix}
-1 & 4 & -1 & 4 & 0 & 0 & -1 & 0 & -4 \\
0 & -1 & 4 & -1 & 0 & -1 & 0 & -4 & 0 \\
0 & 0 & -1 & 0 & 0 & 4 & -4 & 0 & -1 \\
0 & 0 & 0 & -1 & 0 & -4 & 4 & -1 & 0 \\
0 & 0 & 0 & 0 & 0 & 0 & 0 & 0 & 0 \\
0 & 0 & 0 & 0 & 0 & -1 & 0 & -1 & 4 \\
0 & 0 & 0 & 0 & 0 & 0 & -1 & 4 & -1 \\
0 & 0 & 0 & 0 & 0 & 0 & 0 & -1 & 4 \\
0 & 0 & 0 & 0 & 0 & 0 & 0 & 0 & -1
\end{bmatrix}.
$$

This filter works on a 3×3-pixel plane of support, and the coefficients are arranged according to the lexicographical ordering described in Chapter 4. It is left as an exercise for the reader to show that a single impulse present in the 3×3-pixel block under consideration at any specific location is eliminated by this filter at the output. We also leave it to the reader to show that ordinary edges present in this block will result in high output values. The combination of these two properties provide us with the desired characteristics of edge extraction and impulsive noise reduction.

Figure 10.14*a* displays the image of a boat used in the experiments described in this example. This image was corrupted by a 5% impulsive noise. The noisy image is shown in Figure 10.14*b*. The results of processing these two images with the Sobel operator and the operator $\mathbf{H}_2$ are shown in Figures 10.14*c*–10.4*f*. These images are bilevel images obtained by thresholding the output of the two operators so that the pixel takes the value 0 whenever the corresponding input pixel is above a threshold τ, and the pixel takes the value 255 whenever the input pixel is below τ. We can clearly see that while both systems work reasonably well on the clean input image, the operator $\mathbf{H}_2$ significantly outperforms the Sobel operator when the input image is corrupted by impulsive noise.

10.6.2 Image Enhancement

Digital image enhancement is the process by which perceptually more pleasing images are generated from given input images using digital image processing techniques. Examples of image enhancement applications include edge sharpening, and noise removal without blurring the edges. Although linear filters are useful in many applications of image enhancement, there are several reasons why nonlinear filters are attractive in such problems. The human visual system is intrinsically nonlinear, and any algorithm that is perceptually tuned will almost certainly be nonlinear. Furthermore, images are formed using a multiplicative process involving the intensity of the incident light and the reflection coefficient of the objects from which the light is reflected. Finally, images are often distorted by multiplicative or frequency-dependent noises, requiring the use of nonlinear filters to reduce such distortions. The design procedures for many of the algorithms discussed in this section are strictly problem-dependent, and only recently have some generalizations

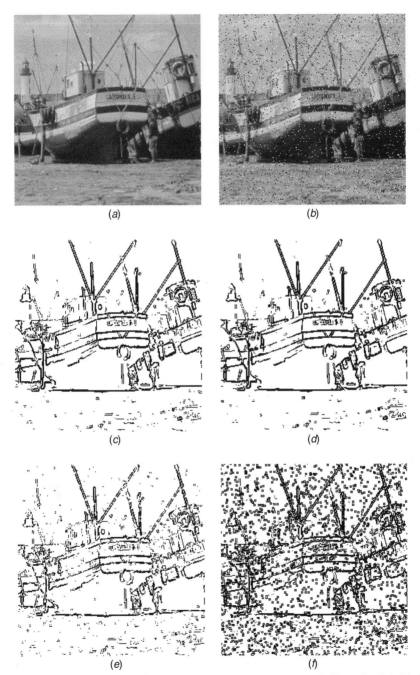

Figure 10.14 Edge maps obtained after filtering and thresholding as in Example 10.4: (*a*) the original input image; (*b*) noisy input image available for processing; (*c*) response of the quadratic operator $\mathbf{H}_2$ to the clean input image; *(d)* response of the Sobel operator to the clean input image; (*e*) response of the quadratic operator $\mathbf{H}_2$ to the noisy input image; (*f*) response of the Sobel operator to the noisy input image.

been introduced for solving well-performance defined classes of problems. In the following subsections, we consider several specific problems of image enhancement.

10.6.2.1 Edge-Preserving Noise Smoothing

The objective of *edge-preserving noise smoothing* is the removal of noise from an image without causing deterioration of small details and contours. Linear filters cannot offer satisfactory results in such problems because there exists a direct tradeoff between noise cancellation and detail preservation capabilities of linear filters. Nonlinear operators based on rank-order operators such as median filters have been traditionally employed for this problem. Median filters and their variations are useful when the images contain noise distortions with a long-tailed distribution such as the double exponential distribution or impulsive noises. However, order-statistic filters often fail to cancel distortions such as those caused by additive Gaussian noise. Quadratic filters have been successfully employed in such situations. Quadratic filters are especially effective in enhancing noisy input images with low contrast that result from difficult lighting environments. In particular, isotropic quadratic operators defined on a 3 × 3-pixel plane of support have been designed for this purpose using numerical optimization techniques and decision-directed approaches [238,239]. An example of a quadratic filter so designed and its performance was discussed in Example 4.6. Experimental results obtained from images having only 16 contiguous input gray levels out of the possible 256 levels showed that the most commonly used techniques yield relatively poor results. In particular, a simple linear gray-level expansion amplifies the noise and renders it more visible. Linear smoothing cancels noise at the expense of the image details. Median filtering does not provide satisfactory results because of the *patchwork* effect it produces due to the reduced number of gray levels in the input image. Quadratic filters achieve the detail preservation typical of the median filter together with the smoothing effect of linear smoothing filters, while minimizing the drawbacks of both.

10.6.2.2 Unsharp Masking

Another classic problem in the field of image enhancement is to find operators that are capable of sharpening the details of an image, but are reasonably insensitive to noise. A simple approach that has long been used for the enhancement of slightly blurred or imperfectly contrasted images is *linear unsharp masking*. Its basic idea, depicted in Figure 10.15, is to add a scaled highpass component derived from the original image to the image itself. This

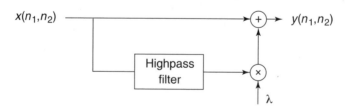

Figure 10.15   Linear unsharp masking structure.

process usually results in visually pleasing output images. The sharpening effect can also be tuned by choosing the scaling factor appropriately. However, due to the presence of the highpass filter in the linear unsharp masking system, it is very sensitive to noise. In the following, we describe two methods for modifying the linear unsharp masking scheme by incorporating nonlinear filters into its structure.

Algorithms Based on Teager's Operator. Teager's algorithm and its variations have been applied in image enhancement problems by replacing the highpass filter of the linear unsharp masking scheme with a nonlinear operator as shown in Figure 10.16. In most cases, the nonlinear operators are simple quadratic or cubic filters.

One example of a two-dimensional quadratic filter employed for nonlinear unsharp masking is obtained by applying the filtering operation of (10.6) along both the vertical and horizontal directions, which results in the following relationship:

$$y(n_1, n_2) = 2x^2(n_1, n_2) - x(n_1 - 1, n_2)x(n_1 + 1, n_2) - x(n_1, n_2 - 1)x(n_1, n_2 + 1). \tag{10.15}$$

Another operator is obtained by applying the filtering operation along the diagonal directions, which gives

$$\begin{aligned} y(n_1, n_2) = {} & 2x^2(n_1, n_2) - x(n_1 - 1, n_2 + 1)x(n_1 + 1, n_2 - 1) \\ & - x(n_1 - 1, n_2 - 1)x(n_1 + 1, n_2 + 1). \end{aligned} \tag{10.16}$$

When the average value of the input signal is large compared to its standard deviation, one can show that these operators behave like local-mean weighted adaptive highpass filters [181,182]. The derivation is based on a direct extension of the following approximation used for a one-dimensional Teager algorithm to two dimensions:

$$y(n) \approx l_m(n)[2x(n) - x(n - 1) - x(n + 1)], \tag{10.17}$$

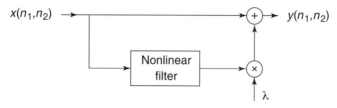

Figure 10.16 Nonlinear unsharp masking structure.

where $l_m(n)$ is an estimate of the local mean value given by

$$l_m(n) = \frac{x(n-1) + x(n) + x(n+1)}{3} \tag{10.18}$$

and the second term in (10.17) describes a *Laplacian filter* that has highpass characteristics. The corresponding approximations for the two-dimensional case are

$$y(n_1, n_2) \approx l_m(n_1, n_2) \times [4x(n_1, n_2) - x(n_1, n_2 - 1) - x(n_1, n_2 + 1)$$
$$- x(n_1 + 1, n_2) - x(n_1 - 1, n_2)], \tag{10.19}$$

and

$$y(n_1, n_2) \approx l_m(n_1, n_2) \times [4x(n_1, n_2) - x(n_1 - 1, n_2 - 1) - x(n_1 + 1, n_2 + 1)$$
$$- x(n_1 - 1, n_2 + 1) - x(n_1 + 1, n_2 - 1)], \tag{10.20}$$

respectively, where

$$l_m(n_1, n_2) = \frac{\sum_{i_1=-1}^{1} \sum_{i_2=-1}^{1} x(n_1 - i_1, n_2 - i_2)}{9}. \tag{10.21}$$

The local-mean adaptive nature of Teager's operators is important since it is tuned to the response of the human visual system and renders the enhancement perceptually uniform according to the Weber's law. As briefly described in Chapter 4, Weber's law states that the just-noticeable brightness difference is proportional to the average background brightness. Because of the characterization of the output of the nonlinear operator as a product of the local mean value and the output of a highpass filter, the overall effect is that of yielding a smaller output in darker areas and therefore reducing the perceivable noise with respect to the highpass filters employed in the linear unsharp masking schemes. However, the direct use of such operators in unsharp masking schemes may still introduce some visible noise depending on the scaling factor λ chosen. This effect can be alleviated by resorting to higher-order polynomial filters.

Higher-Order Polynomial Operators. In order to improve the performances of the conventional unsharp masking scheme, it is necessary to condition the behavior of the highpass filter in order to emphasize only local luminance changes that are due to the true details of the image, avoiding the amplification of variations due to the noise. Higher-order polynomial operators can be used to achieve this result according to two different strategies. The first approach [246,320] is based on a modification of the classic unsharp masking scheme in which a polynomial filter is connected in cascade to the linear highpass filter, as shown in Figure 10.17. In [246], the nonlinear filter is formed as a parallel connection of a linear lowpass filter and a

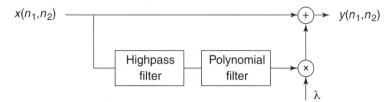

Figure 10.17 A modification to the linear unsharp masking structure using cascade nonlinear operators.

nonlinear operator that takes the form of the Hammerstein model that contains a cubic or an absolute-value quadratic nonlinearity. Since the output of the nonlinear operator is amplitude-dependent, small amplitude variations in the highpass filtered data, which can be attributed with reasonable justification to noise, are smoothed by the lowpass linear component, while large variations, which correspond to relevant spatial edges or details, are further amplified by the polynomial sharpener. The choice of cubic or absolute-value quadratic nonlinearities in the system preserves the sign information during transitions, and thus allows processing transitions from dark to bright and from bright to dark in the same manner.

Remark 10.3 In general, an absolute-value quadratic filter includes terms of the type $x(n_1, n_2)|x(n_1 - m_1, n_2 - m_2)|$. Ramponi and Sicuranza [246] use only products of the type $x(n_1, n_2)|x(n_1, n_2)|$. Such filters can be considered as extensions of the usual Volterra filters in which one or more input samples appear with their absolute value in the expression given for the generic kernel of order p. We refer to these nonlinear filters as *pseudopolynomial* filters.

The second approach [249,252] is based on the use of a control signal obtained from an *edge sensor* that multiplies the output of the highpass filter, as shown in Figure 10.18. When the edge sensor has classified the present filtering mask position as belonging to the neighborhood of an edge, the highpass filter is activated and the unsharp masking operation is performed. On the contrary, when the filtering mask

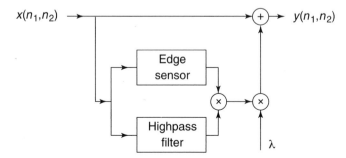

Figure 10.18 A modification to the linear unsharp masking structure using an edge sensor.

covers a uniform part of the image, the unsharp masking action is not activated. Both the highpass filter and the edge sensor can be elementary operators, and act on very small planes of support. In the one-dimensional case, the sharpening component can be defined as

$$z(n) = [x(n-1) - x(n+1)]^2 [2x(n) - x(n-1) - x(n+1)], \qquad (10.22)$$

where the edge sensor is given by the first factor on the right-hand side, and the highpass filter is the Laplacian operator. It is clear that the output of the edge sensor will be large only when the difference $x(n-1) - x(n+1)$ is large enough, that is, when the filtering mask overlaps two objects having different luminances. Moreover, the squaring operation prevents interpreting small luminance variations due to noise as true image details. A two-dimensional expression of the sharpening component is obtained by applying the filtering operation of (10.22) along both vertical and horizontal directions, which results in the following relationship:

$$\begin{aligned}
z(n_1, n_2) = {}& [x(n_1 - 1, n_2) - x(n_1 + 1, n_2)]^2 \\
& \times [2x(n_1, n_2) - x(n_1 - 1, n_2) - x(n_1 + 1, n_2)] \\
& + [x(n_1, n_2 - 1) - x(n_1, n_2 + 1)]^2 \\
& \times [2x(n_1, n_2) - x(n_1, n_2 - 1) - x(n_1, n_2 + 1)] \qquad (10.23)
\end{aligned}$$

It is easy to obtain a corresponding expression of the 3×3 operator involving the diagonal directions.

Several variations of the ideas described above are available in the literature for nonlinear unsharp masking, in particular for further reducing the noise sensitivity. For example, a way of making unsharp masking more robust to noise is to use a more effective edge sensor, such as the Sobel filter and/or insert a lowpass filter in the direct path [250,252].

Example 10.5 An application of the polynomial unsharp masking technique is given here. In Figure 10.19a, a 256×256-pixel portion of the original 8-bit/pixel image *Lena* is shown. The effect of conventional unsharp masking is seen in Figure 10.19b. The linear highpass filter is the 3×3 Laplacian operator

$$\begin{aligned}
z(n_1, n_2) = {}& [4x(n_1, n_2) - x(n_1, n_2 - 1) - x(n_1, n_2 + 1) \\
& - x(n_1 + 1, n_2) - x(n_1 - 1, n_2)]
\end{aligned}$$

and the scaling factor λ is equal to 0.6. The image is much sharper, but noise is clearly visible in uniform areas. With the algorithm based on the Teager's operator of (10.19) the result of Figure 10.19c is obtained. Because of the varying behavior of this operator in dark and bright areas, the effect of noise is less visible where the average luminance is low, while it is quite apparent in bright areas such as in Lena's forehead, cheek, and shoulder. The scaling factor used was $\lambda = \frac{1}{256}$. Finally, Figure

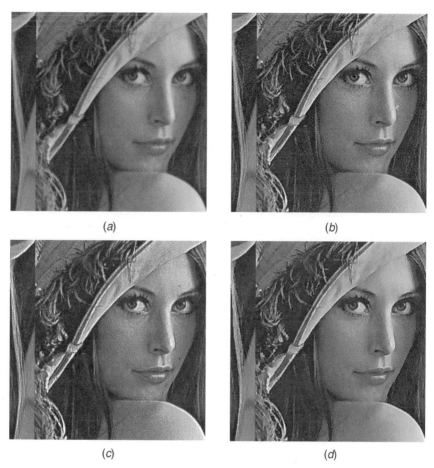

Figure 10.19 Image enhancement with linear and polynomial filters: (*a*) original image; (*b*) image processed with the linear unsharp masking technique; (*c*) image processed with the algorithm based on the Teager operator; (*d*) image processed with the polynomial filter including an edge sensor. (Courtesy of G. Ramponi. Copyright © 1996 SPIE *Journal of Electronic Imaging.*)

10.19*d* shows the result of the unsharp masking using the filter of (10.23), with $\lambda = 0.001$. In this image the enhancement of details is achieved, while the noise effects are negligible.

To test the noise robustness of these operators, the same original image has been spoiled by an additive Gaussian noise of variance 50. The new test image is shown in Figure 10.20*a*. The unsharp masking with the Laplacian filter ($\lambda = 0.55$) and the Teager operator ($\lambda = \frac{1}{256}$) give the images of Figures 10.20*b* and 10.20*c*, respectively. Both images are well sharpened, but noise is appreciable. The result of the unsharp masking using the filter of (10.23), with $\lambda = 0.001$, shown in Figure 10.20*d*, appears to be overall less noisy.

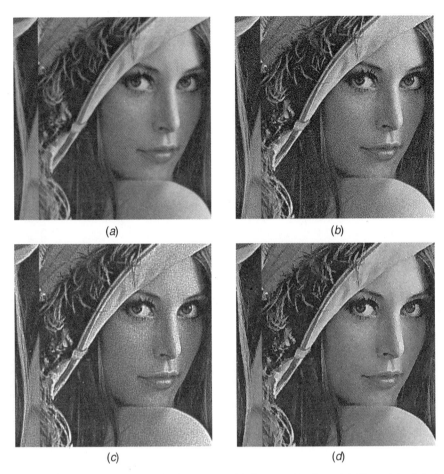

(a)   (b)

(c)   (d)

Figure 10.20 Enhancement of a noisy image with linear and polynomial filters: (a) original noisy image; (b) image processed with the linear unsharp masking technique; (c) image processed with the algorithm based on the Teager operator; (d) image processed with the polynomial filter including an edge sensor. (Courtesy of G. Ramponi. Copyright © 1996 SPIE *Journal of Electronic, Imaging.*)

10.6.3   Processing of Document Images

Translating documents in electronic form is a preliminary task in many activities such as office automation, manipulation of text files on a computer terminal, recognition of handwritten texts, and validation of check signatures. All these topics are also relevant in the area of multimedia applications. A document image is a peculiar image since it is usually acquired using a gray-level scale even though the documents themselves may be binary images. Because the original documents may not be in good condition and the scanning process may often be performed in difficult environments, the digitized images are often noisy and have low contrast.

Most character recognition systems first create two-level images from the gray-scale document images and apply the recognition algorithms to the resulting binary images. Because of the poor quality of the input images, direct binarization using even an appropriately selected threshold function often results in blurred or connected character edges, and this, in turn, results in poor recognition of the characters. The following example describes a quadratic filter introduced by Ramponi and Fontanot [247] for preprocessing document images. Mo and Mathews presented an adaptive version of this preprocessor in [183].

Example 10.6 The input image employed in this example is a digitized version of a typewritten mail address image, and is shown in Figure 10.21*a*. This image was binarized using a thresholding scheme described by Otsu [210], and the resulting

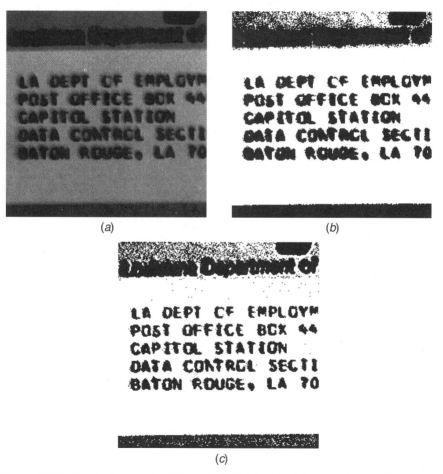

Figure 10.21 Binarization results of Example 10.6: (*a*) original document image; (*b*) result of direct binarization; (*c*) result of quadratic preprocessing and binarization. (Courtesy of S. Mo. Copyright © 1998 IEEE.)

image is shown in Figure 10.21b. The generic quadratic filter employed in the experiments processed nine input samples at each time to find its output as

$$y(n_1, n_2) = \sum_{k_1=-1}^{1} \sum_{k_2=-1}^{1} h_1(k_1, k_2)x(n_1 - dk_1, n_2 - dk_2) + \sum_{k_1=-1}^{1} \sum_{k_2=-1}^{1} \sum_{k_3=-1}^{1} \sum_{k_4=-1}^{1}$$
$$\times h_2(k_1, k_2, k_3, k_4)x(n_1 - dk_1, n_2 - dk_2)x(n_1 - dk_3, n_2 - dk_4),$$

where d is a positive integer number. By choosing d to be 1, 2 or 3, the plane of support of the filter becomes 3×3, 5×5 or 7×7 pixels, respectively, without increasing the complexity of the filter. The choice of d is made based on the thickness of the characters present in the input image. For character sizes similar to those in the input image, $d = 3$ is a good choice. For thinner characters, a smaller value of d should be selected.

Ramponi and Fontanot [247] first normalized the input image so that the normalized image $x_n(n_1, n_2)$ took values in the range [0,1] and then employed the symmetry and isometry conditions as well as additional simplifications in the design of the filter to obtain an easy-to-realize system with input–output relationship

$$y(n_1, n_2) = x_0(n_1, n_2) + \vec{\mathbf{H}}_s^T(n_1, n_2)\mathbf{X}_s(n_1, n_2),$$

where

$$x_0(n_1, n_2) = x_n^2(n_1, n_2) + \frac{1}{8} \sum_{k_1=-1}^{1} \sum_{\substack{k_2=-1 \\ [k_1, k_2] \neq [0,0]}}^{1}$$
$$\times [x_n(n_1 - dk_1, n_2 - dk_2) - x_n^2(n_1 - dk_1, n_2 - dk_2)],$$

and $\mathbf{X}_s(n_1, n_2)$ is a two-element input vector given by

$$\mathbf{X}_s(n_1, n_2) = [x_1(n_1, n_2) \quad x_2(n_1, n_2)]^T,$$

with $x_1(n_1, n_2)$ and $x_2(n_1, n_2)$ defined as

$$x_1(n_1, n_2) = x_n(n_1, n_2) - x_n^2(n_1, n_2) - \frac{1}{8} \sum_{k_1=-1}^{1} \sum_{\substack{k_2=-1 \\ [k_1, k_2] \neq [0,0]}}^{1}$$
$$\times [x_n(n_1 - dk_1, n_2 - dk_2) - x_n^2(n_1 - dk_1, n_2 - dk_2)]$$

and

$$x_2(n_1, n_2) = x_n(n_1, n_2) \sum_{k_1=-1}^{1} \sum_{\substack{k_2=-1 \\ [k_1,k_2] \neq [0,0]}}^{1} x_n(n_1 - dk_1, n_2 - dk_2)$$

$$- [x_n(n_1 - d, n_2) + x_n(n_1 + d, n_2)][x_n(n_1, n_2 - d) + x_n(n_1, n_2 + d)]$$

$$- \tfrac{1}{2}x_n(n_1 - d, n_2 - d)[x_n(n_1 - d, n_2) + x_n(n_1, n_2 - d)]$$

$$- \tfrac{1}{2}x_n(n_1 - d, n_2 + d)[x_n(n_1 - d, n_2) + x_n(n_1, n_2 + d)]$$

$$- \tfrac{1}{2}x_n(n_1 + d, n_2 + d)[x_n(n_1 + d, n_2) + x_n(n_1, n_2 + d)]$$

$$- \tfrac{1}{2}x_n(n_1 + d, n_2 - d)[x_n(n_1 + d, n_2) + x_n(n_1, n_2 - d)],$$

respectively. The coefficient vector $\vec{\mathbf{H}}_s(n_1, n_2)$ also contains two elements denoted by $h_{s,1}(n_1, n_2)$ and $h_{s,2}(n_1, n_2)$ so that

$$\vec{\mathbf{H}}_s(n_1, n_2) = [h_{s,1}(n_1, n_2) \quad h_{s,2}(n_1, n_2)]^T.$$

Ramponi and Fontanot showed [247] that this filter contains a component with lowpass characteristics that reduces the noise present in the input images and another component that serves to enhance the edges in the input images. Figure 10.21c shows the result of binarizing the output of the quadratic filter. We observe that the character edges are much better defined and separated in this figure, implying that character recognition on the image that was preprocessed using the quadratic filter should perform significantly better than recognition performed on the image obtained through the direct binarization process. However, the quadratic processing did not perform as well in the top portion of the image as in the other areas. This was primarily because the statistics of the input image in this region were significantly different from that in other areas, and therefore a spatially invariant preprocessor is not able to perform equally well in all areas of the image. Adaptive algorithms similar to the one described by Mo and Mathews [183] may be required for satisfactory performance on a larger class of images.

10.7 POLYNOMIAL PROCESSING OF IMAGE SEQUENCES

The reasons given for using polynomial filters in two-dimensional image processing applications are valid for applications involving image sequences also. In this section, we describe the applications of polynomial filters in spatial and temporal prediction and interpolation of image sequences.

10.7.1 Prediction of Image Sequences

Differential pulse code modulation and hybrid transform coding are effective methods for reducing the bit rate for transmission and storage of digital image sequences. Both methods use a temporal predictor and subsequent coding of the prediction errors. However, the efficiency of fixed linear predictors is limited because most images deviate significantly from the stationarity and Gaussianity required for optimal performance of fixed linear predictors. Nonlinear predictors for image sequences were introduced in [236,291] to overcome this problem. It was shown experimentally that simple quadratic predictors outperformed linear predictors, especially in areas where motion is present.

The approach in [236] consists of subdividing each frame of the image sequence into small blocks that are then classified as *active* or *inactive* on the basis of the mean-square difference between corresponding blocks in adjacent frames. Each inactive block is transmitted only once. For each active block the coefficients of the quadratic predictor are computed using the least-squares criterion and then the prediction errors are computed and transmitted after coding.

A pixel-by-pixel adaptive quadratic predictor for image sequences is described in [291]. In this method, a normalized LMS algorithm was employed to update the coefficients. As was described for the case of predicting speech signals, the prediction is based on the reconstructed signals in image compression applications also. This is done so that the images can be reconstructed at the receiver without having to transmit the adaptive filter coefficients. Sicuranza and Ramponi [291] also employed a switching strategy between linear and nonlinear predictors. Each pixel was classified as belonging to a smooth area or a detail area such as one containing edges using a measure of the activity in a block surrounding the pixel. The nonlinear predictor was used only in the detail area. In this scheme, additional overhead information regarding which predictor is employed for each pixel must be transmitted to the receiver.

Example 10.7 We consider a three-dimensional predictor with the five-sample support as depicted in Figure 10.22. The test input sequence *Judy* contained 64 frames with gray-level values ranging from 0 to 255 and pictured a woman moving her head actively against a detailed background. The complexity of the predictor is that associated with implementing five linear coefficients and 15 quadratic coefficients. The coefficients were evaluated using the least-squares criterion for contiguous blocks of 16×16 pixels each. Figure 10.23 shows the active part of one of the frames of the input sequence along with scaled maps of the prediction errors between the original and the predicted frame using a fixed linear predictor, the optimum least-squares linear predictor, and the optimum least-squares quadratic predictor. Less detail of the original frame is recognizable in the error map obtained using the quadratic predictor, indicating the ability of the quadratic predictor to better model the image sequence. Experiments described in [236,291] indicate that the nonlinear predictors are especially effective in the moving areas of the image sequence.

The computational complexity is different for the linear and nonlinear predictors in this example. We can choose a linear predictor with a larger support to render its complexity comparable to that of the nonlinear predictor. However, experiments have shown that this does not provide significant improvement in performance over the linear predictor with a smaller support.

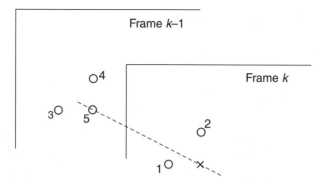

Figure 10.22   Support of the predictor used in Example 10.7.

10.7.2 Interpolation of Image Sequences

Skipping frames at the transmitter and interpolating the skipped frames at the receiver is an attractive method for coding image sequences at low bit rates. Since linear interpolation by temporal filtering performs poorly in moving areas, we need alternate approaches for compensating for the motion of the objects in the scenes during interpolation. One such method is based on the use of a quadratic interpolation filter. The rationale for using nonlinear interpolators is similar to that employed for nonlinear processing for prediction and enhancement of images. A typical approach for interpolation of image sequences is as follows [83]. Assume that the lth frame must be interpolated from the $(l-1)$th and $(l+1)$th frames of the reconstructed image sequence at the receiver. Each frame of the image sequence is subdivided into small blocks and the mean-square error between each block in the lth frame and the corresponding block in the $(l-1)$th frame is computed. If the computed measure of the difference image is smaller than a given threshold, the block does not contain any significant changes. Consequently, no interpolation is required, and the block in the lth frame can be replaced during interpolation with the reconstructed block in the previous frame. If the threshold is exceeded, the optimum least-squares quadratic estimator is computed from the corresponding blocks in the $(l-1)$th and $(l+1)$th frames. The results obtained on typical videoconference sequences in which every other frame is skipped have confirmed the ability of the polynomial interpolators to outperform linear interpolators. Experiments have shown that quadratic interpolators are able to compensate in some sense for motion present in the scenes.

Example 10.8   Figure 10.24 displays the support of the quadratic filter employed in this example for interpolation of an image sequence. The pixels in the skipped frame l are interpolated using two pixels chosen from the same line in the $(l-1)$th frame and two pixels from the same column in the $(l+1)$th frame. Therefore, the nonlinear interpolator has four coefficients in the linear component and 10 coefficients in the homogeneous quadratic component. The test sequence used is the progressive TV

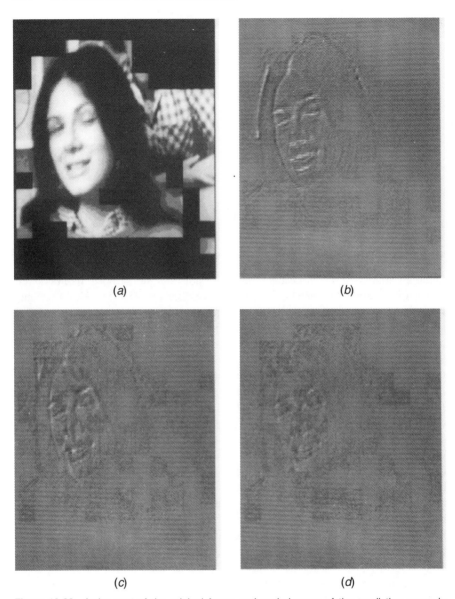

(a) (b)

(c) (d)

Figure 10.23 Active part of the original frame and scaled maps of the prediction errors in Example 10.7: (a) active part of the original frame; (b) fixed linear prediction; (c) optimum least-squares linear prediction; (d) optimum least-squares quadratic prediction. (Copyright © 1987 IEE *Electronic Letters.*)

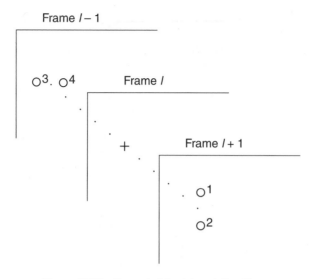

Figure 10.24   Support of the interpolation filters.

sequence *Judy.* Each image of 256×256 pixels is subdivided into blocks of 16×16 pixels each. Figure 10.25 shows the original frame that was interpolated, along with the appropriately amplified versions of the interpolation errors when three different filters were employed for the interpolation. The interpolators employed were a heuristic fixed interpolator that simply averaged the four input pixels to the interpolator, the optimum block least-squares linear interpolator and the optimum block least-squares quadratic interpolator. The least-squares filters were evaluated and used only in active blocks. The error maps of the fixed and linear interpolators contain information on the moving edges of the original image, while the linear–quadratic filter gives a smoother error map. The mean-square errors for the interpolated frame are equal to 20.4, 12.3, and 8.4 for the fixed, linear, and quadratic interpolators, respectively. These facts demonstrate the superiority of the nonlinear approach when moderate motion is present in the scene.

10.8   A BRIEF OVERVIEW OF SOME MORE APPLICATIONS

In this section, we overview a few more applications of polynomial systems briefly. Because of the nature of our descriptions, no experimental results are provided, but appropriate references are made to publications available in the literature.

10.8.1   Adaptive Noise Cancellation

There are many applications in which a signal of interest is corrupted by undesirable interference signals. Examples include biomedical signals such as electroencephalo-

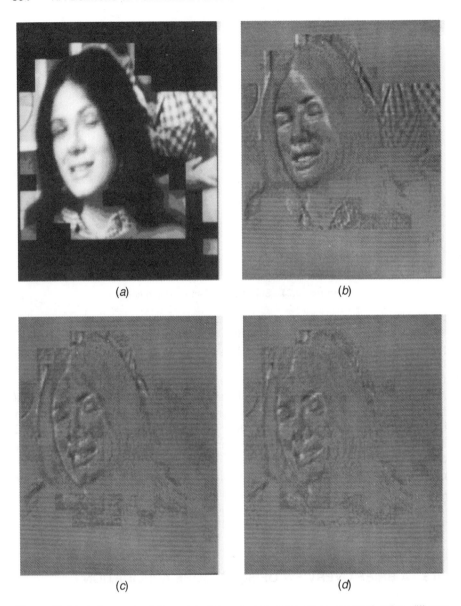

Figure 10.25   Active part of the original image and error maps obtained using different interpolators in Example 10.8: (*a*) active part of the original frame to be interpolated; (*b*) fixed linear interpolator with four coefficients equal to 0.25; (*c*) optimum least-squares linear interpolator; (*d*) optimum least-squares quadratic interpolator. (Copyright © 1987 IEE *Electronic Letters.*)

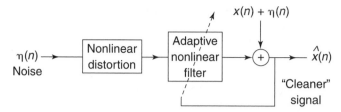

Figure 10.26 An adaptive nonlinear noise cancellation system.

gram (EEG) and electrocardiogram (EKG) that are corrupted by line voltage due to poor isolation, and speech signals that may be corrupted by aircraft or automobile engine noise. In many such situations, a reference signal that is related to the interfering signal is available, and it can be used to reduce the distortions in the signal of interest. Adaptive noise cancellation is a technique that uses an adaptive filter to process the reference noise in order to estimate the actual noise that has corrupted the signal. The estimated noise is then subtracted from the noisy signal to cancel the noise from the signal. Figure 10.26 shows the block diagram of an adaptive noise canceller. This system employs the adaptive filter to estimate the noisy signal using the reference noise. If the noise is independent of the signal of interest, the output of the adaptive filter will be an estimate of the distortions introduced into the signal. Consequently, the estimation error in the system is a cleaner version of the signal of interest. In many situations, the noise corrupts the signal after undergoing some nonlinear transformation. In other situations, the reference signal available to the adaptive filter may have been subject to some nonlinear transformations. In both these situations, it is appropriate to employ nonlinear system models in the adaptive filter.

Stapleton and Bass describe an example of adaptive noise cancellation for speech signals corrupted by engine noise in an aircraft cockpit in [297]. The nonlinearities in the problem arose from a moderately hard-clipped circuit that distorts the reference signal available to the noise canceller. The Hammerstein model was used in this work to model the nonlinearity. An iterative technique was employed to estimate the coefficients of the noise cancellation system. The nonlinear noise canceller produced improvements up to 5 dB in the signal-to-noise ratio over a linear noise canceller in this work. Another example in which the Volterra series representation was used to perform adaptive nonlinear noise cancellation is given in [58].

10.8.2 Stabilization of Moored Vessels in Random Sea Waves

Modeling the nonlinear dynamic behavior of offshore structures subject to random sea wave excitations [111,116] is a typical example of a situation in which the quadratic model is appropriate, while the linear model is not. The relationship between the random sea wave excitation and the corresponding response of a moored vessel is characterized by the presence of the so-called low-frequency drift oscillations (LFDO). The LFDO in response to sea wave fluctuations can be explained with the presence of a second-order force proportional to the square of the instantaneous wave heights. Due to this square-law relationship, the drift oscillations generate sum and difference frequencies in response to the spectral components of the incident sea wave. This process may result in the generation of low-frequency components within the resonant bandwidth of the vessel-mooring system. These components will be amplified, and therefore the second-order force may produce the dominant component in the response of the moored vessel. The physics behind the generation of the LFDO indicates that it can be accurately modeled using a second-order Volterra system. A simplified model for the LFDO is given by a square-law device followed by a linear filter [117]. This model corresponds to the Hammerstein model described in Chapter 2. Techniques for stabilization of moored vessels employing quadratic system models have been described in [116,117].

10.8.3 Signal Detection

The detection problem applied to signals consists, in general, of discriminating between the following two hypotheses:

1. The vector **X** contains the useful part (the *signal*) as well as the undesired part (the *noise*).
2. The vector **X** contains noise only.

A number of criteria have been applied to the design of optimal filters for detection. In this context, quadratic filters have been investigated by Balakrishnan [11] and Baker [9], respectively, in problems related to the noise-in-noise detection for sonar applications. Other contributions to the detection and estimation problem using quadratic or linear–quadratic filters can be found in [51,77,226,227]. Taft conducted an analysis of the linear–quadratic filters for improving signal detection, and its extension to more than one dimension and reported on applications to image segmentation and texture discrimination [308]. A first result of this analysis is that for Gaussian input signal probability distributions the quadratic and linear terms could be considered decoupled for optimization under the mostly used criteria. Moreover, solutions for non-Gaussian input distributions can be obtained by solving systems of linear equations. Simulations showed that for equal means but separated variances on the two hypotheses (1 and 2), the homogeneous quadratic filter performed best, while for the case of separated means and separated variances, the linear–quadratic filter performed best.

10.8.3.1 Energy Detection Energy detectors are typically employed to detect the presence of components in the received signals with energy patterns significantly different From those in the noise in which they are embedded.

A conventional linear energy detector consists of a linear, time-invariant filter followed by a magnitude-square operator and possibly a smoothing filter [6,7]. Such systems are commonly employed in both time-varying and time-invariant operating environments. The output of such a system without the smoothing operator can be expressed as

$$y_L(n) = \left| \sum_{m=0}^{M-1} h_L(m)x(n - m) \right|^2, \tag{10.24}$$

where $x(n)$ is the input signal and $h_L(n)$ is the unit impulse response signal of the linear filter. If both the input signal and the impulse response are real, (10.24) represents a quadratic filter given in the form of the Wiener model with input–output relationship

$$y_L(n) = \sum_{m_1=0}^{M-1} \sum_{m_2=0}^{M-1} x(n - m_1)x(n - m_2)h_L(m_1)h_L(m_2), \tag{10.25}$$

where the product $h_L(m_1)h_L(m_2)$ indicates that the quadratic kernel is separable.

Other simple versions of energy detectors have been derived from a general quadratic kernel of the form [7]

$$y(n) = \sum_{m=0}^{M-1} h(m)x(n + m)x(n - m). \tag{10.26}$$

This detector is a sparse quadratic operator. This system may also be considered as an extension of the Teager operator. It has been studied and used for nonlinear analysis of speech [7], and useful results have been obtained in problems such as preprocessing of speech for pitch tracking and burst recognition. In another application [6], quadratic detectors with 128 coefficients were designed using standard FIR filter design techniques to provide a bank of bandpass nonlinear filters to detect energy components in various frequency bands.

10.8.3.2 *Texture Discrimination* Another application of truncated Volterra filters involves image processing for texture discrimination. According to the well-known work of Julesz [106], textures that have identical second-order statistics but different higher-order statistics cannot be easily discriminated by a human observer unless some recognizable clusters or lines are present in the image. Polynomial operators process their input signals on the basis of higher-order statistics of the signals, and consequently, they can be employed to separate a mixture of two textures. Example 4.8 provided a demonstration of the ability of a simple quadratic filter to discriminate between two synthetic textures. Ramponi described the design of such a system [245]. A related work is that due to Ramponi and Sicuranza [243].

Several other formal approaches based on the exploitation of higher-order statistics of the input signals are available in the literature. Nonminimum phase linear modeling of non-Gaussian random fields using higher-order statistics has been investigated [305,311,315,318]. Second- and third-order statistics of random fields have been used in [87] to characterize real and synthetic texture images using AR and ARMA models. The problem of image texture analysis in the presence of noise has been examined using higher-order statistics in [299,300].

10.9   EXERCISES

10.1   One difficulty with the predistortion technique for loudspeaker linearization is that the estimate of the inverse filter will be biased if the measured loudspeaker output contains noise. Another approach [76] for linearization of loudspeaker nonlinearities is shown in Figure 10.27. Suppose that the predistorter employs a homogeneous quadratic system model with N-sample memory and we wish to adapt the coefficients of the predistorter using a stochastic gradient method that attempts to reduce the mean-square value of the estimation error defined as the difference between the measured value of the loudspeaker output and a delayed version of the input to the predistorter. To accomplish this, we require knowledge of the input–output relationship of the loudspeaker, which may be estimated using an adaptive filter as shown in the figure. Model the input–output relationship using a homogeneous quadratic filter with M-sample memory. Assuming slow variations of the coefficients of the adaptive filters, derive a set of update equations for the coefficients of the predistortion filter. You may substitute the estimated values of the parameters of the loudspeaker characteristics in place of the true values of the parameters.

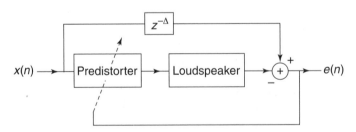

Figure 10.27   An approach for adaptive linearization of loudspeaker nonlinearities.

10.2   *Computing Assignment*: Consider a nonlinear echo canceler for a digital subscriber loop modem. Assume that the echo path is affected by a nonlinearity so that its input–output relation is described by the Wiener model

$$y(n) = w(n) + \beta w^2(n),$$

where $w(n)$ is the output of a linear system described by the impulse response

$$h(i) = \exp(-0.8i)$$

for $0 \leq i \leq N - 1$. Fix $N = 8$ and assume an input signal described by two bits. Using the LMS adaptation algorithm, find the residual error of the echo canceler as a function of the number of iterations for the following FIR configurations:

1. Linear taps and $\beta = 0$
2. Linear and nonlinear taps and $\beta = 0.1$
3. Linear and nonlinear taps and $\beta = 1.0$.

Consider the two cases where the received signal power from the far-end is zero or -30 dB. Comment on the results obtained.

10.3 Show that the quadratic filter defined by the coefficient matrix $\mathbf{H}_2$ in Example 10.4 eliminates a single impulse present inside the 3×3-pixel block under consideration at each spatial location.

10.4 *Computing Assignment:* Verify by computer simulations that the homogeneous quadratic filter

$$
\mathbf{H}_2 =
\begin{bmatrix}
1 & \frac{1}{2} & 0 & \frac{1}{2} & 0 & -\frac{1}{2} & 0 & -\frac{1}{2} & -1 \\
\frac{1}{2} & 0 & \frac{1}{2} & 0 & 0 & 0 & -\frac{1}{2} & 0 & -\frac{1}{2} \\
0 & \frac{1}{2} & 1 & -\frac{1}{2} & 0 & \frac{1}{2} & -1 & -\frac{1}{2} & 0 \\
\frac{1}{2} & 0 & -\frac{1}{2} & 0 & 0 & 0 & \frac{1}{2} & 0 & -\frac{1}{2} \\
0 & 0 & 0 & 0 & 0 & 0 & 0 & 0 & 0 \\
-\frac{1}{2} & 0 & \frac{1}{2} & 0 & 0 & 0 & -\frac{1}{2} & 0 & \frac{1}{2} \\
0 & -\frac{1}{2} & -1 & \frac{1}{2} & 0 & -\frac{1}{2} & 1 & \frac{1}{2} & 0 \\
-\frac{1}{2} & 0 & -\frac{1}{2} & 0 & 0 & 0 & \frac{1}{2} & 0 & \frac{1}{2} \\
-1 & -\frac{1}{2} & 0 & -\frac{1}{2} & 0 & \frac{1}{2} & 0 & \frac{1}{2} & 1
\end{bmatrix}
$$

is able to perform edge extraction and Gaussian noise reduction.

10.5 An extension of the two-dimensional Teager algorithms is given by the following expression:

$$
y(n_1, n_2) = 2x^2(n_1, n_2) - x(n_1 - 2, n_2)x(n_1 + 2, n_2)
$$
$$
- x(n_1, n_2 - 2)x(n_1, n_2 + 2).
$$

Assume that the average value of the input signal is large compared to its standard deviation, and show that this filter approximates a local-mean weighted two-dimensional bandpass filter

$$y(n_1, n_2) = l_m(n_1, n_2)[4x(n_1, n_2) - x(n_1, n_2 - 2) - x(n_1, n_2 + 2)$$
$$- x(n_1 + 2, n_2) - x(n_1 - 2, n_2)],$$

where $l_m(n_1, n_2)$ is an estimate of the local mean.

10.6 *Computing Assignment:* A way of making unsharp masking more robust to noise is to use a more effective edge sensor in the scheme of Figure 10.18. Show by suitable computer simulations that beneficial effects can be obtained using as edge sensor the Sobel operator. Repeat the simulations for different variances of an additive Gaussian noise.

10.7 *Computing Assignment:* Another way to reduce the noise effects in the unsharp masking techniques is to insert a lowpass filter in the direct input–output branch in Figure 10.18. Show, by suitable computer simulations, the effects obtained using lowpass filters with different bandwiths for different levels of additive Gaussian noises.

10.8 *Computing Assignment:* The quality of the results of the enhancement techniques is usually subjectively evaluated. An objective measurement can be derived by defining a *detail variance* and a *background variance* for each pixel of the processed image [252]. The procedure to define such measurements is as follows:

(**a**) Compute the local variance on a small mask, typically 3×3 pixels, centered on each pixel.

(**b**) Form two classes of pixels by defining a threshold that separates the pixel considered as belonging to the background from those belonging to details or contours of the image.

(**c**) Compute the detail variance (DV) by averaging the local variances of all the pixels representing a detail according to the previous classification. Similarly, compute the background variance (BV). Then, a high-quality enhanced image should be characterized by high values of DV and low values of BV.

Show by experiments that

(**a**) The absolute values of DV and BV depend on the threshold chosen, but their relative values can be used to perform comparisons among images obtained with different operators.

(**b**) Characterize with the DV–BV metric the results obtained on a noisy test image using the unsharp masking schemes, including the pure Laplacian filter and the Laplacian filter triggered by an edge sensor.

(**c**) Repeat the measurements with a lowpass filter inserted in the direct branch of the unsharp masking scheme.

10.9 This exercise considers the derivation of an adaptive preprocessor for document image processing [183]. Figure 10.28 shows a model for

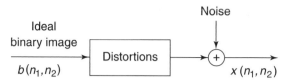

Figure 10.28 A model for the generation of scanned document images.

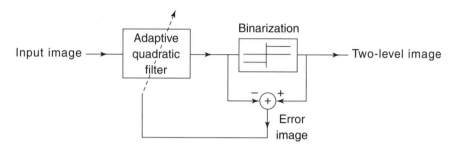

Figure 10.29 Adaptive preprocessor for document images.

generating the scanned document image starting from an ideal document image that takes on only one of two values. This model is similar to models of binary communication channels. A method to compensate for the distortions caused during the scanning process is to input the scanned image $x(n_1, n_2)$ into an adaptive filter as shown in Figure 10.29. The output $y(n_1, n_2)$ of the adaptive filter is mapped into one of two values 0 or 2α, depending on whether $y(n_1, n_2)$ is smaller than or larger than a positive threshold value α. The objective of the adaptive filter is to make its output as close to 0 or 2α as possible. Let $y'(n_1, n_2)$ denote the output of the binarization process described above. Develop a stochastic gradient adaptive filter that attempts to reduce the squared error $(y(n_1, n_2) - y'(n_1, n_2))^2$ at each location. Assume a quadratic system model for the adaptive filter.

10.10 *Computing Assignment*: Design a linear-quadratic predictor using a pixel-by-pixel adaptation algorithm based on the normalized LMS criterion. Choose the three-dimensional causal support, shown in Figure 10.30, formed by the three previous pixels in the spatial and temporal directions. The predictor is formed with three linear coefficients and six quadratic coefficients.

(**a**) Test the performances on a sequence that contains edges, details, and relevant motion.

(**b**) Compare the results with those obtained with an adaptive linear predictor working on the same support by computing the corresponding PSNRs and some significant prediction error map.

(**c**) Compare the results also from a subjective point of view, by analyzing the part of the sequence containing spatial details and/or moving objects.

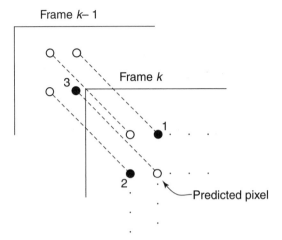

Figure 10.30   Support of the prediction filter.

(**d**) On the basis of the previous results, design a linear predictor for smooth areas and a linear–quadratic predictor for moving or detailed areas. You will need to incorporate in the adaptive predictor a method for classifying the pixels into those that belong to smooth or detail areas.

10.11   *Computing Assignment:* Consider the interpolation of image sequences and assume the same environment as in Example 10.8. Your task involves reconstructing a skipped image using the corresponding blocks in the previous and subsequent frames. Assume that each image is subdivided into blocks of 16×16 pixels. Compare the results obtained using the optimum block least-squares linear interpolator and the optimum block least-squares quadratic interpolator on the support shown in Figure 10.24.

11

SOME RELATED TOPICS AND
RECENT DEVELOPMENTS

The previous chapters described in detail the fundamentals of analysis, design, and realization of polynomial systems, and several applications of polynomial signal processing. In this concluding chapter, we describe some related or ongoing work at the time this book was written. In many ways, the ideas discussed in the following sections involve somewhat nontraditional approaches to the analysis of nonlinear systems, and applications in some nontraditional areas. It is clear that many more such developments will occur in the near future, in terms of both understanding the characteristics of nonlinear systems and new uses for such systems. The different sections of this chapter are connected only by the themes of nonlinearity in the system models and the novelty of the ideas discussed in them. The discussions are necessarily brief because of the emerging nature of most of these ideas. Even so, it is our hope that these discussions will provide the reader with a feel for the directions the research and applications in this field will take in the near future.

11.1 NONLINEAR CLASSIFIERS

Applications of polynomial functions are common in the field of pattern recognition. Linear classifiers have limited capabilities since they work adequately well only in linearly separable forms of pattern discrimination problems. In general, the discriminator boundaries in most pattern classification problems are hyperplanes in multidimensional spaces. An example of the inadequacy of linear discriminant functions can be seen by considering very simple problems such as the two-bit parity checking problem. In the two-bit parity checking problem, the discriminant function should be able to identify if the two bits present at its input are identical or of different polarity. No single straight line that can achieve this separation of the input patterns exists on the two-dimensional plane. The discriminator boundaries for this problem assume hyperbolic or elliptical shapes. Complex decision boundaries such as those for this example can be achieved by using nonlinear discriminant functions.

A practical approach for designing nonlinear discriminants consists of using a preprocessing stage that computes an appropriate nonlinear function of the input samples. Very often, the nonlinearity employed is a polynomial function. An example of the application of polynomial discriminant functions in classification problems is described by Widrow and Lehr [330]. In their work, a polynomial preprocessor computes the actual inputs to an adaptive linear combiner called *Adaline*. The block diagram of a system that contains appropriately weighted linear and quadratic terms for the two-bit parity check problem is shown in Figure 11.1. The thresholding condition that discriminates between the two parities occurs when the output s of this system is equal to zero. The thresholding condition for the circuit of Figure 11.1 is given by

$$s = w_0 + w_1 x_1 + w_2 x_2 + w_{11} x_1^2 + w_{12} x_1 x_2 + w_{22} x_2^2 = 0, \qquad (11.1)$$

where w_i and $w_{i,j}$ are the weights applied to the input signals. With proper choice of these weights, the separating boundaries in the pattern space can be established so that the two-bit parity problem can be solved.

The objective of the circuit described above was to identify the boundaries of the region that separate the different classifications. An alternative approach is to perform a nonlinear transformation of the input samples into a domain in which the classification can be achieved using linear discriminants. A schematic block diagram of a structure for accomplishing this objective is shown in Figure 11.2. In this case also, the nonlinearity is often described by a polynomial expansion of the elements of the input samples [255]. For a generic problem involving N samples and up to p-th order of nonlinearities, the nonlinearity is described using a Volterra series-like expansion given by

$$y = w_0 + \sum_{m_1=1}^{N} w_{m_1} x_{m_1} + \sum_{m_1=1}^{N} \sum_{m_2=1}^{N} w_{m_1 m_2} x_{m_1} x_{m_2} + \cdots$$

$$+ \sum_{m_1=1}^{N} \sum_{m_2=1}^{N} \cdots \sum_{m_p=1}^{N} w_{m_1 m_2 \cdots m_p} x_{m_1} x_{m_2} \cdots x_{m_p}. \qquad (11.2)$$

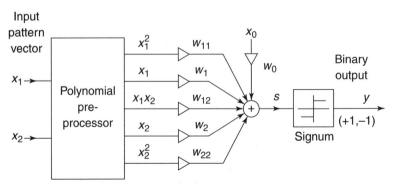

Figure 11.1   A linear combiner with a polynomial preprocessor.

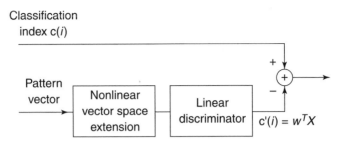

Figure 11.2 Classification using a nonlinear transformation and a linear discriminator.

The optimum weights can be derived by applying any of the commonly employed numerical search procedures or an adaptive training procedure. The main advantage of this approach is that the error surface for the optimization problem is a convex function of the coefficients in most practical situations, and therefore, there exists a single global minimum for this problem.

11.1.1 Higher-Order Neural Networks

Artificial neural networks are often used to solve problems in the field of nonlinear classification. A typical example of an artificial neural network is the multilayer perceptron network in which each layer is formed by forward connections of *neurons* consisting of a set of linear adaptive combiners followed by a memoryless nonlinearity. The network is then trained using an adaptation technique such as the backpropagation technique [264]. An interesting area of recent research involves a generalized backpropagation technique that can be used to train higher-order neural networks that incorporate a polynomial preprocessor for each neuron [82]. A single-layer perceptron in this class can be represented by the input–output relationship

$$y_i = \theta\left(w_{i,0} + \sum_{m_1=1}^{N} w_{i,m_1} x_{m_1} + \sum_{m_1=1}^{N}\sum_{m_2=1}^{N} w_{i,m_1 m_2} x_{m_1} x_{m_2} + \cdots \right.$$
$$\left. + \sum_{m_1=1}^{N}\sum_{m_2=1}^{N}\sum_{m_3=1}^{N} w_{i,m_1 m_2 m_3} x_{m_1} x_{m_2} x_{m_3} + \cdots \right),$$
(11.3)

where $\theta(\cdot)$ is a memoryless nonlinearity. Perceptrons with nonlinear links of this form as shown in Figure 11.3 are simple examples of higher-order neural networks. Such networks are useful in pattern recognition problems where the invariance of the output to changes in the position and the size of the input patterns is required [108,222,296]. The invariance can be achieved by applying suitable constraints to the weights w of a perceptron of the type shown in Figure 11.3 [129].

Neural networks are also extensively used in the fields of controls and communications. It has been shown that functions realized by hidden layer nodes in a multilayer network can be realized or approximated by a single-layer network,

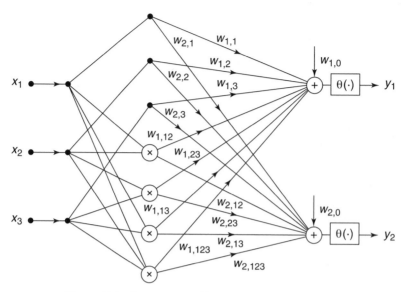

Figure 11.3 A perceptron with a polynomial preprocessor.

provided a sufficient number of nonlinear combinations of the input elements are incorporated into the nodes [217]. Polynomial combinations are often used in such networks. Functional nonlinearities that are different from polynomial nonlinearities may also be used in such systems. Higher-order neural networks have been applied in various problems from the identification and control of nonlinear systems to the equalization of nonlinear transmission channels [48,49,266].

11.1.2 Polynomial Invariants

The location, orientation, and size of an object in an image are irrelevant in pattern recognition problems that attempt to recognize the presence of that object in the input image. Therefore, the pattern recognition systems should ideally be invariant to transformations involving spatial and temporal translation, rotation, and scaling. One method for accomplishing this objective is to design systems that operate on invariant features in the input images. Invariants to a particular transformation are image characteristics that remain unchanged when an object in the scene is subjected to that transformation. As we saw in Chapter 10, many image processing tasks require nonlinear processing. Polynomial filters and transforms designed to extract invariants for group actions such as translations and rotations have been recently developed [36,79].

The characterization of systems through equivalence classes of the input signals is the basis for deriving polynomial invariants. Let us consider a system $\mathbf{S}$ that performs a vector-valued mapping of input vectors $\mathbf{x}$ belonging to an appropriate vector space to a space of real or complex-valued M-dimensional vectors; that is, the operator $\mathbf{S}$ is described by M scalar function denoted by

$$\mathbf{S}(\mathbf{x}) = [s_1(\mathbf{x}), \ldots, s_M(\mathbf{x})]^T. \tag{11.4}$$

Two vectors $\mathbf{x}_1$ and $\mathbf{x}_2$ are *equivalent* inputs for $\mathbf{S}$ if they are mapped by the system to the same output, i.e., $\mathbf{S}(\mathbf{x}_1) = \mathbf{S}(\mathbf{x}_2)$. Equivalence classes of input vectors can be generated by an operator g for which $\mathbf{x}_1 = g(\mathbf{x}_2)$ implies that $\mathbf{S}(\mathbf{x}_1) = \mathbf{S}(\mathbf{x}_2)$. In other words, the system $\mathbf{S}$ is invariant with respect to the transformation g of the input data. Let G define a set of transformations, and let g be a member of this set. We assume that G is a finite set of order $|G|$, where $|G|$ is the number of elements in the set. The basic idea for constructing polynomial operators that are invariant to the transformations in G is to start with a monomial of the form

$$f(\mathbf{x}) = x(0)^{b_0} x(1)^{b_1} \cdots x(N-1)^{b_{N-1}}, \tag{11.5}$$

where $b_0 + b_1 + \cdots + b_{N-1} \leq |G|$, and then to average $f(g(\mathbf{x}))$ over all possible members of the set G to get

$$A[f](\mathbf{x}) = \frac{1}{|G|} \sum_{g \in G} f(g(\mathbf{x})). \tag{11.6}$$

This process, called *group averaging*, is described in detail by Schulz-Mirbach [276].

When $f(\mathbf{x})$ is a monomial as in (11.5) and g is a linear transformation, the group average $A[f](\mathbf{x})$ is also a polynomial and can therefore be written as

$$A[f](\mathbf{x}) = \sum_i \mathbf{H}_i[\mathbf{x}], \tag{11.7}$$

where the operator $\mathbf{H}_i$ is the component of the output corresponding to the polynomial of degree i, i.e., $\mathbf{H}_i[\mathbf{x}]$ takes the form

$$\mathbf{H}_i[\mathbf{x}] = \sum_{m_1} \cdots \sum_{m_i} H_i(m_1, \ldots, m_i) \mathbf{x}(m_1) \cdots \mathbf{x}(m_i), \tag{11.8}$$

where $h_i(m_1, \ldots, m_i)$ denotes the coefficients of the ith-order term of the polynomial expansion in (11.7). These derivations suggest that we can interpret the expression in (11.7) as a truncated Volterra series expansion of $A[f](\mathbf{x})$, implying that we have a method for designing truncated Volterra filters that are invariant to certain features in the input images. Burkhardt and Schulz-Mirbach have described details of the design methodology, properties of the transformations so designed, and some applications [36].

11.2 EVALUATION OF LOCAL INTRINSIC DIMENSIONALITY

The local intrinsic dimensionality of a signal is a measure of the degrees of freedom used by the signal in a neighborhood around the location of interest. This concept is closely related to investigation of the information reduction strategies employed in biological vision systems, and to the function of specialized neurons in the mammalian visual cortex [209,340]. For two-dimensional images, it is possible to distinguish among intrinsically zero-dimensional signals (I0D signals), namely constant signals, intrinsically one-dimensional signals (I1D signals) such as straight lines and edges, and intrinsically two-dimensional signals (I2D signals) such as corners, line ends, junctions, and spots. Figure 11.4 illustrates the basic types of intrinsic dimensionality that occur in two-dimensional images. The image is segmented into I0D, I1D, and I2D signals in this figure. We can observe in this example that there is a gradual decrease in the probability of occurrence of the I0D signals to the probability of occurence of the I2D signals. There also exists a hierarchy of intrinsic dimensionality in terms of redundancy, since knowledge about the I2D signals can be used to predict the I1D and I0D signals. This type of statistical hierarchy is an inherent property of natural images. From an information theoretic point of view, these properties imply that the information content of a local image feature increases systematically with the corresponding degree of intrinsic dimensionality. Moreover, a substantial portion of the information contained in the

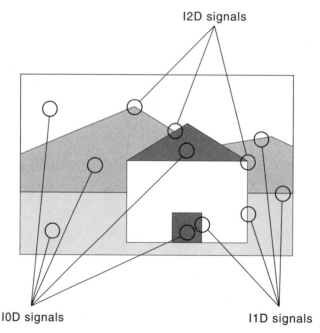

Figure 11.4 Basic types of intrinsic dimensionality in a simple image. (Adapted from Krieger and Zetzsche [128].)

image can be recovered from the knowledge of its I2D regions. The basic ideas related to the concept of local intrinsic dimensionality are described by Krieger and Zetsche in [128] and references cited therein. The authors of [128] also show that the I2D signals can be detected using quadratic operators.

The detectors for I2D signals can be designed on the basis of the ability of certain quadratic operators to be insensitive to I0D and I1D signals, and thus respond exclusively to I2D signals. A large class of such I2D operators can be obtained using structures similar to the one showed in Figure 11.5. Each branch of the structure is formed by a linear differentiator in the horizontal or vertical directions, connected in cascade with a linear two-dimensional filter with frequency response determined by the nature of the I2D signal to be detected. The outputs of the two branches are multiplied together and the difference signal of the two outputs are evaluated. The rationale for using quadratic Volterra filters is that it is possible to design quadratic filters with varying levels of sensitivity to I2D signals with different orientations and at the same time able to attenuate or eliminate I0D and I1D signals present in the filter's plane of support. Because of the orientation properties of such filters, it is possible to derive different types of isotropic and anisotropic I2D operators that are able to extract local I2D features from images. While linear filters can typically separate low frequencies from high frequencies, quadratic filters can discriminate between I1D and I2D signals. These polynomial filters enable selective processing of the intrinsic dimensionality and may be viewed as nonlinear operators that exploit the redundancies present in two-dimensional signals.

The intrinsic two-dimensional operators are closely related to certain classes of visual neurons. In particular, they are closely related to neurons with orientation-selectivity properties. Although the basic phenomenon of I2D selectivity has been known for more than three decades in neurophysiology, modeling of these cells is a

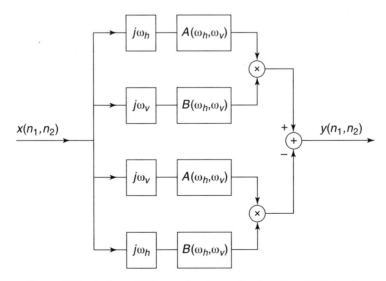

Figure 11.5 A class of I2D operators insensitive to I0D and I1D signals.

recent development. The approach in [128] suggests a framework for signal modeling using the concept of intrinsic dimensionality, and also offers a strategy for the systematic neurophysiological measurement of the properties of I2D selective neurons. This approach may lead to the development of improved criteria for classifying different types of visual neurons.

The basic principle of local intrinsic dimensionality can be extended to higher-dimensional signals. For example, in the processing of image sequences, nonlinear intrinsically three-dimensional operators can be used to identify significant motion discontinuities. This and related issues are open research problems at present.

11.3 RATIONAL FILTERS

A major advantage of Volterra filters is that they are nonlinear with respect to their input but linear with respect to their coefficients. Consequently, they possess many interesting properties that make them attractive in several applications. However, as we have seen in the previous chapters, the computational complexity associated with higher-order Volterra systems is often very high. Furthermore, since the output of the Volterra systems tends to become large as the magnitudes of the input signal become large, the utility of truncated Volterra filters is limited to a finite dynamic range of input signals. Rational functions have recently been proposed as an alternative to truncated Volterra filters in several signal processing applications including detection and estimation [89] and image processing [248]. The input–output relationship of a rational filter is the ratio of two polynomials in its input signal. Rational filters have at least the same approximation capabilities as do truncated Volterra filters since the latter can be considered as a special case of the former. In many applications, the rational filters are able to approximate nonlinear systems to a desired level of accuracy with a lower complexity than truncated Volterra filters. They also often possess better extrapolation capabilities. Furthermore, it has been shown [89] that it is possible to estimate the parameters of a rational filter by simple extensions of linear adaptive algorithms.

11.3.1 Estimation of the Parameters of a Rational Filter

Let the input–output relationship of a rational filter have the form

$$y(n) = \frac{a_0 + \sum\limits_{m_1=0}^{N-1} a_{1,m_1} x(n-m_1) + \sum\limits_{m_1=0}^{N-1}\sum\limits_{m_2=0}^{N-1} a_{2,m_1,m_2} x(n-m_1)x(n-m_2) + \cdots}{b_0 + \sum\limits_{m_1=0}^{N-1} b_{1,m_1} x(n-m_1) + \sum\limits_{m_1=0}^{N-1}\sum\limits_{m_2=0}^{N-1} b_{2,m_1,m_2} x(n-m_1)x(n-m_2) + \cdots}.$$

$$(11.9)$$

Assuming that we have symmetric kernels for the numerator and denominator polynomials, the representation described in (11.9) is unique up to a constant factor in the numerator and denominator coefficients. Without loss of generality, we may fix the value of a specific coefficient in the system model to be one in (11.9). We will assume in the sequel that $b_0 = 1$. To be useful in practical applications, the rational filter must be formed with finite polynomials of low orders. To see how the coefficients of a rational filter can be estimated using extensions of procedures for estimating the parameters of a linear system model, let us denote the input–output relationship of the system model by $N_\alpha[x(n)]/(1 + D_\beta[x(n)])$, where $N_\alpha[x(n)]$ and $1 + D_\beta[x(n)]$ are numerator and denominator polynomials with orders α and β, respectively. Also, let $d(n)$ represent the desired response signal of the system. We formulate the estimation problem using a least-squares cost function as that of minimizing

$$\mathcal{E} = \sum_{i=0}^{M} \left| d(i) - \frac{N_\alpha[x(i)]}{1 + D_\beta[x(i)]} \right|^2, \tag{11.10}$$

where M is the total number of input samples available for the minimization of the global error $\mathcal{E}$. We can solve this optimization problem by evaluating the derivatives of (11.10) with respect to the coefficients of the filter, equating them to zero, and then solving for the unknown coefficients to obtain the desired solution. Unfortunately, this formulation requires the solution of a system of nonlinear equations. An alternative is to change the error function $\mathcal{E}$ using an approach similar to the equation error method for adaptive IIR filters. Assuming that the noise contamination in the desired response signal is small, we can approximate the estimation error as

$$e(i) = d(i) - (N_\alpha[x(i)] - d(i)D_\beta[(x(i)]). \tag{11.11}$$

Substituting this approximation in (11.10) results in an optimization problem that is linear in the coefficients. The coefficients can be estimated using a batch process or a variation of algorithms such as the LMS and RLS adaptive filters. The details of such procedures are given in the work by Leung and Haykin [89], where the adaptive rational filter is then applied to the estimation of direction of arrival in array processing, and to a radar detection problem.

11.3.2 Rational Filters for Image Processing

Rational filters have recently been proposed as an alternative to truncated Volterra filters for solving many image processing tasks. In this subsection, we discuss the derivation of a rational filter for noise smoothing. We will see that this system is relatively simple to implement, and that it has the capability of providing reasonably robust performance. We describe the rational filter starting from a simple symmetric, one-dimensional, linear filter with three coefficients. Let the output $y(n)$ of this system be given by

$$y(n) = w[x(n-1) + x(n+1)] + (1 - 2w)x(n). \tag{11.12}$$

This operator has unit gain at zero frequency. For values of the parameter w in the range $0 < w \leq \frac{1}{4}$, the system exhibits lowpass characteristics with monotonically decreasing magnitude response with increasing frequency. For $w = \frac{1}{4}$, the gain is zero at π radians per sample, and for higher values of w the zero-gain frequency moves below π radians per sample. This type of filter, with an appropriately chosen value for w, can be used to remove high-frequency noise components from a given signal. However, as we saw in Chapters 4 and 10, linear lowpass filters create unacceptable blurring effects in many image processing applications. We can reduce such effects by adapting the parameter w in such a way that the lowpass action occurs primarily in smooth areas of the input image and little or no smoothing occurs in edge areas. Such adaptation can be achieved by using suitable edge sensors such as a Laplacian operator at each position of the filter mask. One of the simplest edge sensors computes the quantity $[x(n-1) - x(n+1)]^2$ at each value of n. We can devise a rational filter that incorporates this edge sensor to control the smoothing action due to the filter in (11.12). Several different formulations of the rational filter have been described in the literature [248,251]. One such formulation results in a rational filter with input–output relationship

$$y(n) = \frac{w[x(n-1) + x(n+1)]}{wk[x(n-1) - x(n+1)]^2 + 1} + \left(1 - \frac{2w}{wk[x(n-1) - x(n+1)]^2 + 1}\right)x(n).$$

(11.13)

Combining the two terms in this representation, we can express the input–output relationship as the ratio of a third-order polynomial and a second-order polynomial as given by

$$y(n) = \frac{x(n-1) + x(n+1) + x(n)\{k[x(n-1) - x(n+1)]^2 + 1/w - 2\}}{k[x(n-1) - x(n+1)]^2 + 1/w}.$$

(11.14)

The parameter k is used to control the characteristics of the filter, and always takes positive values. For $k = 0$, we obtain the linear filter in (11.12). As $k \to \infty$, the filter becomes an identity system, and has no effect on the input images. For intermediate values of k, the term $k[x(n-1) - x(n+1)]^2$ in the denominator controls the smoothing action of the filter as a function of the dynamics of the input signal. In the presence of an edge, $k[x(n-1) - x(n+1)]^2$ takes larger values and therefore the filter will have reduced smoothing effect. On the other hand, in smooth areas, this term takes small values, allowing the filter to provide significant smoothing action at such locations.

Two-dimensional operators can be derived by applying the operator in (11.14) on a 3×3 mask in the $0°$, $45°$, $90°$, and $135°$ directions in order to provide orientation sensitivity to the smoothing action of the filter. Such operators have been successfully applied in detail preserving smoothing of noisy images [248,251]. The rational filter approach can also be used to derive other operators for edge-sensitive image interpolation [44], interpolation of the DC components of JPEG [Joint Photographic

Expert Group (ISO)]-coded images [253], and contrast enhancement [254]. Extension of rational filters to three dimensions and its applications to image sequence processing can also be devised.

The approach to image processing by means of rational filters shares many characteristics with recently proposed techniques based on the concept of *anisotropic diffusion* [223]. These relations could offer new insights for the design of efficient rational filters for various image processing tasks. This and related topics are subjects of active research at this time.

11.4 BLIND IDENTIFICATION AND EQUALIZATION OF NONLINEAR CHANNELS

Most of the existing work in the field of the nonlinear system modeling involves identification methods based on the availability of the input *and* the output signals. Blind identification and equalization of nonlinear systems, where the input signal to the system to be identified is not directly available, is an emerging area of advanced research. Blind equalization approaches are of interest in digital communications since no training input and no interruption of the transmission are necessary to equalize the channel. Therefore, for channels exhibiting multipath phenomena, time-varying nonlinearities, or high data rates, blind methods are attractive. An example of a blind Volterra system identification problem is the method described by Prakryia and Hatzinakos [231]. This method constrains the system model to be an LNL interconnection of two linear subsystems separated by a memoryless nonlinearity. In addition, the input sequence is required to be circularly symmetric, and the first subsystem can be fully identified only if it is of minimum phase, and the identification of the linear subsystem is based on the higher-order polyspectrum.

A general approach for blind equalization and identification of nonlinear single-input multiple-output truncated Volterra filters has been presented in [80]. While impossible with a single output, multiple outputs make it possible to blindly deconvolve several truncated Volterra channels simultaneously. This approach requires only that the input sequence satisfy a persistent excitation condition [200] and that the channel transfer matrix have full row rank. The input is allowed to be deterministic or random with unknown distribution. Unlike the method described by Prakryia and Hatzinakos [231], the approach described by Giannakis and Serpedin [80] does not require knowledge of the higher-order statistics of the input to the equalizer for its implementation. Interestingly, the nonlinear channels are equalized with linear FIR filters in this method. This fact can be justified intuitively since the vector equalizer can be seen as a beamformer that, thanks to its diversity, is capable of nulling the nonlinearities and equalizing the linear part. The results of several simulations involving experiments in the equalization of magnetic recording channels, and real and complex communication channels have been presented in [80]. However, a number of questions about the capabilities of the method are still open research issues. Such issues include analytic performance evaluation, the selection of the optimum equalizer delay, the determination of the order of the

Volterra model from output data only, and the capabilities of the algorithm in the presence of significant amounts of channel noise. The complexity of the algorithm is relatively high, and therefore, additional work on efficient on-line versions are also required before this method can be used in practice.

11.5   NEW MATHEMATICAL FORMULATIONS

Several recent developments have introduced new ways of representing and analyzing polynomial systems. In what follows, we describe two approaches that have the potential to make the design, analysis, or realization of polynomial systems easier.

11.5.1   A Diagonal Coordinate Representation for Volterra Filters

A diagonal coordinate representation for Volterra filters was developed by Raz and Van Veen in [256]. This representation allows a truncated Volterra system to be characterized as a bank of linear filters whose coefficients are defined by the "diagonal" entries of its kernels. Such a characterization allows the derivation of efficient realizations of Volterra systems using fast convolution techniques. The diagonal coordinate representation also offers insights into the relationship between the characteristics of the output signal in the frequency domain and the filter parameters.

Let the output of a causal, time-invariant, finite-memory, homogeneous pth-order Volterra filter be given by

$$y_p(n) = \sum_{m_1=0}^{N-1} \sum_{m_2=m_1}^{N-1} \cdots \sum_{m_p=m_{p-1}}^{N-1} h_p(m_1, m_2, \ldots, m_p) \prod_{i=1}^{p} x(n - m_i). \qquad (11.15)$$

Let us introduce the following change of coordinates into this expression:

$$s = m_1$$
$$s + r_{i-1} = m_i; \quad i = 2, \ldots, p. \qquad (11.16)$$

Using the definitions given above, (11.15) can be written as

$$y_p(n) = \sum_{s=0}^{N-1} \sum_{r_1=0}^{N-1-s} \sum_{r_2=r_1}^{N-1-s} \cdots$$
$$\sum_{r_{p-1}=r_{p-2}}^{N-1-s} h_p(s, s + r_1, \ldots, s + r_{p-1}) x(n - s) \prod_{i=1}^{p-1} x(n - s - r_i). \qquad (11.17)$$

If the pth-order kernel is represented as a sampled hypercube of the same order, the coordinates corresponding to the indices m_i are the Cartesian coordinates while the coordinates corresponding to the indices r_i and s are aligned along the diagonals of the hypercube. We define a diagonal to be any line that is parallel to the main diagonal of the hypercube. The coordinate value r_i denotes the distance of the first element of the corresponding diagonal from the first element of the main diagonal along the ith Cartesian axis. Consequently, the index r_i in the diagonal coordinate system defined by the preceding transformation defines the diagonal on which a particular entry is located. Similarly, the index s specifies the position of this entry along that diagonal. By changing the order of summations in (11.17) so that the summation over s is performed first, we get

$$y_p(n) = \sum_{r_1=0}^{N-1} \sum_{r_2=r_1}^{N-1} \cdots$$

$$\sum_{r_{p-1}=r_{p-2}}^{N-1} \sum_{s=0}^{N-1-r_{p-1}} h_p(s, s+r_1, \ldots, s+r_{p-1}) x(n-s) \prod_{i=1}^{p-1} x(n-s-r_i). \quad (11.18)$$

Let us now define a set of new signals as

$$v_{r_1,\ldots,r_{p-1}}(n) = x(n) \prod_{i=1}^{p-1} x(n-r_i). \quad (11.19)$$

Let us also define linear filters corresponding to the preceding input signals with the impulse response signals given by

$$g_{r_1,\ldots,r_{p-1}}(s) = h_p(s, s+r_1, \ldots, s+r_{p-1}). \quad (11.20)$$

Substituting (11.19) and (11.20) in (11.18) transforms the input–output relationship for the Volterra system as

$$y(n) = \sum_{r_1=0}^{N-1} \sum_{r_2=r_1}^{N-1} \cdots \sum_{r_{p-1}=r_{p-2}}^{N-1} \sum_{s=0}^{N-1-r_{p-1}} g_{r_1,\ldots,r_{p-1}}(s) v_{r_1,\ldots,r_{p-1}}(n-s). \quad (11.21)$$

Therefore, the output of the homogeneous pth-order Volterra filter can be expressed as the sum of one-dimensional convolutions involving the sequences $v_{r_1,\ldots,r_{p-1}}(n)$. Furthermore, we can obtain a one-dimensional frequency-domain representation for the Volterra system by taking the discrete-time Fourier transform of (11.21). This procedure yields

$$Y(\omega) = \sum_{r_1=0}^{N-1} \sum_{r_2=r_1}^{N-1} \cdots \sum_{r_{p-1}=r_{p-2}}^{N-1} V_{r_1,\ldots,r_{p-1}}(\omega) G_{r_1,\ldots,r_{p-1}}(\omega), \quad (11.22)$$

where $V_{r_1,\ldots,r_{p-1}}(\omega)$ and $G_{r_1,\ldots,r_{p-1}}(\omega)$ are the discrete-time Fourier transforms of $v_{r_1,\ldots,r_{p-1}}(n)$ and $g_{r_1,\ldots,r_{p-1}}(n)$, respectively. This representation is particularly useful since the frequency content of the output signal is directly related to the frequency response of the diagonal elements of the kernel. Efficient implementations of Volterra filters for processing carrier-based input signals using the diagonal coordinate systems were derived using fast convolution techniques [256].

11.5.2 *V*-Vector Algebra for Adaptive Volterra Filters

Almost all of the adaptive filtering algorithms employing polynomial system models described in Chapter 7 are derived by extending the conventional adaptive linear filtering algorithms with the help of a linear, multichannel interpretation of the system model. Such extensions are straightforward for the derivation of the LMS and conventional RLS adaptive filters. The adaptive algorithms for polynomial filters are obtained in these cases by simply replacing the input vector for the linear system models with the corresponding input vector for the polynomial model. However, derivation of computationally efficient RLS adaptive filters such as the fast RLS Volterra filter derived in Chapter 7 requires additional operations involving manipulations of intermediate variables in the algorithms. Such manipulations are not always trivial. A novel approach called *V-vector algebra* that is suitable for the development of fast adaptive algorithms equipped with polynomial system models as direct extensions of linear adaptive techniques was presented in [43]. In this approach, the input vector and associated variables are replaced by an entity called the *V*-vector that is a nonrectangular matrix. For example, the *V*-vector formed with the input samples of a homogeneous second-order Volterra filter with *N*-sample memory is an array in which the nonlinear combinations of the input samples are arranged as

$$
\begin{matrix}
x^2(n) & x^2(n-1) & \cdots & & x^2(n-N+1) \\
 & x(n)x(n-1) & \cdots & & x(n-N+2)x(n-N+1) \\
 & & \vdots & & \\
 & & x(n)x(n-N+1) & &
\end{matrix}
\quad . \tag{11.23}
$$

The advantage of this structure of the input "vector" is that the new samples in the *V*-vector at time *n* enter from the left, and the samples that are discarded appear along the right end of the *V*-vector. Consequently, no additional manipulations are required on the input vector to identify the samples that are discarded at each time. Similar to the definition of the *V*-vector, it is possible to introduce the *V-matrix* which is a *V*-vector whose elements are again *V*-vectors. The relevant definitions, properties, and operations with *V*-vectors and *V*-matrices have been described in [43]. Using such a formalism, it is possible to derive fast and numerically stable adaptive polynomial filters in a relatively direct manner.

11.6 AN APPLICATION OF VOLTERRA FILTERS IN COMPUTER ENGINEERING

Memory caches are employed in designs of computer architechtures and large programs to facilitate faster operation of computer systems and programs. The basic idea is that the computer keeps track of the use of the pages in its memory addresses by a program, and saves in a cache the most frequently used pages. In this way, accessing the most frequently used pages can be performed efficiently. A memory management algorithm is employed to predict the sequences of memory pages addressed by the process under execution, and to move the pages through various levels of memory hierarchy. The sequence of page numbers addressed by the process may be regarded as a signal or time series, and proper memory management involves accurate prediction of this time series. Recent experimental work [191] has shown that this signal can be modeled as a nonlinear time series. Consequently, it is possible to predict the access rate of pages in the computer memory with the help of a nonlinear predictor. This information can then be used to design the replacement and allocation policies for the memory cache. A memory management algorithm that utilizes polynomial models to predict which pages will be requested in the near future has been described in [191]. The predicted information is combined with a demand paging policy, thus leading to an effective virtual memory management system. The experimental results provided by Mumolo and Bernardis [191] showed that the actual form of the predictor can be described as a sparse linear–quadratic filter. This work also showed that the use of the nonlinear predictor reduced the occurrences of "faults," or instances when the desired page was not in the cache, by about 10% over conventional memory management algorithms.

Appendix A

PROPERTIES OF
KRONECKER PRODUCTS

A *Kronecker product,* also known as a *direct product* or a *tensor product,* is a concept having its origin in group theory with successful applications in various fields of matrix theory.

The Kronecker product $\mathbf{A} \otimes \mathbf{B}$ of an $(L_1 \times M_1)$-element matrix $\mathbf{A}$ with an $(L_2 \times M_2)$-element matrix $\mathbf{B}$ is an $(L_1 L_2 \times M_1 M_2)$-element matrix given by

$$
\mathbf{A} \otimes \mathbf{B} = \begin{bmatrix}
a_{0,0}\mathbf{B} & a_{0,1}\mathbf{B} & \cdots & a_{0,M_1-1}\mathbf{B} \\
a_{1,0}\mathbf{B} & a_{1,1}\mathbf{B} & \cdots & a_{1,M_1-1}\mathbf{B} \\
\vdots & \vdots & \vdots & \vdots \\
a_{L_1-1,0}\mathbf{B} & a_{L_1-1,1}\mathbf{B} & \cdots & a_{L_1-1,M_1-1}\mathbf{B}
\end{bmatrix}. \tag{A.1}
$$

It has $L_1 M_1$ blocks, and the (i,j)th block is the matrix $a_{i,j}\mathbf{B}$ formed with $L_2 \times M_2$ elements. For example, let

$$
\mathbf{A} = \begin{bmatrix} a_{0,0} & a_{0,1} \\ a_{1,0} & a_{1,1} \end{bmatrix} \tag{A.2}
$$

and

$$
\mathbf{B} = \begin{bmatrix} b_{0,0} & b_{0,1} \\ b_{1,0} & b_{1,1} \end{bmatrix}. \tag{A.3}
$$

Then

$$
\mathbf{A} \otimes \mathbf{B} = \begin{bmatrix}
a_{0,0}b_{0,0} & a_{0,0}b_{0,1} & a_{0,1}b_{0,0} & a_{0,1}b_{0,1} \\
a_{0,0}b_{1,0} & a_{0,0}b_{1,1} & a_{0,1}b_{1,0} & a_{0,1}b_{1,1} \\
a_{1,0}b_{0,0} & a_{1,0}b_{0,1} & a_{1,1}b_{0,0} & a_{1,1}b_{0,1} \\
a_{1,0}b_{1,0} & a_{1,0}b_{1,1} & a_{1,1}b_{1,0} & a_{1,1}b_{1,1}
\end{bmatrix}. \tag{A.4}
$$

We briefly review some properties and rules for Kronecker products. The proofs and further discussions can be found, for example, in Brewer [27] and Graham [85].

419

Multiplication by a Scalar. For every scalar constant α

$$\mathbf{A} \otimes \alpha\mathbf{B} = \alpha(\mathbf{A} \otimes \mathbf{B}). \tag{A.5}$$

Distributive Property. The Kronecker product is distributive with respect to addition:

$$(\mathbf{A} + \mathbf{B}) \otimes \mathbf{C} = \mathbf{A} \otimes \mathbf{C} + \mathbf{B} \otimes \mathbf{C} \tag{A.6}$$

and

$$\mathbf{A} \otimes (\mathbf{B} + \mathbf{C}) = \mathbf{A} \otimes \mathbf{B} + \mathbf{A} \otimes \mathbf{C}. \tag{A.7}$$

Associative Property. The Kronecker product is associative:

$$\mathbf{A} \otimes (\mathbf{B} \otimes \mathbf{C}) = (\mathbf{A} \otimes \mathbf{B}) \otimes \mathbf{C}. \tag{A.8}$$

Transposition Rule

$$(\mathbf{A} \otimes \mathbf{B})^T = \mathbf{A}^T \otimes \mathbf{B}^T. \tag{A.9}$$

Mixed-Product Rule

$$(\mathbf{A} \otimes \mathbf{B})(\mathbf{C} \otimes \mathbf{D}) = \mathbf{A}\mathbf{C} \otimes \mathbf{B}\mathbf{D} \tag{A.10}$$

provided that the dimensions of the matrices are compatible.

Inversion Rule. If $\mathbf{A}^{-1}$ and $\mathbf{B}^{-1}$ exist, then

$$(\mathbf{A} \otimes \mathbf{B})^{-1} = \mathbf{A}^{-1} \otimes \mathbf{B}^{-1} \tag{A.11}$$

This list is not exhaustive. For a compendium of rules and properties of the Kronecker products, see the article by Brewer [27] and Graham's book [85].

Appendix B

LU DECOMPOSITION OF SQUARE MATRICES

Consider a symmetric $N \times N$ matrix $\mathbf{H}$ of rank $r \leq N$ and whose first r minors are different from zero, i.e.,

$$D_k = \det \begin{bmatrix} h(1,1) & h(1,2) & \cdots & h(1,k) \\ h(2,1) & h(2,2) & \cdots & h(2,k) \\ \vdots & \vdots & & \vdots \\ h(k,1) & h(k,2) & \cdots & h(k,k) \end{bmatrix} \neq 0 \tag{B.1}$$

for $k = 1, 2, \ldots, r$. The condition on $\mathbf{H}$ expressed in (B.1) can always be satisfied using appropriate permutations of its rows and columns. The LU decomposition of the matrix $\mathbf{H}$ involves expressing it in the form

$$\mathbf{H} = \sum_{i=1}^{r} d_i \mathbf{L}_i \mathbf{L}_i^T, \tag{B.2}$$

where the vector $\mathbf{L}_i$ contains N elements and the scalar multipliers d_i are real numbers.

The vectors $\mathbf{L}_i$ are given by the following expression:

$$\mathbf{L}_i = [0 \quad \cdots \quad 0 \quad 1 \quad l_i(i+1) \quad l_i(i+2) \quad \cdots \quad l_i(N)]^T; \quad i = 1, 2, \ldots, r \tag{B.3}$$

where

$$l_k(q) = \frac{H\begin{pmatrix} 1 & 2 & \cdots & k-1 & q \\ 1 & 2 & \cdots & k-1 & k \end{pmatrix}}{D_k} \tag{B.4}$$

for $q = k+1, \ldots, N$ and $k = 1, 2, \ldots, r$. In (B.4), $H\begin{pmatrix} i_1 & i_2 & \cdots & i_r \\ k_1 & k_2 & \cdots & k_r \end{pmatrix}$ is the determinant of a minor of order r of $\mathbf{H}$ defined as

$$H\begin{pmatrix} i_1 & i_2 & \cdots & i_r \\ k_1 & k_2 & \cdots & k_r \end{pmatrix} = \det \begin{bmatrix} h(i_1, k_1) & \cdots & h(i_1, k_r) \\ \vdots & & \vdots \\ h(i_r, k_1) & \cdots & h(i_r, k_r) \end{bmatrix} \tag{B.5}$$

for $1 \le i_1 < i_2 < \cdots < i_r$ and $1 \le k_1 < k_2 < \cdots < k_r$. Finally, the scalar multipliers are given by the expressions

$$d_1 = D_1; \qquad d_2 = \frac{D_2}{D_1}; \qquad \cdots \qquad d_r = \frac{D_r}{D_{r-1}}, \tag{B.6}$$

thus completing the computation of the parameters of the LU decomposition of the matrix $\mathbf{H}$.

Appendix C

LINEAR LATTICE PREDICTORS

Figure C.1 displays the basic structure of a linear lattice predictor of order three. Given a wide sense stationary input signal $x(n)$, a lattice predictor of order N computes the backward prediction error signals $b_0(n), b_1(n), \ldots, b_N(n)$ and the forward prediction error signals $f_0(n), f_1(n), \ldots, f_N(n)$ in an order-recursive manner. In other words, the lattice predictor computes the kth-order forward and backward prediction errors using the $(k-1)$th-order forward and backward prediction errors, and $\rho(k)$, the coefficient of the lattice predictor at the kth stage, as

$$f_k(n) = f_{k-1}(n) - \rho(k)b_{k-1}(n-1) \tag{C.1}$$

and

$$b_k(n) = b_{k-1}(n-1) - \rho(k)f_{k-1}(n), \tag{C.2}$$

respectively, where the prediction error signals at the zeroth stage are defined to be $b_0(n) = x(n)$ and $f_0(n) = x(n)$.

For any appropriate choice of optimization criterion, the kth-order forward prediction error is defined as the optimum estimation error for the problem of linearly predicting $x(n)$ using $x(n-1), x(n-2), \ldots, x(n-k)$. In the same manner, the kth-order backward prediction error is defined as the optimum estimation error

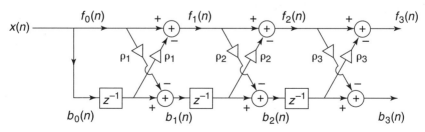

Figure C.1 A linear lattice predictor.

for the problem of linearly predicting $x(n - k)$ using $x(n), x(n - 1), \ldots,$ $x(n - k + 1)$. Let $a_{l,k}$, for $l = 1, 2, \ldots, k$ represent the optimum direct form coefficients of the kth-order forward predictor. Similarly, let $s_{l,k}, l = 0,$ $1, \ldots, k - 1$ represent the optimum direct-form coefficients of the kth-order backward predictor. For wide sense stationary input signals and the minimum mean-square error optimization criterion, it can be shown that the forward and backward predictor coefficients satisfy the mirror image symmetry $a_{l,k} = s_{k-l,k}$. The lattice predictor shown in Figure C.1 is an alternate realization that implements both the forward and backward predictors in one structure. The coefficient $\rho(i)$ is known as the *reflection coefficient* of the ith stage. It is relatively straightforward to show that the backward prediction error signals are mutually orthogonal, and that the set $\{b_0(n), b_1(n), \ldots, b_N(n)\}$ form an orthogonal basis set for the set $\{x(n),$ $x(n - 1), \ldots, x(n - N)\}$.

Given the $(N + 1) \times (N + 1)$-element autocorrelation matrix $\mathbf{R}_{xx}$ of the signal $x(n)$, we can efficiently evaluate the reflection coefficients using the Levinson–Durbin algorithm [90]. This algorithm is tabulated in Table C.1. We recall that wide sense stationarity of the input signal implies that the autocorrelation matrix is Toeplitz. In Table C.1, the variable $r_{xx}(k)$ denotes the autocorrelation of $x(k)$ at lag k. The variable σ_k^2 in the table denotes the mean-square value of the forward prediction error at the output of the kth stage. Using the symmetry of the forward and backward predictor coefficients, we can show that the mean-square value of the kth-order backward prediction error signal is also given by σ_k^2. Procedures for converting from the direct-form to the lattice predictor structure and vice versa can also be derived

TABLE C.1   The Levinson–Durbin Algorithm

Given $\mathbf{R}_{xx}$: $(N + 1) \times (N + 1)$ autocorrelation matrix (Toeplitz);
   Nth-order of prediction

Initialization

$$\rho_1 = \frac{r_{xx}(1)}{r_{xx}(0)}$$

$$a_{1,1} = \rho_1$$

$$\sigma_1^2 = r_{xx}(0)(1 - \rho_1^2)$$

Main iteration
   Do for $k = 2, 3, \ldots, N$

$$\rho_k = \frac{r_{xx}(k) - \sum_{l=1}^{k-1} a_{l,k-1} r_{xx}(k - l)}{\sigma_{k-1}^2}$$

$$a_{k,k} = \rho_k$$

$$a_{l,k} = a_{l,k-1} - \rho_k a_{k-l,k-1}; \qquad l = 1, 2, \ldots, k - 1$$

$$\sigma_k^2 = \sigma_{k-1}^2(1 - \rho_k^2)$$

---

from the Levinson–Durbin algorithm. The equations describing the conversion procedures are given in Tables C.2 and C.3.

TABLE C.2 Transformation of Reflection Coefficients to Direct-Form Predictor Coefficients

---

Given ρ_l; $l = 1, 2, \ldots, N$; Reflection Coefficients of order up to N

Initialization

$$a_{1,1} = \rho_1$$

Main iteration

  Do for $k = 2, 3, \ldots, N$

  $a_{k,k} = \rho_k$

  $a_{l,k} = a_{l,k-1} - \rho_k a_{k-l,k-1};$ $l = 1, 2, \ldots, k-1$

TABLE C.3 Transformation of Direct-Form Predictor Coefficients to Reflection Coefficients

---

Given $a_{l,L}$; $l = 1, 2, \ldots, N$; Nth-order predictor coefficients

Initialization

$$a_{1,1} = \rho_1$$

Main iteration

  Do for $k = N, N-1, \ldots, 2$

  $a_{l,k-1} = \dfrac{a_{l,k} + \rho_k a_{k-l,k}}{(1 - \rho_k^2)};$ $l = 1, 2, \ldots, k-1$

  $\rho_{k-1} = a_{k-1,k-l}$

REFERENCES

1. L. Agarossi, S. Bellini, C. Beoni, and P. Migliorati, "Nonlinear effects in high density optical recording," *Proc. Int. Workshop on HDTV-96,* Los Angeles, California, October 8–9, 1996.

2. O. Agazzi, D. G. Messerschmitt, and D. A. Hodges, "Nonlinear echo cancellation of data signals," *IEEE Trans. Commun.,* vol. COM-30, pp. 2421–2433, November 1982.

3. A. Alper, "A consideration of the discrete Volterra series," *IEEE Trans. Automatic Control,* vol. AC-10, no. 3, pp. 322–327, July 1965.

4. J. Amorocho and A. Brandsetter, "Determination of nonlinear functional response functions in rainfall-runoff processes," *Water Resources Research,* vol. 7, no. 5, pp. 1087–1101, 1971.

5. J. Astola and P. Kuosmanen, *Fundamentals of Nonlinear Digital Filtering,* CRC Press, Boca Raton, New York, 1997.

6. L. Atlas, J. Fang, P. Loughlin, and W. Music, "Resolution advantages of quadratic signal processing," *Proc. SPIE Int. Symp., Conf. 1566,* San Diego, California, July 21–26, 1991.

7. L. Atlas and J. Fang, "Quadratic detectors for general nonlinear analysis of speech," *Proc. IEEE Int. Conf. Acoustics, Speech and Signal Processing,* San Francisco, California, March 1992.

8. H. K. Baik and V. J. Mathews, "Adaptive bilinear lattice filters," *IEEE Trans. Signal Process.,* vol. 41, no. 6, pp. 2033–2046, June 1993.

9. C. R. Baker, "Optimum quadratic detection of a random vector in Gaussian noise," *IEEE Trans. Commun. Technol.,* vol. 14, pp. 802–805, December 1966.

10. A. V. Balakrishnan, "On a characterization of processes for which optimal mean-square systems are of specified form," *IEEE Trans. Inform. Theory,* vol. IT-6, September 1960.

11. A. V. Balakrishnan, "Computing algorithms for detection problems," *Proc. Symp. Computer Processing in Communications,* Polytechnic Press, New York, pp. 603–615, 1969.

12. H. Barker and S. Ambati, "Nonlinear sampled-data system analysis by multidimensional z-transform," *Proc. IEE,* vol. 119, pp. 1407–1413, 1972.

13. J. Barrett, "The use of functionals in the analysis of nonlinear physical systems," *J. Electron. and Control,* vol. 15, pp. 567–615, 1963.

14. J. S. Bendat, *Nonlinear System Analysis and Identification,* Wiley-Interscience, New York, 1990.

15. S. Benedetto, E. Biglieri, and S. Daffara, "Modeling and performance evaluation of nonlinear satellite links— a Volterra series approach," *IEEE Trans. Aerospace Electron. Syst.,* vol. AES-15, pp. 494–507, 1979.

16. S. Benedetto and E. Biglieri, "Nonlinear equalization of digital satellite channels," *IEEE J. Select. Areas Commun.,* vol. SAC-1, pp. 57–62, January 1983.

17. S. Benedetto, E. Biglieri, and V. Castellani, *Digital Transmission Theory,* Prentice Hall, Englewood Cliffs, New Jersey, 1987.

18. E. Biglieri, "Theory of Volterra processors and some applications," *Proc. IEEE Int. Conf. Acoustics, Speech and Signal Processing,* Paris, France, May 1982, pp. 294–297.

19. E. Biglieri, A. Gersho, R. D. Gitlin, and T. L. Lim, "Adaptive cancellation of nonlinear inter-symbol interference for voiceband data transmission," *IEEE J. Select. Areas Commun.,* vol. SAC-2, pp. 765–777, September 1984.

20. E. Biglieri, S. Barberis, and M. Catena, "Analysis and compensation of nonlinearities in digital transmission systems," *IEEE J. Select. Areas Commun.,* vol. SAC-6, pp. 42–51, January 1988.

21. E. Biglieri, E. Chiaberto, G. P. Maccone, and E. Viterbo, "Compensation of nonlinearities in high-density magnetic recording channels," *IEEE Trans. Magn.,* vol. 30, no. 6, pp. 5079–5086, November 1994.

22. S. A. Billings, "Identification of nonlinear systems—a survey," *IEE Proc.,* vol. 127, pt. D, no. 6, pp. 272–285, November 1980.

23. P. Bondon, M. Benidir, and B. Picinbono, Sur l'inversibilité, de la matrice de corrélation d'un vecteur aléatoire à composantes polynomiales, *Colloque GRETSI,* Juan-les-Pins, France, pp. 13–16, 1991.

24. P. Bondon, "Polynomial estimation of the amplitude of a signal," *IEEE Trans. Info. Theory,* vol. 40, no. 3, pp. 960–965, May 1994.

25. T. Bose and M.-Q. Chen, "BIBO stability of the discrete bilinear system," *Digital Signal Process.: A Review J.,* vol. 5, no. 3, 1995.

26. T. Bose and M.-Q. Chen, "Conjugate gradient method in adaptive bilinear filtering," *IEEE Trans. Signal Process.,* vol. 43, no. 6, pp. 1503–1508, June 1995.

27. J. W. Brewer, "Kronecker products and Matrix calculus in system theory," *IEEE Trans. Circuits Syst.,* vol. CAS-25, pp. 772–781, 1978.

28. M. B. Brilliant, *Theory of the Analysis of Nonlinear Systems,* Technical Report No. 345, Research Laboratory of Electronics, MIT., Cambridge, MA, 1958.

29. D. R. Brillinger, "An introduction to polyspectra," *Ann. Math. Statist.,* vol. 36, pp. 1351–1374, 1965.

30. D. R. Brillinger, "The identification of polynomial systems by means of higher order spectra," *J. Sound Vibrat.,* vol. 12, pp. 301–313, 1970.

31. D. R. Brillinger, J. Guckenheimer, P. E. Guttorp, and G. Oster, "Empirical modelling of population time series data: the case of age and density dependent vital rates," *Lect. Math Life Sci., Am. Math. Soc.,* vol. 13, pp. 65–90, 1980.

32. R. W. Brockett, "Volterra series and geometric control theory," *Automatica,* vol. 12, pp. 167–176, 1976.

33. C. Bruni, G. Di Pillo, and G. Koch, "On the mathematical models of bilinear systems," *Ricerche di Automatica,* vol. 2, no. 1, pp. 11–26, 1971.

34. C. Bruni, G. Di Pillo, and G. Koch, "Bilinear systems: An appealing class of 'nearly linear systems' in theory and applications," *IEEE Trans. Aut. Control,* vol. AC-19, no. 4, pp. 334–348, August 1974.

35. B. K. Burcher, W. G. Price, and R. E. D. Bishop, "Uses of functional analysis in ship dynamics," *Proc. Royal Soc., Series A,* vol. 332, no. 15, pp. 23–35, 1973.

36. H. Burkhardt and H. Schulz-Mirbach, "A contribution to nonlinear system theory," *Proc. IEEE Workshop on Nonlinear Signal and Image Processing,* Halkidiki, Greece, June 1995.

37. A. Bush, *Some Techniques for the Synthesis of Nonlinear Systems,* Technical Report No. 441, Research Laboratory of Electronics, MIT, Cambridge, MA, 1966.

38. G. Carayannis, D. Manolakis, and N. Kalouptsidis, "A fast sequential algorithm for least-squares filtering and prediction," *IEEE Trans. Acoustics, Speech and Signal Processing,* vol. ASSP-31, no. 6, p. 1394–1402, December 1983.

39. A. Carini and E. Mumolo, "Adaptive stabilization of recursive second order polynomial filters by means of a stability test," *Proc. IEEE Workshop on Nonlinear Signal and Image Processing,* Halkidiki, Greece, pp. 939–942, June 1995.

40. A. Carini, G. L. Sicuranza, and V. J. Mathews, "On the exact inverse and the pth order inverse of certain nonlinear systems," *Proc. 1997 IEEE Workshop on Nonlinear Signal and Image Processing,* Mackinac Island, Michigan, United States of America, September, 7–11, 1997.

41. A. Carini, G. L. Sicuranza, and V. J. Mathews, "On the inversion of certain nonlinear systems," *IEEE Signal Proc. Letters,* Vol. 4, no.12, pp. 334–336, December 1997.

42. A. Carini, G. L. Sicuranza, and V. J. Mathews, "Equalization and linearization of nonlinear systems," *Proc. IEEE Int. Conf. Acoust., Speech, and Signal Processing,* Seattle, WA, May 1998.

43. A. Carini, E. Mumolo, and G. L. Sicuranza, "V-vector algebra and its application to Volterra adaptive filtering", *IEEE Trans. Circuits Syst. II: Analog Digital Signal Process.* vol. 46, no. 5, May 1999.

44. S. Carrato, G. Ramponi, and S. Marsi, "A simple edge-sensitive image interpolation filter," *Proc. Third IEEE Int. Conf. Image Processing,* ICIP-96, Lausanne, CH, September 16–19, 1996.

45. J. R. Casar-Corredera, M. Garcia-Otero, and A. R. Figueiras-Vidal, "Data echo nonlinear cancellation," *Proc. IEEE Int. Conf. Acoust., Speech, and Signal Processing,* Tampa, Florida, pp. 32.4.1–32.4.4, March 1985.

46. H. H. Chang, C. L. Nikias, and A. N. Venetsanopoulos, "Reconfigurable systolic array implementation of quadratic filters," *IEEE Trans. Circuits Syst.,* vol. CAS-33, pp. 845–847, August 1986.

47. H. H. Chang, C. L. Nikias, and A. N. Venetsanopoulos, "Efficient implementations of quadratic filters," *IEEE Trans. Acoust., Speech, Signal Process.,* vol. ASSP-34, pp. 1511–1528, December 1986.

48. C. H. Chang, S. Siu, and C. H. Wei, "A polynomial-perceptron based decision feedback equalizer with a robust learning algorithm," *Signal Process.,* vol. 47, no. 2, pp. 145–158, November 1995.

49. S. Chen, G. J. Gibson, and C. F. N. Cowan, "Adaptive channel equalization using a polynomial perceptron structure," *IEE Proc.*, pt. I, vol. 137, no. 5, pp. 257–264, 1990.

50. D. A. Chesler, "Optimum multiple-input nonlinear systems with Gaussian inputs," *IRE Trans. Inform. Theory,* vol. 8, no. 5, pp. 237–245, September 1962.

51. P. Chevalier and B. Picinbono, "Optimal linear-quadratic array for detection," *Proc. IEEE Int. Conf. Acoustics, and Signal Processing,* Glasgow, Scotland, May 1989.

52. L. O. Chua and N. Ng, "Frequency-domain analysis of nonlinear systems: general theory," *IEE J. Electron. Circuits Syst.,* vol. 3, pp. 165–185, 1979.

53. L. O. Chua and N. Ng, "Frequency-domain analysis of nonlinear systems: formulation of transfer functions," *IEE J. Electron. Circuits Syst.,* vol. 3, pp. 257–269, 1979.

54. J. M. Cioffi and T. Kailath, "Fast recursive least squares transversal filters for adaptive filtering," *IEEE Trans. Acoust. Speech Signal Process.,* vol. ASSP-32, no. 2, pp. 304–337, March 1984.

55. J. M. Cioffi, "Limited precision effects in adaptive filtering," *IEEE Trans. Circuits Syst.,* vol. CAS-34, no. 7, pp. 821–833, July 1987.

56. S. J. Clancy and W. J. Rugh, "A note on the identification of discrete-time polynomial systems," *IEEE Trans. Automat. Control,* vol. AC-24, pp. 975–978, December 1979.

57. G. A. Clark, S. R. Parker, and S. K. Mitra, "A uniform approach to time- and frequency-domain realization of FIR adaptive digital filters," *IEEE Trans. Acoust. Speech Signal Process.,* vol. ASSP-31, no. 5, pp. 1073–1083, October 1983.

58. M. J. Coker and D. N. Simkins, "A nonlinear adaptive noise canceller," *Proc. IEEE Int. Conf. Acoustics, Speech and Signal Processing,* Denver, Colorado, pp. 470–473, 1980.

59. C. F. N. Cowan and P. F. Adams, "Non-linear system modeling: concept and application," *Proc. IEEE Int. Conference Acoustics, Speech and Signal Processing,* San Diego, California, pp. 45.6-1–45.6-4, March 1984.

60. H. Cramér, *Mathematical Methods of Statistics,* Princeton University Press, Princeton, New Jersey, 1946.

61. P. d'Alessandro, A. Isidori, and A. Ruberti, "Realizations and structure theory of bilinear dynamical systems," *SIAM J. Control,* vol. 12, no. 3, pp. 517–535, 1974.

62. S. Dasgupta, Y. Shrivastava, and G. Krener, "Persistent excitation in bilinear systems," *IEEE Trans. Automat. Control,* vol. 36, no. 2, pp. 305–313, March 1991.

63. C. E. Davila, A. J. Welch, and H. G. Rylander, III, "A second-order adaptive Volterra filter with rapid convergence," *IEEE Trans. Acoust. Speech Signal Process.,* vol. ASSP-35, no. 9, pp. 1259–1263, September 1987.

64. J. Dieudonné, *Foundations of Modern Analysis,* Springer, Berlin, Heidelberg, New York, 1976.

65. C. Eun and E. J. Powers, "A new Volterra predistorter based on the indirect learning architecture," *IEEE Trans. Signal Process.,* vol. 45, no. 1, pp. 223–227, January 1997.

66. D. D. Falconer, "Adaptive equalization of channel nonlinearities in QAM data transmission systems," *Bell Syst. Tech. J.,* vol. 57, pp. 2589–2611, September 1978.

67. S. Y. Fakhouri, "Identification of the Volterra kernels of nonlinear systems," *Proc. IEE,* vol. 127, pt. D, pp. 296–304, November 1980.

68. P. L. Feintuch, "An adaptive recursive LMS filter," *Proc. IEEE,* vol. 64, no. 11, pp. 1622–1624, November 1976.

69. F. Fnaiech and L. Ljung, "Recursive identification of bilinear systems," *Int. J. Control,* vol. 45, no. 2, pp. 453–470, 1987.

70. W. A. Frank, "An efficient approximation to the quadratic Volterra filter and its application in real-time loudspeaker linearization," *Signal Process.,* vol. 45, no. 1, pp. 97–114, July 1995.

71. W. A. Frank, "Sampling requirements for Volterra system identification," *IEEE Signal Process. Lett.,* vol. 3, no. 9, pp. 266–268, September 1996.

72. M. Fréchet, "Sur les fonctionelles continues," *Annales Scientifiques de L'Ecole Normale Superieure,* 3rd Ser., vol. 27, pp. 193–216, May 1910.

73. A. S. French and E. G. Butz, "Measuring Volterra kernels of a nonlinear system using the fast Fourier transform algorithm," *Int. J. Control,* vol. 17, pp. 529–539, 1973.

74. A. S. French, "Practical nonlinear system analysis by Wiener kernel estimation in the frequency domain," *Biol. Cybernet.,* vol. 24, pp. 111–119, 1976.

75. F. R. Gantmacher, *The Theory of Matrices,* Chelsea, New York, 1960.

76. F. X. Y. Gao and W. M. Snelgrove, "Adaptive linearization of a loudspeaker," *Proc. IEEE Int. Conf. Acoustics, Speech and Signal Processing,* Toronto, Canada, pp. 3589–3592, May 1991.

77. W. A. Gardner, "Structurally constrained receivers for signal detection and estimation," *IEEE Trans. Commun,* vol. COM-24, pp. 578–592, June 1976.

78. D. A. George, *Continuous Nonlinear Systems,* Technical Report No. 355, Research Laboratory of Electronics, MIT, Cambridge, Massachusetts, 1959.

79. G. Gheen, "Distortion invariant Volterra filters," *Pattern Recogn.,* vol. 27, no. 4, pp. 517–523, October 1980.

80. G. G. Giannakis and E. Serpedin, "Linear multichannel blind equalizers of nonlinear FIR Volterra channels," *IEEE Trans. Signal Process.,* vol. 45, no. 1, pp. 67–81, January 1997.

81. E. Gilbert, "Functional expansions for the response of nonlinear differential systems," *IEEE Trans. Automat. Control,* vol. AC-22, pp. 909–921, 1977.

82. C. L. Giles, R. D. Griffin, and T. Maxwell, "Encoding geometric invariances in higher order neural networks," in *Neural Information Processing Systems,* D. Z. Anderson, ed., American Institute of Physics, New York, 1988, pp. 301–309.

83. R. Glavina, S. Cucchi, and G. Sicuranza, "Nonlinear interpolation of TV image sequences," *Electron. Lett.,* vol. 23, no. 15, pp. 778–780, July 16, 1987.

84. J. Goldman, "A Volterra series description of crosstalk interference in communication systems," *Bell Syst. Tech. J.,* vol. 52, no. 5, pp. 649–668, 1973.

85. A. Graham, *Kronecker products and matrix calculus with applications.* Ellis Horwood Limited, Chichester, England, 1981.

86. C. W. J. Granger and A. P. Andersen, *An Introduction to Bilinear Time Series Models,* Vandenhoeck and Ruprecht, Gottingen, 1978.

87. T. E. Hall and G. B. Giannakis, "Bispectral analysis and model validation of texture images," *IEEE Trans. Image Process.,* vol. 4, no. 7, pp. 996–1009, July 1995.

88. S. Haykin, *Neural Networks,* Macmillan, 1993.

89. H. Leung and S. Haykin, "Detection and estimation using an adaptive rational function filter," *IEEE Trans. Signal Process.,* vol. 42, no. 12, pp. 3366–3376, December 1994.

90. S. Haykin, *Adaptive Filter Theory,* Third Edition, Prentice-Hall, Englewood Cliffs, New Jersey, 1996.

91. R. Hermann, "Volterra modeling of digital magnetic saturation recording channels," *IEEE Trans. Magn.,* vol. 26, no. 5, pp. 2125–2127, September 1990.

92. M. J. Hinich and D. M. Patterson, "Identification of the coefficients in a non-linear time series of the quadratic type," *J. Econometrics,* vol. 30, pp. 269–288, 1985.

93. M. L. Honig and D. Messerschmitt, *Adaptive Filters: Structures, Algorithms, and Applications,* Kluwer Academic Publishers, Boston, 1984.

94. I. W. Hunter and M. J. Korenberg, "The identification of nonlinear biological systems: Wiener and Hammerstein cascade models," *Biol. Cybernet.,* vol. 55, pp. 135–144, 1986.

95. S. Im and E. J. Powers, "A third-order frequency-domain adaptive Volterra filter," *Proc. IEEE Workshop on Nonlinear Signal and Image Processing,* Halkidiki, Greece, pp. 931–934, June 1995.

96. S. Im and E. J. Powers, "A third-order frequency-domain adaptive Volterra filter," *IEEE Signal Process. Lett.,* vol. 4, no. 3, pp. 75–78, March 1997.

97. S. L. S. Jacobi, J. S. Kowalik, and J. T. Pizzo, *Iterative Methods for Nonlinear Optimization Problems,* Prentice-Hall, Englewood Cliffs, New Jersey, 1972.

98. N. C. Jagan and D. C. Reddy, "Evaluation of response of nonlinear discrete systems using multidimensional z-transforms," *Proc. IEE,* vol. 119, no. 10, pp. 1521–1525, October 1972.

99. A. K. Jain, *Fundamentals of Digital Image Processing,* Prentice-Hall, Englewood Cliffs, New Jersey, 1989.

100. A. Javed, B. A. Syrett, and P. A. Goud, "Intermodulation distortion analysis of reflection-type IMPATT amplifiers using Volterra series representation," *IEEE Trans. Microwave Theory Tech.,* vol. MTT-25, no. 9, pp. 729–733, September 1977.

101. N. S. Jayant and P. Noll, *Digital Coding of Waveforms,* Prentice-Hall, Englewood Cliffs, New Jersey, 1984.

102. N. S. Jayant, J. D. Johnston, and R. J. Safranek, "Signal compression based on models of human perception," *Proc. IEEE,* vol. 81, no. 10, pp. 1385–1424, October 1993.

103. A. H. Jazwinsky, *Stochastic Processes and Filtering Theory,* Academic Press, New York, 1970.

104. C. R. Johnson, Jr., "Adaptive IIR filtering: current results and open issues," *IEEE Trans. Inform. Theory,* vol. IT-30, no. 2, pp. 237–250, March 1984.

105. K. S. Joo and T. Bose, "A fast conjugate gradient algorithm for 2-D nonlinear adaptive filtering," *Proc. Int. Conf. Digital Signal Processing,* Limasol, Cyprus, pp. 314–319, 1995.

106. B. Julesz, "Visual patterns discrimination," *IRE Trans. Inform. Theory,* vol. IT-8, pp. 84–92, February 1962.

107. J. Kaiser, "On a simple algorithm to calculate the 'energy' of a signal," in *Proc. IEEE Int. Conf. Acoustics, Speech and Signal Processing,* Albuquerque, New Mexico, pp. 381–384, 1990.

108. T. Kanaoka, R. Chellappa, M. Yoshitaka, and S. Tomita, "A higher-order neural network for distortion invariant pattern recognition," *Pattern Recogn. Letters,* pp. 837–841, December 1992.

109. G. Karam and H. Sari, "Analysis of predistortion, equalization, and ISI cancellation techniques in digital radio systems with nonlinear transmit amplifier," *IEEE Trans. Commun.* vol. 37, pp. 1245–1253, December 1989.

110. K. I. Kim and E. J. Powers, "Digital bispectrum analysis and its applications to nonlinear wave interaction," *IEEE Trans. Plasma Sci.,* vol. PS-7, pp. 120–131, June 1979.

111. K. I. Kim, E. J. Powers, C. P. Ritz, R. W. Miksad, and F. J. Fischer, "Modeling of nonlinear systems in offshore engineering for non-Gaussian inputs," in *Offshore Engineering,* vol. 5, F. L. L. B. Carneiro, R. C. Batista, and A. J. Ferrante, eds., Pentech, London, 1986, pp. 405–419.

112. K. I. Kim and E. J. Powers, "A digital method of modeling quadratically nonlinear systems with a general random input," *IEEE Trans. Acoust. Speech Signal Process.,* vol. ASSP-36, pp. 1758–1769, November 1988.

113. S. Kirkpatrick, C. D. Gelatt, and M. P. Vecchi, "Optimization by simulated annealing," *Science,* vol. 220, no. 4598, May 1983.

114. W. Klippel, "Dynamic measurement and interpretation of the nonlinear parameters of electrodynamic loudspeakers," *J. Audio Eng. Soc.,* vol. 38, pp. 944–955, December 1990.

115. T. Koh and E. J. Powers, "An adaptive nonlinear digital filter with lattice orthogonalization," *IEEE Int. Conf. Acoustics, Speech and Signal Processing,* Boston, Massachusetts, pp. 37–40, April 1983.

116. T. Koh, E. J. Powers, R. W. Miksad, and F. J. Fischer, "Application of nonlinear digital filter to modeling low-frequency drift oscillations of moored vessels in random sea," *Proc. 1984 Offshore Technology Conf.,* pp. 309–314, 1984.

117. T. Koh and E. J. Powers, "Second-order Volterra filtering and its application to nonlinear system identification," *IEEE Trans. Acoust. Speech Signal Process.,* vol. ASSP-33, no. 6, pp. 1445–1455, December 1985.

118. M. J. Korenberg, "Identification of biological cascades of linear and static nonlinear systems," *Proc. Sixteenth Midwest Symp. Circuit Theory,* pp. 1–9, 1973.

119. M. J. Korenberg, "Crosscorrelation analysis of neural cascades," *Proc. Tenth Annu. Rocky Mountain Bioengineering Symp.,* pp. 47–52, 1973.

120. M. J. Korenberg, "Statistical identification of parallel cascades of linear and nonlinear systems," *Proc. IFAC Symp. Identification and System Parameter Estimation,* Washington DC, pp. 669–674, 1983.

121. M. J. Korenberg and I. W. Hunter, "The identification of nonlinear biological systems: LNL cascade models," *Biological Cybernet.,* vol. 55, pp. 125–134, 1986.

122. M. J. Korenberg, "A rapid and accurate method for estimating the kernels of a nonlinear system with lengthy memory," *15th Biennial Symp. Comm.,* Dept. E.E., Queen's University, Kingston, Ontario, Canada, pp. 57–60, June 3–6, 1990.

123. M. J. Korenberg, "Parallel cascade identification and kernel estimation for nonlinear systems," *Ann. Biomed. Eng.,* vol. 19, 1991.

124. St. Kotsios and N. Kalouptsidis, "BIBO stability criteria for a certain class of discrete nonlinear systems," *Int. J. Control,* vol. 58, No. 3, pp. 707–730, September 1993.

125. P. Koukoulas and N. Kalouptsidis, "Nonlinear system identification using Gaussian inputs," *IEEE Trans. Signal Process.,* vol. 43, no. 8, pp. 1831–1841, August 1995.

126. A. J. Krener, "Linearization and bilinearization of control systems," *Proc. 1974 Allerton Conf. Circuits and System Theory,* Urbana, Illinois, pp. 834–843, 1974.

127. H. Kressel (ed), *Semiconductor Devices for Optical Communications* (Topics in Applied Physics, vol. 39), Springer Verlag, New York, 1980.

128. G. Krieger and C. Zetzsche, "Nonlinear image operator for the evaluation of local intrinsic dimensionality," *IEEE Trans. Image Process.*, vol. 5, no. 6, pp. 1026–1042, June 1996.

129. S. Kröner, "A structural neural network invariant to cyclic shifts and rotations," *Proc. Int. Conf. Computer Analysis of Images and Patterns,* Kiel, Germany, pp. 384–391, September 1997.

130. Y. H. Ku and A. A. Wolf, "Volterra-Wiener functionals for the analysis of nonlinear systems," *J. Franklin Inst.*, pp. 9–26, 1966.

131. A. Lavi and S. Narayanan, "Analysis of a class of nonlinear discrete systems using multidimensional modified z-transform," *IEEE Trans. Automat. Control,* vol. AC-13, no. 1, pp. 90–93, February 1968.

132. S. Lawson, "Direct approach to design of PCAS filters with combined gain and phase specification," *IEE Proc. Vis. Image Signal Process.*, vol. 141, pp. 161–167, June 1994.

133. Y. W. Lee and M. Schetzen, "Measurement of the Wiener kernels of nonlinear system by crosscorrelation," *Int. J. Control,* vol. 2, no. 3, pp. 237–254, September 1965.

134. J. Lee and V. J. Mathews, "Output Error Adaptive Bilinear Filters," *Proc. Twenty Fifth Asilomar Conf. Signals, Systems and Computers,* Pacific Grove, California, pp. 207–211, November 1991.

135. J. Lee, *Adaptive Polynomial Filtering Algorithms,* Ph. D. Dissertation, University of Utah, 1992.

136. J. Lee and V. J. Mathews, "A fast recursive least squares adaptive second order Volterra filter and its performance analysis," *IEEE Trans. Signal Process.*, vol. 41, no. 3, pp. 1087–1102, March 1993.

137. J. Lee and V. J. Mathews, "Adaptive bilinear predictors," *Proc. IEEE Int. Conf. Acoustics, Speech and Signal Processing,* Adelaide, Australia, April 1994.

138. J. Lee and V. J. Mathews, "A stability theorem for bilinear systems," *IEEE Trans. Signal Process.*, vol. 41, no. 7, pp. 1871–1873, July 1994.

139. J. Lee and V. J. Mathews, "A stability result for RLS adaptive bilinear filters," *IEEE Signal Process. Lett.*, vol. 1, no. 12, December 1994.

140. J. Lee and V. J. Mathews, "Output error LMS bilinear filters with stability monitoring," *Proc. IEEE Int. Conf. Acoustics, Speech and Signal Processing,* Detroit, Michigan, May 1995.

141. P. J. Lenk and S. R. Parker, "Nonlinear modeling by discrete orthogonal lattice structures," *Proc. IEEE Int. Symp. Circuits and Syst.,* San Jose, California, pp. 486–489, 1986.

142. V. P. Leonov and A. N. Shiryaev, "On a method of calculation of semi-invariants," *Theory Probab. Appl.,* vol. IV, no. 3, pp. 319–328, 1959.

143. C. Lesiak and A. J. Krener, "The existence and uniqueness of Volterra series for nonlinear systems," *IEEE Trans. Automat. Control,* vol. AC-23, no. 6, pp. 1090–1095, December 1978.

144. D. W. Lin, "On digital implementation of the fast Kalman algorithm," *IEEE Trans. Acoust. Speech Signal Process.*, vol. ASSP-32, no. 5, pp. 998–1005, October 1984.

145. F. Ling and J. G. Proakis, "A generalized multichannel least squares lattice algorithm based on sequential processing stages," *IEEE Trans. Acoust. Speech Signal Process.*, vol. ASSP-32, no. 2, pp. 381–390, April 1984.

146. R. P. Lippmann, "An introduction to computing with neural nets," *IEEE ASSP Magazine*, vol. 4, no. 2, pp. 4–22, April 1987.

147. L. Ljung, D. Falconer, and M. Morf, "Fast calculation of gain matrices for recursive estimation schemes," *Int. J. Control,* vol. 27, pp. 1–19, January 1978.

148. L. Ljung and S. Ljung, "Error propagation properties of recursive least-squares adaptation algorithms," *Automatica,* vol. 21, no. 2, pp. 157–167, 1985.

149. Y. Lou, C. L. Nikias, and A. N. Venetsanopoulos, "Efficient VLSI array processing structures for adaptive quadratic digital filters," *J. Circuits Systems Signal Process.,* vol. 7, no. 2, pp. 253–273, 1988.

150. R. W. Lucky, "Modulation and detection for data transmission on the telephone channel," in *New Directions in Signal Processing in Communication and Control,* J. K. Skwirzynski, ed., Noordhoff, Leiden, Holland, 1975.

151. G. K. Ma, J. Lee, and V. J. Mathews, "A fast RLS bilinear filter for channel equalization," *Proc. IEEE Int. Conf. Acoustics, Speech and Signal Processing,* Adelaide, Australia, pp. III-257–III-260, April 1994.

152. D. Mansour and A. H. Gray, "Frequency-domain nonlinear adaptive filter," *Proc. IEEE Int. Conf. Acoust., Speech Signal Process.,* Toronto, Canada, pp. 550–553, 1991.

153. P. Maragos and R. W. Schafer, "Morphological skeleton representation and coding of binary images," *IEEE Trans. Acoust. Speech Signal Process.,* vol. ASSP-34, no. 5, pp. 1228–1244, October 1986.

154. P. Maragos and R. W. Schafer, "Morphological filters, Part I: their set theoretic analysis and relations to linear shift invariant filters," *IEEE Trans. Acoust. Speech Signal Process.,* vol. ASSP-35, no. 8, pp. 1153–1169, August 1987.

155. P. Maragos, T. F. Quatieri, and J. F. Kaiser, "Speech nonlinearities, modulations, and energy operators", *Proc. IEEE Int. Conf. Acoustics, Speech and Signal Processing,* Toronto, Canada, pp. 421–424, 1991.

156. P. Maragos, J. F. Kaiser, and T. F. Quatieri, "On amplitude and frequency demodulation using energy operators," *IEEE Trans. Signal Process.,* vol. 41, pp. 1532–1550, April 1993.

157. P. Maragos, J. F. Kaiser, and T. F. Quatieri, "Energy separation in signal modulations with application to speech analysis," *IEEE Trans. Signal Process.,* vol. 41, pp. 3024–3051, October 1993.

158. P. Maragos, A. Potamianos, and B. Santhanam, "Instantaneous energy operators: applications to speech processing and communications", *Proc. IEEE Workshop on Nonlinear Signal and Image Processing,* Halkidiki, Greece, pp. 955–958, June 1995.

159. P. Maragos and A. Potamianos, "Higher-order differential energy operators," *IEEE Signal Process. Lett.,* vol. 2, pp. 152–154, August, 1994.

160. P. Z. Marmarelis and K.-I. Naka, "White-noise analysis of a neuron chain: an application of the Wiener theory," *Science,* vol. 175, no. 4027, pp. 1276–1278, March 17, 1972.

161. P. Z. Marmarelis and V. Z. Marmarelis, *Analysis of Physiological Systems,* Plenum Press, New York, 1978.

162. S. L. Marple, *Digital Spectral Analysis With Applications,* Prentice-Hall, Englewood Cliffs, New Jersey, 1987.

163. S. Marsi and G. L. Sicuranza, "On reduced-complexity and approximation of quadratic filters," *Proc. 27th Asilomar Conf. Signals, Systems and Computers,* Pacific Grove, California, November 1993.

164. V. J. Mathews and J. Lee, "A fast recursive least-squares second-order Volterra filter," *Proc. IEEE Int. Conf. Acoustics, Speech and Signal Processing,* New York, pp. 1383–1386, April 1988.

165. V. J. Mathews, "Adaptive polynomial filters" *IEEE Signal Process. Magazine,* vol. 8, no. 3, pp. 10–26, July 1991.

166. V. J. Mathews and Z. Xie, "Stochastic gradient adaptive filters with gradient adaptive step sizes," *Proc. IEEE Int. Conf. Acoustics, Speech, and Signal Processing,* pp. 1385–1388, Albuquerque, New Mexico, April 3–6, 1990.

167. V. J. Mathews and T. K. Moon, "Parameter estimation for a bilinear time series model," *Proc. IEEE Int. Conf. Acoustics, Speech, and Signal Processing,* Toronto, Canada, pp. 3513–3516, May 1991.

168. V. J. Mathews and Z. Xie, "Stochastic gradient adaptive filters with gradient adaptive step sizes," *IEEE Trans. Signal Process.,* vol. 41, no. 6, pp. 2075–2087, June 1993.

169. V. J. Mathews and J. Lee, "Techniques for bilinear time series analysis," *Proc. Twenty Seventh Annu. Asilomar Conf. Signals, Systems and Computers,* Pacific Grove, California, November 1993.

170. V. J. Mathews, "Orthogonalization of correlated Gaussian signals for Volterra system identification," *IEEE Signal Process. Lett.,* vol. 2, no. 10, pp. 188–190, October 1995.

171. V. J. Mathews, "Adaptive Volterra filters using orthogonal structures," *IEEE Signal Process. Lett.,* vol. 3, no. 12, pp. 307–309, December 1996.

172. V. J. Mathews and S. C. Douglas, *Adaptive Filters,* Prentice-Hall, Englewood Cliffs, New Jersey, 2000.

173. J. E. Mazo, "On the independence theory of equalizer convergence," *Bell Syst. Tech. J.,* vol. 58, no. 5, pp. 963–993, May–June 1979.

174. T. E. McCannon, N. C. Gallagher, D. Minoo-Hamedani, and G. L. Wise, "On the design of nonlinear discrete-time predictors", *IEEE Trans. Inform. Theory,* vol. IT-28, pp. 366–371, March 1982.

175. J. M. Mendel, "Tutorial on high order statistics (spectra) in signal processing and system theory: theoretical results and some applications," *Proc. IEEE,* vol. 79, no. 3, pp. 278–305, March 1991.

176. J. M. Mendel, *Lessons in Digital Estimation Theory,* Prentice-Hall, Englewood-Cliffs, New Jersey, 1995.

177. B. G. Mertzios, G. Sicuranza, and A. N. Venetsanopoulos, "Efficient realizations of two-dimensional quadratic digital filters," *IEEE Trans. Acoust. Speech Signal Process.,* vol. ASSP-37, no. 5, pp. 765–768, May 1989.

178. B. G. Mertzios, "Parallel modeling and structure of nonlinear Volterra discrete systems," *IEEE Trans. Circuits Syst. —I:Fund. Theory Appl.,* vol. 41, no. 5, pp. 359–371, May 1994.

179. R. G. Meyer, M. J. Shensa, and R. Eschenbach, "Crossmodulation and intermodulation in amplifiers at high frequencies," *IEEE J Solid State Circuits,* vol. 7, no. 1, pp. 16–23, February 1972.

180. K. Mio, E. Moisan, and P. O. Amblard, "Nonlinear noise cancellation: bilinear filtering of a sonar experiment," *Proc. IEEE Workshop on Nonlinear Signal and Image Processing,* Halkidiki, Greece, pp. 416–419, June 1995.

181. S. K. Mitra, H. Li, I. Lin, and T. Yu, "A new class of nonlinear filters for image enhancement," *Proc. Int. Conf. IEEE Acoustics, Speech, and Signal Processing,* Toronto, Canada, pp. 2525–2528, May 1991.

182. S. K. Mitra, S. Thurnhofer, M. Lighstone, and N. Strobel, "Two-dimensional Teager operators and their image processing applications," *Proc. IEEE Workshop on Nonlinear Signal and Image Processing,* Halkidiki, Greece, pp. 959–962, June 1995.

183. S. Mo and V. J. Mathews, "Adaptive quadratic preprocessing of document images for binarization," *IEEE Trans. Image Process.,* vol. 7, no. 7, pp. 992–999, July 1998.

184. R. R. Mohler and C. N. Shen, *Optimal Control of Nuclear Reactors,* Academic Press, New York, 1972.

185. R. R. Mohler, *Bilinear, Control Processes,* Academic Press, New York, 1973.

186. R. R. Mohler and W. J. Kolodziej, "An overview of bilinear system theory and applications," *IEEE Trans. Syst., Man Cybernet.,* vol. SMC-10, no. 10, pp. 683–688, October 1980.

187. R. R. Mohler, C. Bruni, and A. Gandolphi, "A system approach to immunology," *Proc. IEEE,* vol. 68, no. 8, pp. 964–990, August 1980.

188. B. Mulgrew and C. F. N. Cowan, "Equalisation techniques using non-linear adaptive filters", *Proc. COST 229 Workshop on Adaptive Algorithms: Applications and Non Classical Schemes,* Vigo, Spain, pp. 1–19, March 1991.

189. E. Mumolo and D. Francescato, "Adaptive predictive coding of speech by means of Volterra predictors," *Proc. IEEE Winter Workshop on Nonlinear Digital Signal Processing,* Tampere, Finland, pp. 2.1.4.1–2.1.4.4, January 1993.

190. E. Mumolo and A. Carini, "A stability condition for adaptive recursive second-order polynomial filters," *Signal Process.,* vol. 54, no. 1, pp. 85–90, October 1996.

191. E. Mumolo and G. Bernardis, "A novel demand prefetching algorithm based on Volterra adaptive prediction for virtual memory management," *Proc. Int. Conf. System Science,* Wailea, Hawaii, January 7–10, 1997.

192. S. W. Nam and E. J. Powers, "Application of higher order spectral analysis to cubically nonlinear system identification," *IEEE Trans. Signal Process.,* vol. 42, no. 7, pp. 1746–1765, July 1994.

193. S. Narayanan, "Transistor distortion analysis using Volterra series representation," *Bell Syst. Tech. J.,* vol. 46, pp. 991–1024, May–June 1967.

194. K. S. Narendra and P. G. Gallman, "An iterative method for the identification of nonlinear systems using a Hammerstein model," *IEEE Trans. Automat. Control,* vol. AC-11, pp. 546–550, July 1966.

195. A. N. Netravali and R. J. P. de Figuereido, "On the identification of nonlinear dynamical systems", *IEEE Trans. Automat. Control,* vol. AC-16, pp. 28–36, February 1971.

196. C. L. Nikias and M. R. Raghuveer, "Bispectrum estimation: a digital signal processing framework," Proc. IEEE, vol. 75, no. 7, pp. 869–891, July 1987.

197. C. L. Nikias and A. P. Petropulu, *Higher-Order Spectral Analysis: A Nonlinear Signal Processing Framework,* Prentice-Hall, Englewood Cliffs, New Jersey, 1993.

198. C. L. Nikias and J. M. Mendel, "Signal processing with higher-order spectra," *IEEE ASSP Magazine,* vol. 10, no. 3, pp. 10–37, July 1993.

199. R. D. Nowak and B. D. Van Veen, "Volterra filtering with spectral constraints," *Proc. IEEE Int. Conf. Acoustics, Speech and Signal Processing*, Adelaide, South Australia, vol. IV, pp. 137–140, April 1994.

200. R. D. Nowak and B. D. Van Veen, "Random and pseudorandom inputs for Volterra system identification," *IEEE Trans. Signal Process.,* vol. 42, no. 8, pp. 2124–2135, August 1994.

201. R. Nowak and B. Van Veen, "Invertibility of higher order moment matrices," *IEEE Trans. Signal Process.,* vol. 43, no. 3, pp. 705–708, March 1995.

202. R. Nowak and B. Van Veen, "Tensor product basis approximations for Volterra filters," *IEEE Trans. Signal Process.,* vol. 44, no. 1, pp. 36–50, January 1996.

203. R. Nowak and B. Van Veen, "Volterra filter equalization: a fixed point approach," *IEEE Trans. Signal Process.,* vol. 45, no. 2, pp. 377–388, February 1997.

204. R. Nowak and R. Baraniuk, "Wavelet-based transformations for nonlinear signal processing," *IEEE Trans. Signal Process.,* vol. 47, no. 7, pp. 1852–1865, July 1999.

205. H. Ogura, "Orthogonal functionals of Poisson processes," *IEEE Trans. Inform. Theory,* vol. IT-18, no. 4, pp. 473–481, July 1972.

206. A. V. Oppenheim, R. W. Schafer, and T. G. Stockham, Jr., "Nonlinear filtering of multiplied and convolved signals," *Proc. IEEE,* vol. 56, no. 8, pp. 1264–1291, August 1968.

207. A. V. Oppenheim, A. S. Willsky, and I. T. Young, *Signals and Systems,* Prentice-Hall, Englewood Cliffs, New Jersey, 1983.

208. A. V. Oppenheim and R. Schafer, *Discrete-Time Signal Processing,* Prentice Hall, Englewood Cliffs, New Jersey, 1989.

209. G. A. Orban, *Neuronal Operations in the Visual Cortex,* Springer, Heidelberg, Germany, 1984.

210. N. Otsu, "A threshold selection method from gray-level histograms," *IEEE Trans. Syst., Man Cybernet.,* vol. SMC-9, no. 1, pp. 62–66, January 1979.

211. S. Özgünel, A. N. Kayran, and E. Panayirci, "Nonlinear channel equalization using multichannel adaptive lattice algorithms," *IEEE Int. Symp. Circuits and Systems,* Singapore, pp. 2826–2829, June 1991.

212. G. Palm and T. Poggio, "The Volterra representation and the Wiener expansion: validity and pitfalls," *SIAM J. Appl. Math.,* vol. 33, pp. 195–216, 1977.

213. G. Palm, "On the representation and approximation of nonlinear systems, Part II: Discrete time," *Biol. Cybernet.* vol. 34, pp. 49–52, 1979.

214. T. M. Panicker and V. J. Mathews, "Parallel-cascade realizations of higher-order truncated Volterra filters," *Proc. IEEE Int. Conf. Acoustics, Speech and Signal Processing,* Atlanta, May 1996.

215. T. M. Panicker and V. J. Mathews, "Parallel-cascade realizations and approximations of truncated Volterra systems," *IEEE Trans. Signal Process.* vol. 46, no. 10, pp. 2829–2832, October 1998.

216. T. M. Panicker, V. J. Mathews, and G. L. Sicuranza, "Adaptive parallel-cascade truncated Volterra filters," *IEEE Trans. Signal Process.,* vol. 46, no. 10, pp. 2664–2673, October 1998.

217. Y. H. Pao, *Adaptive Pattern Recognition and Neural Networks,* Addison-Wesley, Reading, Massachusetts, 1989.

218. A. Papoulis, *Probability, Random Variables and Stochastic Processes,* McGraw-Hill, New York, 1984.

219. R. Parente, "Nonlinear differential equations and analytic system theory," *SIAM J. Appl. Math.,* vol. 18, pp. 41–66, 1970.

220. S. R. Parker and F. A. Perry, "A discrete ARMA model for nonlinear system identification," *IEEE Trans. Circuits Systems,* vol. CAS-28, pp. 224–233, March 1981.

221. A. Peled and B. Liu, "A new hardware realization of digital filters," *IEEE Trans. Acoust. Speech Signal Process.,* vol. ASSP-22, pp. 456–462, December 1974.

222. S. J. Perantonis and P. J. G. Lisboa, "Translation, rotation, and scale invariant pattern recognition by high-order neural networks and moment classifiers," *IEEE Trans. Neural Networks,* vol. 3, no. 2, pp. 241–251, March 1992.

223. P. Perona and J. Malik, "Scale-space and edge detection using anisotropic diffusion," *IEEE Trans. Pattern Anal. Machine Intell.,* vol. 12, no. 7, pp. 629–639, July 1990.

224. A. L. Pflug, G. E. Ioup, J. W. Ioup, and R. L. Field, "Properties of higher-order correlations and spectra for band-limited, deterministic transients," *J. Acoust. Soc. Am.,* vol. 91, no. 2, pp. 975–988, February 1992.

225. B. Picinbono, "Quadratic filters," *Proc. IEEE Int. Conf. Acoustics, Speech and Signal Processing,* Paris, France, pp. 298–301, May 1982.

226. B. Picinbono and P. Duvaut, "Optimal linear-quadratic systems for detection and estimation," *IEEE Trans. Inform. Theory,* vol. IT-34, pp. 304–311, March 1988.

227. B. Picinbono, "Higher-order statistical signal processing with Volterra filters," *Proc. Workshop on Higher-Order Spectral Analysis,* Vail, Colorado, pp. 62–67, June 1989.

228. I. Pitas and A. N. Venetsanopoulos, *Nonlinear Digital Filters: Priciples and Applications,* Kluwer Academic Publishers, Boston, Massachusetts, 1990.

229. E. J. Powers, J. Y. Hong, and Y. C. Kim, "Cross sections and radar equations for nonlinear scatterers," *IEEE Trans. Aerospace Electron. Syst.,* vol. AES-17, pp. 602–605, July 1981.

230. E. J. Powers, R. W. Miksad, C. P. Ritz, and R. S. Solis, "Application of digital processing techniques to measure nonlinear dynamics of transition to turbulence," *Proc. AIAA 10th Aeroacoust. Conf.,* Seattle, WA, July 1986.

231. S. Prakryia and D. Hatzinakos, "Blind identification of LTI-ZMNL-LTI nonlinear channel models," *IEEE Trans. Signal Process.,* vol. 43, no. 12, pp. 3007–3013, December 1995.

232. J. G. Proakis, *Digital Communications,* Second Edition, McGraw-Hill, New York, 1989.

233. S. Pupolin and L. J. Greenstein, "Performance analysis of digital radio links with nonlinear transmit amplifier," *IEEE J. Select. Areas Commun.,* vol. SAC-5, pp. 534–546, April 1987.

234. G. Ramponi, "Edge extraction by a class of second-order nonlinear filters," *Electron. Lett.,* vol. 22, no. 9, pp. 482–484, April 24, 1986.

235. G. Ramponi, G. L. Sicuranza, and W. Ukovich, "An optimization approach to the design of nonlinear Volterra filters," *Proc. Third Eur. Signal Processing Conf.,* The Hague, The Netherlands, 1986.

236. G. Ramponi, G. L. Sicuranza, and S. Cucchi, "2- and 3-D nonlinear predictors," *Proc. IEEE Int. Conf. Acoustics, Speech and Signal Processing,* Dallas, Texas, pp. 1079–1082, April 1987.

237. G. Ramponi and W. Ukovich, "Quadratic 2-D filter design by optimization techniques," *Proc. Int. Conf. Digital Signal Processing,* Firenze, Italy, pp. 59–63, September 1987.

238. G. Ramponi and G. L. Sicuranza, "Decision-directed nonlinear filter for image processing," *Electron. Lett.,* vol. 23, no. 23, pp. 1218–1219, November 5, 1987.

239. G. Ramponi and G. L. Sicuranza, "Quadratic digital filters for image processing," *IEEE Trans. Acoust. Speech Signal Process.,* vol. ASSP-36, no. 6, pp. 937–939, June 1988.

240. G. Ramponi, "Enhancement of low-contrast images by nonlinear operators," *Alta Frequenza,* vol. LVII, no. 7, pp. 451–455, September 1988.

241. G. Ramponi, G. L. Sicuranza, and W. Ukovich, "A computational method for the design of 2-D nonlinear Volterra filters," *IEEE Trans. Circuits Systems,* vol. CAS-35, no. 9, pp. 1095–1102, September 1988.

242. G. Ramponi, "Quadratic filters for image enhancement," in *Proc. Eur. Signal Processing Conf.,* Grenoble, France, September 1988.

243. G. Ramponi and G. L. Sicuranza, "Texture discrimination via higher-order statistics," *Proc. Workshop on Higher-Order Spectral Analysis,* Vail, Colorado, pp. 100–105, June 1989.

244. G. Ramponi, "Design of 2-D quadratic filters using their bi-impulse response," *Proc. Eur. Conf. Circuit Theory and Design,* Brighton, United Kingdom, pp. 415–419, September 1989.

245. G. Ramponi, "Bi-impulse response design of isotropic quadratic filters," *IEEE Proc.,* vol. 78, no. 4, pp. 665–677, April 1990.

246. G. Ramponi and G. L. Sicuranza, "Image sharpening using a polynomial operator," *Proc. Eur. Conf. Circuit Theory and Design,* Davos, Switzerland, pp. 1431–1436, 1993.

247. G. Ramponi and P. Fontanot, "Enhancing document images with a quadratic filter," *Signal Process.,* vol. 33, pp. 23–34, July 1993.

248. G. Ramponi, "Detail-preserving filter for noisy images," *Electron. Lett.,* vol. 31, no. 11, pp. 865–866, May 25, 1995.

249. G. Ramponi, "A simple cubic operator for sharpening an image," *Proc. IEEE Workshop on Nonlinear Signal and Image Processing,* Halkidiki, Greece, pp. 963–966, June 1995.

250. G. Ramponi and G. L. Sicuranza, "Sobel-Laplacian image sharpening for noisy data," *Proc. Int. Conf. Digital Signal Processing,* Limassol, Cyprus, pp. 296–301, 1995.

251. G. Ramponi, "The rational filter for image smoothing," *IEEE Signal Proc. Lett.* vol. 3, no. 3, pp. 63–65, March 1996.

252. G. Ramponi, N. Strobel, S. K. Mitra, T. H. Yu, "Nonlinear unsharp masking methods for image contrast enhancement," *Journal of Electronic Imaging,* Vol. 5, pp. 353–366, July 1996.

253. G. Ramponi and S. Carrato, "Interpolation of the DC components of coded images using a rational filter," *Proc. Fourth IEEE Int. Conf. on Image Processing,* Santa Barbara, CA, October 26–29, 1997.

254. G. Ramponi and A. Polesel, "A rational unsharp masking technique," *J. Electron. Imaging,* vol. 7, no. 2, April 1998.

255. P. J. W. Rayner and M. R. Lynch, "A new connectionist model based on a non-linear adaptive filter," *Proc. IEEE Int. Conf. Acoustics, Speech and Signal Processing,* Glasgow, Scotland, pp. 1191–1194, May 23–26, 1989.

256. G. M. Raz and B. D. Van Veen, "Baseband Volterra filters for implementing carrier based nonlinearities," *Proc. 1997 IEEE Workshop on Nonlinear Signal and Image Processing,* Mackinac Island, Michigan, September 7–11, 1997.

257. P. A. Regalia, *Adaptive IIR Filtering in Signal Processing and Control,* Mercel Dekker, New York, 1995.

258. C. P. Ritz, E. J. Powers, D. L. Brower, T. Rhodes, R. D. Bengstone, N. C. Luhmann, and W. A. Peebles, "Digital spectral analysis—a diagnostic for fusion research," *Proc. 11th Symp. Fusion Engineering,* Austin, Texas, pp. 1185–1188, November 1985.

259. C. P. Ritz and E. J. Powers, "Estimation of nonlinear transfer functions for fully developed turbulence," *Physica,* 20D, pp. 320–334, 1986.

260. M. Rosenblatt, *Stationary Sequences and Random Fields,* Birkhauser, Boston, Massachusetts, 1985.

261. S. Ross, *A First Course in Probability,* Third Edition, Macmillan, New York, 1988.

262. R. J. Roy and J. Sherman, "A learning technique for Volterra series representation," *IEEE Trans. Automat. Control,* vol. AC-12, December 1967.

263. E. Roy, R. W. Stewart, and T. S. Durrani, "Higher order system identification with second order Volterra IIR filters," *IEEE Signal Process. Lett.,* vol. 3, no. 10, pp. 276–279, October 1996.

264. W. J. Rugh, *Nonlinear System Theory. The Volterra-Wiener Approach,* Johns Hopkins University Press, Baltimore, Maryland, 1981.

265. D. E. Rummelhart and J. L. McClelland (eds.), *Parallel Distributed Processing,* MIT Press, Cambridge, Massachusetts, 1986.

266. N. Sadegh, "A preceptron network for functional identification and control of nonlinear systems," *IEEE Trans. Neural Networks,* vol. 4, no. 6, pp. 982–988, November 1993.

267. A. Sandberg and L. Stark, "Wiener *G*-functional analysis as an approach to nonlinear characteristics of human pupil light reflex," *Brain Research,* vol. 11, no. 1, pp. 194–211, October 1968.

268. I. W. Sandberg, "Signal distortion in nonlinear feedback systems," *Bell Systems Tech. J.,* vol. 42, pp. 2533–2550, November 1963.

269. I. W. Sandberg, "Some results in the theory of physical systems governed by nonlinear functional equations," *Bell Systems Tech. J.,* vol. 44, no. 5, pp. 871–898, May–June 1965.

270. I. W. Sandberg, "The mathematical foundations of associated expansions for mildly nonlinear systems," *IEEE Trans. Circuits Syst.,* vol. CAS-30, no. 7, pp. 441–455, 1983.

271. I. W. Sandberg, "Uniform approximation with doubly finite Volterra series," *IEEE Trans. Signal Process.,* vol. 40, no. 6, pp. 1438–1442, June 1992.

272. A. Sarti and S. Pupolin, "Recursive techniques for the synthesis of a pth - order inverse of a Volterra system," *Eur. Trans. Telecommun.,* vol. 3, no. 4, pp. 315–322, August 1992.

273. M. Schetzen, "A theory of nonlinear system identification," *Int. J. Control,* vol. 20, no. 4, pp. 577–592, October 1974.

274. M. Schetzen, "Nonlinear system modeling based on the Wiener theory," *Proc. IEEE,* vol. 69, no. 12, pp. 1557–1573, December 1993.

275. M. Schetzen, *The Volterra and Wiener Theories of Nonlinear Systems,* Reprint Edition, R. E. Krieger Publishing Company, Malabar, Florida, 1989.

276. H. Schulz-Mirbach, "Constructing invariant features by averaging techniques," *Proc. 12'th ICPR,* vol. II, pp. 178–182, 1994.

277. A. Segall and T. Kailath, "Orthogonal functionals of independent-increment processes," *IEEE Trans. Inform. Theory,* vol. IT-22, no. 3, pp. 287–298, May 1976.

278. J. Serra, *Image Analysis and Mathematical Morphology,* Academic Press, 1983.

279. J. Serra, "Introduction to mathematical morphology," *Computer Vision, Graphics Image Process.,* vol. 35, pp. 283–305, 1986.

280. S. A. O. Sesay and T. Subba Rao, "Difference equations for higher-order moments and cumulants for bilinear time series model BP $(p, 0, p, 1)$," *J. Time Series Anal.,* vol. 12, no. 2, pp. 159–177, February 1991.

281. J. J. Shynk, "Adaptive IIR filtering," *IEEE ASSP Magazine,* vol. 6, no. 2, pp. 4–21, April 1989.

282. J. J. Shynk, "Frequency-domain and multirate adaptive filtering," *IEEE Signal Process. Magazine,* vol. 9, no. 1, pp. 14–37, January 1992.

283. G. L. Sicuranza, A. Bucconi, and P. Mitri, "Adaptive echo cancellation with nonlinear digital filters," *Proc. IEEE Int. Conf. Acoustics, Speech and Signal Processing,* San Diego, California, pp. 3.10.1–3.10.4, March 1984.

284. G. L. Sicuranza, "Theory and realization of nonlinear digital filters," *Proc. IEEE Int. Symp. Circuit and Systems,* Montreal, Canada, pp. 242–245, May 1984.

285. G. L. Sicuranza and G. Ramponi, "Distributed arithmetic implementation of nonlinear echo cancellers," *Proc. IEEE Int. Conf. Acoustics, Speech and Signal Processing,* Tampa, Florida, pp. 1617–1620, March 1985.

286. G. L. Sicuranza, "Nonlinear digital filter realization by distributed arithmetic," *IEEE Trans. Acoust. Speech Signal Process.,* vol. ASSP-33, no. 4, pp. 939–945, August 1985.

287. G. L. Sicuranza and G. Ramponi, "Theory and realization of M-D nonlinear digital filters," *Proc. IEEE Int. Conf. Acoustics, Speech and Signal Processing,* Tokyo, Japan, pp. 1061–1064, April 1986.

288. G. L. Sicuranza and G. Ramponi, "Adaptive nonlinear digital filters using distributed arithmetic," *IEEE Trans. Acoust, Speech, Signal Process.* vol. ASSP-34, pp. 518–526, June 1986.

289. G. L. Sicuranza and G. Ramponi, "A variable-step adaptation algorithm for memory-oriented Volterra filters," *IEEE Trans. Acoust. Speech Signal Process.,* vol. ASSP-35, pp. 1492–1494, October 1987.

290. G. L. Sicuranza, "Quadratic filters for signal processing," *Proc. IEEE,* vol. 80, August 1992.

291. G. L. Sicuranza and G. Ramponi, "Adaptive nonlinear prediction of TV image sequences," *Electron. Lett.,* vol. 25, no. 8, pp. 526–527, April 13, 1989.

292. A. Siegel, T. Imamura, and W. C. Meecham, "Wiener-Hermite functional expansion in turbulence with the Burgers model," *Phys. Fluids,* vol. 60, no. 10, pp. 1519–1521, October 1963.

293. T. Siu and M. Schetzen, "Convergence of Volterra series representation and BIBO stability of bilinear systems," *Int. J. Syst. Sci.,* vol. 22, no. 12, pp. 2679–2684, 1991.

294. D. T. M. Slock and T. Kailath, "Numerically stable fast transversal filters for recursive least-squares adaptive filtering," *IEEE Trans. Signal Process.,* vol. 39, no. 1, pp. 92–114, January 1991.

295. T. Söderstrom and P. Stoica, *System Identification,* Prentice-Hall, Englewood Cliffs, New Jersey, 1989.

296. L. Spirkovska and M. B. Reid, "Coarse-coded higher-order neural networks for PSRI object recognition," *IEEE Trans. Neural Networks,* vol. 4, no. 2, pp. 276–283, March 1993.

297. J. C. Stapleton and S. C. Bass, "Adaptive noise cancellation for a class of nonlinear, dynamic reference channels," *IEEE Trans. Circuits Syst.,* vol. CAS-32, pp. 143–150, February 1985.

298. D. Starer and A. Nehorai, "Adaptive polynomial factorization by coefficient matching," *IEEE Trans. Signal Process.*, vol. 39, no. 2, pp. 527–530, February 1991.

299. P. T. Stathaki and A. G. Costantinides, "Noisy texture analysis based on higher order statistics and neural network classifiers," *Proc. IEEE Int. Conf. Neural Network Applications to Signal Processing*, Singapore, August 1993.

300. P. T. Stathaki and A. G. Costantinides, "Two-dimensional second order Volterra filters for texture segmentation with application to mammography," *Proc. ATHOS Workshop-93*, Sophia-Antipolis, France, September 1993.

301. T. G. Stockham Jr., "Image processing in the context of a visual model," *Proc. IEEE*, vol. 60, no. 7, pp. 828–842, July 1972.

302. T. Subba Rao, "On the theory of bilinear time series models," *J. Royal Soc. Ser. B*, vol. 43, pp. 244–255, 1981.

303. T. Subba Rao and M. M. Gabr, *An Introduction to Bispectral Analysis and Bilinear Time Series Models*, Springer-Verlag, Berlin, 1984.

304. A. Swamy and J. M. Mendel, "Cumulant-based approach to the harmonic retrieval problems," *Proc. Int. Conf. Acoust., Speech and Signal Processing*, pp. 2264–2267, New York, 1988.

305. A. Swamy, G. B. Giannakis, and J. M. Mendel, "Linear modeling of multidimensional non-Gaussian processes using cumulants," *Mult. Syst. Signal Process.*, vol. 1, pp. 11–37, March 1990.

306. M. A. Syed and V. J. Mathews, "QR-decomposition based algorithms for adaptive Volterra filtering," *IEEE Trans. Circuits Syst.—I: Fund. Theory Appl.*, vol. 40, no. 6, pp. 372–382, June 1993.

307. M. A. Syed and V. J. Mathews, "Lattice algorithms for adaptive Volterra filtering," *IEEE Trans. Circuits Syst.—II: Analog Digital Signal Process.*, vol. 41, no. 3, pp. 202–214, March 1994.

308. J. D. Taft, *Linear and Nonlinear Filters for Signal Detection and Image Segmentation*, Ph.D. Dissertation, Electrical Engineering, University of Pittsburgh, December 1986.

309. H. M. Teager, "Some observations on oral air flow during phonation," *IEEE Trans. Acoust. Speech Signal Process.*, vol. ASSP-28, no. 5, pp. 599–601, October 1980.

310. H. M. Teager and S. M. Teager, "Evidence for nonlinear production mechanism in the vocal tract," *NATO Adv. Study Inst. on Speech Production and Speech Modelling*, Bonas, France, 1989; Kluwer Academic Publishers, 1990.

311. A. M. Tekalp and A. T. Erdem, "Two-dimensional higher-order spectrum factorization with applications in non-Gaussian image modeling," *Proc. Workshop Higher-Order Spectral Anal.*, Vail, Colorado, pp. 186–190, June 1989.

312. E. J. Thomas, "Some considerations on the application of Volterra representation of nonlinear networks to adaptive echo cancellers," *Bell Syst. Tech. J*, vol. 50, pp. 2979–2805, 1971.

313. L. J. Tick, "The estimation of transfer functions of quadratic systems," *Technometrics*, vol. 3, pp. 563–567, November 1961.

314. H. Tong, "Nonlinear time series modelling in population biology: a preliminary case study," in *Nonlinear Time Series and Signal Processing*, R. R. Mohler, ed., Springer-Verlag, Berlin, 1988.

315. M. Tsatsanis and G. B. Giannakis, "Object and texture detection and classification using higher-order statistics," *IEEE Trans. Pattern Anal. Machine Intell.,* vol. 14, no. 7, pp. 733–750, July 1992.

316. J. Tsimbinos and K. V. Lever, "Sampling frequency requirements for identification and compensation of nonlinear systems," *Proc. IEEE Int. Conf. Acoustics, Speech and Signal Processing,* pp. III-513–516, Adelaide, Australia, 1994.

317. Y. Tsividis, "Continuous-time MOSFET-C filters in VLSI," *IEEE J. Solid-State Circuits,* vol. SC-21, pp. 15–29, February 1986.

318. J. K. Tugnait, "Estimation of linear parametric models of non-Gaussian discrete random fields," *Proc. SPIE, Image Processing Algorithms Techniques,* vol. 1452, San Jose, California, pp. 204–215, February 1991.

319. H. Urkowitz, *Signal Theory and Random Processes,* Artech House, Dedham, Massachusetts, 1983.

320. A. Vanzo, G. Ramponi, and G. L. Sicuranza, "An image enhancement technique using polynomial filters," *Proc. IEEE Int. Conf. Image Processing,* Austin, TX, pp. II 477–481, November 1994.

321. L. A. Vassilopoulos, "The applications of statistical theory of non-linear systems to ship performance in random seas," *Int. Ship Building Progress,* vol. 14, no. 150, pp. 54–65, 1967.

322. A. N. Venetsanopoulos, K. M. Ty, and A. C. P. Liou, "High speed architectures for digital image processing," *IEEE Trans. Circuits Syst.,* vol. CAS-34, pp. 887–896, August 1987.

323. M. H. Verhagen, "Round-off error propagation in four generally-applicable, recursive least-squares estimation schemes," *Automatica,* vol. 25, pp. 437–444, 1989.

324. V. Volterra, "Sopra le funzioni che dipendono da altre funzioni," *Rend. Regia Accademia dei Lincei, 2o Sem.,* pp. 97–105, 141–146, 153–158, 1887.

325. V. Volterra, *Leçons sur les functions de lignes,* Gauthier-Villars, Paris, France, 1913.

326. V. Volterra, *Theory of Functionals and of Integral and Integro-Differential Equations,* Dover, New York, 1959.

327. A. Watanabe and L. Stark, "Methods for nonlinear analysis: identification of a biological control system," *Math. Biosci.,* vol. 27, pp. 99–108, 1975.

328. E. T. Whittaker, "Biography of Vito Volterra," in *Obituary Notices of Fellows of The Royal Society,* 1941. Also reproduced in Volterra [326].

329. B. Widrow and S. D. Stearns, *Adaptive Signal Processing,* Prentice-Hall, Englewood Cliffs, New Jersey, 1985.

330. B. Widrow and M. A. Lehr, "30 years of adaptive neural networks: Perceptron, Madaline, and backpropagation," *Proc. IEEE,* vol. 78, no. 9, pp. 1415–1442, September 1990.

331. N. Wiener, *Response of a Nonlinear Device to Noise,"* Report No. 129, Radiation Laboratory, MIT, Cambridge, Massachusetts, April 1942.

332. N. Wiener, *Nonlinear Problems in Random Theory,* The Technology Press, MIT, Cambridge, Massachusetts, and Wiley, New York, 1958.

333. S. J. Wissler, "Parameter estimation for a bilinear time series model," *Proc. Nat. Conf. Undergraduate Research VI,* 1992.

334. L. Yin, R. Yang, M. Gabbouj, and Y. Neuvo, "Weighted median filters: a tutorial," *IEEE Trans. Circuits Syst. II: Analog Digital Signal Process.,* vol. 43, no. 3, pp. 157–192, March 1996.

335. D. H. Youn, K. K. Yu, and V. J. Mathews, "Adaptive nonlinear digital filter with sequential regression algorithm," *Proc. 22nd Annual Allerton Conf. Control, Communication and Computing,* Urbana-Champaign, Illinois, pp. 152–161, October 3–5, 1984.

336. G. D. Zames, *Nonlinear Operators for System Analysis,* Technical Report No. 370, Research Laboratory of Electronics, MIT, Cambridge, Massachusetts, 1960.

337. G. D. Zames, "Functional analysis applied to nonlinear feedback systems," *IEEE Trans. Circuit Theory,* vol. 10, no. 3, pp. 392–404, September 1963.

338. S. M. Zand and J. A. Harder, "Application of nonlinear system identification to the Lower Mekong River, Southeast Asia," *Water Resources Research,* vol. 9, no. 2, pp. 290–297, 1973.

339. J. Zarzycki, *Nonlinear Prediction Ladder Filters for Higher Order Stochastic Sequences,* Springer-Verlag, Berlin, 1985.

340. C. Zetzsche, E. Barth, and B. Wegmann, "The importance of intrinsically two-dimensional image features in biological vision and picture coding," in *Digital Images and Human Vision,* A. B. Watson, ed., MIT Press, Cambridge, Massachusetts, pp. 107–138, 1993.

341. Y.-M. Zhu, "Generalized sampling theorem," *IEEE Trans. Circuits Syst. II: Analog Digital Signal Process.,* vol. 39, no. 8, pp. 587–588, August 1992.

INDEX